DESIGN OF
SPECIAL HAZARD AND
FIRE ALARM SYSTEMS

DESIGN OF SPECIAL HAZARD AND FIRE ALARM SYSTEMS

Second Edition

Robert M. Gagnon, PE, SET, FSFPE

Wake Technical Community Colleg
9101 Fayetteville Road
Raleigh, NC 27603-5696

DELMAR
CENGAGE Learning

Australia • Brazil • Japan • Korea • Mexico • Singapore • Spain • United Kingdom • United States

DELMAR
CENGAGE Learning™

**Design of Special Hazard and Fire Alarm
Systems, Second Edition**
Robert M. Gagnon

Vice President, Technology and Trades ABU:
David Garza

Director of Learning Solutions: Sandy Clark

Managing Editor: Larry Main

Acquisitions Editor: Alison Pace

Product Development Manager: Janet
Maker

Senior Product Manager: Jennifer A. Starr

Marketing Director: Deborah S. Yarnell

Marketing Manager: Erin Coffin

Marketing Coordinator: Patti Garrison

Director of Production: Patty Stephan

Production Manager: Stacy Masucci

Content Project Manager: Jennifer Hanley

Art Director: Benjamin Gleeksman

Technology Project Manager: Kevin Smith

Editorial Assistant: Maria Conto

For product information and technology assistance, contact us at
Cengage Learning Customer & Sales Support, 1-800-354-9706

For permission to use material from this text or product,
submit all requests online at **www.cengage.com/permissions**
Further permissions questions can be emailed to
permissionrequest@cengage.com

Library of Congress Control Number: 2007025847

ISBN-13: 978-1-4180-3950-9

ISBN-10: 1-4180-3950-0

Delmar
Executive Woods
5 Maxwell Drive
Clifton Park, NY 12065
USA

Cengage Learning is a leading provider of customized learning
solutions with office locations around the globe, including Singapore,
the United Kingdom, Australia, Mexico, Brazil, and Japan. Locate your
local office at **international.cengage.com/region**

Cengage Learning products are represented in Canada by
Nelson Education, Ltd.

For your course and learning solutions, visit **delmar.cengage.com**

Visit our corporate website at **cengage.com**

Printed in the United States of America
4 5 6 7 8 9 10 14 13 12 11 10

DEDICATION

Fire is a cruel thief that steals our very young, our elderly, and even the fittest among us. This book is dedicated to fire service professionals, fire protection design professionals, fire science educators, and many others who dedicate their lives to solving problems associated with preventing and overcoming the effects of fire in our society.

Fire service professionals take significant personal risks in performing their duties, and they gain immediate satisfaction in knowing that the risks they take pay off in saving human lives. No reward could be greater.

Fire protection design professionals contribute in a way that is no less significant. They are unsung heroes, responsible for saving the lives of people they may never meet. The design of fire protection systems prevents fires from getting out of control during rescue and extinguishment operations and significantly lessens the risk to fire service professionals.

Fire science educators who are responsible for training fire protection professionals also have a profound effect on the safety and welfare of our society. Their contribution is cumulative and grows exponentially with each graduating class.

All of these esteemed fire protection professionals are working actively to solve problems associated with fire and deserve to be proud of their positive influence on life safety and the protection of property. We are bound by a common goal. We are motivated by a common pride. We are inspired by a common vision.

A special gate to heaven is open to people who dedicate their lives to the safety and welfare of others. May that gate be opened for each of you.

Robert M. Gagnon, PE, SET, FSFPE

CONTENTS

■ HARDWIRED AND ADDRESSABLE (MULTIPLEX) FIRE ALARM
SYSTEMS/471 ■ FIRE ALARM CONTROL UNITS/472 ■
VOLTAGE DROP ON NOTIFICATION APPLIANCE CIRCUITS/479
■ SUMMARY/483 ■ REVIEW QUESTIONS/484 ■ DISCUSSION
QUESTIONS/484 ■ ACTIVITIES/484

FOREWORD TO THE
SECOND EDITION

Despite the widespread use of special hazard systems in the built environment, engineers and technicians have few good references to help them design these systems. The variety of design standards can be difficult to apply by those who do not have a lot of experience in designing special hazard systems. This book bridges the gap between design standards and the expertise needed to design special hazard systems.

Robert M. Gagnon, PE, SET, FSFPE, takes a comprehensive approach in explaining the design of special hazard systems. He begins with design fundamentals—surveys, contract document preparation, and engineering ethics. Then he does an excellent job of explaining the myriad of systems available to protect special hazards and the considerations associated with the different types of system. For each type of system, he provides a step-by-step approach to system design. This approach begins with fundamental principles and expands to state-of-the-art issues.

The author rounds out the book with several chapters on the design of fire alarm systems. He applies the same thoroughness to the choices that must be made when designing fire alarm systems as he provides for special hazard systems.

This second edition significantly expands on the first edition, published in 1998, by addressing suggestions from readers and a diverse review team consisting of engineers, professors, and practitioners. Robert continues to write for an audience of both engineers and engineering technicians. In addition to providing a valuable reference for these two professional fields, his book conveys an understanding of the design decisions that each must make.

The first edition has helped many engineers pass the professional engineering exam, and likewise has helped many engineering technicians complete NICET exams successfully. This second edition is an excellent primer for professionals who are new to the design of special hazard systems and is a superb reference for more experienced practitioners.

Morgan J. Hurley, PE
Technical Director
Society of Fire Protection Engineers

FOREWORD TO THE FIRST EDITION

It is commonly said that there is nothing new under the sun. The idea is simple; almost all of the good ideas seem to have been proposed already, even if they have not been implemented. In the late 1800s, the head of the United States Patent Office proposed that the United States Government eliminate the office because he felt all things that could be invented already had been invented. He felt the agency would soon be useless. Yet, here we are, ending a century that has witnessed a quantum leap in knowledge. In the last 100 years we have seen a technological explosion that is unprecedented in the history of civilization. One cannot help but wonder what the next millennia will bring to fire protection technology.

The writing of a technical text or reference book must serve as a bridge between our past and our future. In my opinion any recently published book is a reflection of the past because it is based upon the theory and practical solutions of the past. *Design of Special Hazard and Fire Alarm Systems* is a harbinger of change because it contains the latest information that is needed to advance students and practitioners. Without a perspective from the past, any field of fire science may lack practicality. Without value being added to a revised version of our body of knowledge, material found in a text or reference book may be visionary, but irrelevant.

Robert M. Gagnon, PE, SET, FSFPE, author of this book, has done an excellent job of keeping these two concepts in mind. The material presented in *Design of Special Hazard and Fire Alarm Systems* achieves the objectives of raising the level of knowledge, but still remains grounded in contemporary practice.

As a writer of texts and journal articles myself, I realize that developing a manuscript is taxing and often frustrating. When a document survives the developmental process it is a commendable act in itself. My congratulations to Robert M. Gagnon, PE, SET, FSFPE for having the discipline to develop *Design of Special Hazard and Fire Alarm Systems,* to the purchasers of this book for a commitment to learning, and to the fire service and aspiring design professionals for their adherence to the further pursuit of excellence in fire science.

Ronny J. Coleman
Chief Deputy Director
and
State Fire Marshal
State of California

PREFACE

I am proud to introduce this second edition of *Design of Special Hazard and Fire Alarm Systems,* which with its direct counterpart, *Design of Water-Based Fire Protection Systems,* comprise a two-book set that provides valuable fire protection system design training for practicing fire protection professionals. Readers of this book are encouraged to obtain *Design of Water-Based Fire Protection Systems,* also published by Delmar Cengage Learning, to complete the story of fire protection system design.

This book provides design advice for students in community colleges, fire academies, and universities and is recommended for use by sprinkler, engineering, plumbing, mechanical, and architectural firms for training entry-level design personnel. All fire protection curricula should require a minimum of two semesters of fire protection system design, at least one for water-based systems, using *Design of Water-Based Fire Protection Systems* as a text, and at least one for special hazard and fire alarm systems, using this book as a text.

Design of Special Hazard and Fire Alarm Systems and its companion text, *Design of Water-Based Fire Protection Systems,* are written so either course may be taken before the other without prerequisite and are structured to permit a student to take only one of these design courses if desired. The books have been written for fire service personnel who encounter fire protection systems in their daily lives and also for practicing or aspiring fire protection technicians and engineers and other fire protection professionals who seek to better understand fire protection systems.

WHY I WROTE THIS BOOK

Design of Special Hazard and Fire Alarm Systems is the second in a two-book set that offers valuable training in fire protection system design training for practicing and aspiring fire protection professionals. The first book, *Design of Water-Based Fire Protection Systems,* is the basis for understanding sprinkler, water spray, water mist, standpipe, and ultra high-speed water spray systems.

This book completes the fire protection design story with its detailed coverage of foam, clean agent, carbon dioxide, dry and wet chemical, explosion suppression, and fire alarm systems. Fire protection is a wonderfully diverse profession, and one book cannot do justice to the broad variety of available suppression systems. Those completing both books will have obtained the minimum requisite tools for designing, specifying, approving, testing, maintaining, and installing the most prevalent fire protection systems encountered in professional practice.

The philosophy that drives both books is the same. To deal effectively with fire protection systems in any capacity, at any level of responsibility, in any field of endeavor, one must thoroughly understand how fire protection systems meet their performance objectives. To discharge one's responsibilities competently with respect to a fire protection system, one must have a firm grasp of the concepts on which that system was created. The only way to understand fire protection systems and their performance objectives is to understand how fire protection systems are designed and calculated. This book is predicated on this fundamental principle.

Some may believe that comprehensive fire protection system design instruction is intended only for those who are directly responsible for the design of fire protection systems. This is a dangerous and potentially lethal position. Fire protection systems are routinely reviewed, approved, maintained, inspected, and tested by fire service personnel, architects, engineers, contractors, and technicians who know significantly less about the systems and their objectives than the person who designed them. This gap in practical design knowledge is the most serious problem facing our profession today, and the gap is widening with the increasing complexity of fire protection technology. This book addresses the dilemma and can be used effectively to narrow the gap.

All who deal with fire protection systems in any capacity have a moral and ethical responsibility to be vigilant in their pursuit of a comprehensive understanding of the art and science of fire protection system design. The only prerequisite for a thorough understanding of the material in this book is a basic understanding of high school mathematics and a commitment to bettering the fire protection profession.

HOW TO USE THIS BOOK

This edition is organized into four logical and progressing units:

Unit 1: Special Hazard and Fire Alarm System Fundamentals

- *Chapter 1* contains essential reference material for creating a drawing and using contract drawings, specifications, and surveys to develop a drawing.

- *Chapter 2* discusses the ethical practice of fire protection system design and layout, a subject as essential to success as knowledge of technical information.

- *Chapter 3* explains how special hazard agents extinguish fires, completing the study of extinguishment introduced in the companion to this book, *Design of Water-Based Fire Protection Systems.*

Unit 2: Water-Based Special Hazard Systems

Chapters 4 through 7 provide a detailed examination of low-expansion and high-expansion foam systems, as well as water mist systems and ultra high-speed water spray systems, using numerous examples, photos, and illustrations to assist readers in understanding these systems. Combined with the exhaustive study available in *Design of Water-Based Fire Protection Systems,* readers can develop a comprehensive appreciation for water as an extinguishing agent.

Unit 3: Gaseous and Particulate Agent Special Hazard Systems

Chapters 8 through 10 use a simplified methodical approach toward the design of clean agent, halon replacement, carbon dioxide, dry chemical, and wet chemical non-aqueous special hazard fire protection systems encountered most commonly.

Unit 4: Special Hazard Detection, Alarm, and Releasing Fire Alarm Systems

Chapters 11 through 15 prepare fire protection professionals thoroughly for the design of fire alarm and detection systems, including a thorough examination of initiating devices, notification appliances, detector placement, fire alarm circuit design, and fire alarm control units.

Most of the chapters in this book that relate to fire protection system design are based on conformance with an accepted national design standard. The nation's principal source for the creation, modification, and publication of codes and standards applicable to fire protection is the National Fire Protection Association (NFPA). Each chapter refers the reader to one or more NFPA standards. Although readers can obtain a good foundation for designing fire protection systems using this book as the sole reference, it is recommended that the applicable NFPA standards be obtained and used in conjunction with this book for optimum understanding of the subject matter.

I have found that the best way to learn fire protection system design is to apply the information at the level of detail presented in this book. Therefore, the reader should complete, as a minimum, the design assignments and activities at the end of each chapter.

Additional Skills

A primary philosophy behind my writing this book is to promote the idea that development of the skill of presenting suppression and detection concepts in a comprehensible drawing enhances the ability to understand and interpret the work of others. The price paid for a lack of design clarity could be the lives of the people who depend on this skill. Some readers may never have done a scaled drawing before reading this book, but understanding the fundamental design concepts and transferring them onto a blank drawing medium is among the most important skills expected of a fire protection professional.

Basic Mathematics Required for This Book

Most readers with a background in introductory high school algebra and geometry should not find the mathematical demands of this book to be a problem. To perform the relatively simple calculations required, the reader should have a calculator with square root, exponent, and inverse keys, because the highest level of math required for this book uses these functions.

The Internet

One of the major technological innovations occurring since the premier edition of this book has been the growth and increased use of the Internet as a source of study and research. It is suggested that readers of this book frequently research areas of interest by entering key words into one or more commonly used search engines. For example, by entering the term "Montreal Protocol" or "Environmental Protection Agency" into an Internet-based search engine, readers can obtain valuable data that will deepen their understanding of the subject matter.

NEW TO THIS EDITION

With great pleasure I introduce to you this second edition of *Design of Special Hazards and Fire Alarm Systems.* I am privileged to have received many appreciative comments on the first edition since its publication and am excited about the many improvements in this new edition.

The second edition includes new material that brings the book up-to-date with the most recent NFPA codes and standards. A unique feature of this book is that it incorporates many helpful suggestions by readers and users of this book who have kept an open and supportive dialogue with me during the past 10 years.

This edition further represents wise counsel from the expert review team that worked closely and cooperatively with me to improve its value and usefulness. Among the many improvements in this edition include organizing the book into logical units and moving the discussion of ethics to the fundamental concepts chapters in Unit 1 to emphasize its importance. Many other technical and pedagogical improvements to this book are attributable to my review team of esteemed professionals and friends, and I am grateful for their many key contributions.

FEATURES OF THIS EDITION

The second edition:

- **Adheres to the new editions of NFPA Standards**, including references and tables and graphics from NFPA 10, 11, 12, 69, 70, 72, 409, 550, 750, and 2001, to ensure that readers receive the most up-to-date information on codes and regulations pertaining to fire protection systems and design

- **Features new technology**, updating readers with the latest information on equipment and system design in the industry

- **Logically organizes the material** into units, each building upon the previous unit in an easily understood presentation of even the most complex and technical content

- **Includes new photos and illustrations** acquired from the most reputable manufacturers in the industry, bringing the content to life and clarifying critical concepts

- **Provides new discussion questions and activities** at the end of each chapter, encouraging students to apply what they have learned in the chapter through classroom or professional discussion and field projects

- **Expands and adds new material in state-of-the-art topics**—wet chemical suppression systems, addressable fire alarm systems, CAD design, and mass communications systems

SUPPLEMENT TO THIS BOOK

New! For this edition we are pleased to offer an Instructor's Guide on CD-ROM, available only to instructors to assist in preparing and delivering a course on fire protection system design. The CD contains the following components:

- **Instructor's Guide** with Lesson Plans and Answers to Review Questions, as well as a suggested course syllabus, a suggested course outline, and a design and research project, to help you prepare and challenge your students in the classroom
- **PowerPoint Presentations** highlighting the essential information within each chapter and correlated with the Lesson Plans in the Instructor's Guide, for a seamless classroom presentation of content
- **Chapter quizzes** in editable Word format, including knowledge-based questions as well as practical problems that will encourage students to apply what they have learned in each chapter. You may edit, add, or delete questions to meet the needs of your course
- **NFPA correlation grid**, which outlines the requirements of several NFPA standards on which the book is based and provides the specific pages in the book where supporting content can be found

Order #: 1-4180-3952-7

NOTE TO THE READERS

I am so pleased with the positive response received from you since the first edition was published. By incorporating your many helpful suggestions, this second edition is a significant improvement over the initial edition. An appreciable amount of new material is devoted to the latest technological advances, along with additional photos and illustrations that bring the material to life and clarify it. The fundamental concepts have been expanded and explained in detail, and the entire book has been reorganized into a logical and cogent presentation.

I sincerely welcome your comments and suggestions for improvement in the next edition. Please contact the author:

Robert M. Gagnon, PE, SET, FSFPE
Gagnon Engineering
RobtGagnon@aol.com
Web site: www.GagnonEngineering.com

ABOUT THE AUTHOR

Robert M. Gagnon, PE, SET, FSFPE is president of Gagnon Engineering and performs fire protection system design and calculation, alarm and detection system design, fire protection engineering, expert witness work, and code consulting. He teaches courses in fire protection and detection system design at the University of Maryland, Department of Fire Protection Engineering, and formerly taught three fire protection and detection system design courses at Montgomery College. After working for 21 years as a fire protection systems designer for "Automatic" Sprinkler Corporation of America and the Fireguard Corporation, he formed Gagnon Engineering. He appears in the 51st edition of *Who's Who in America* and was elected a fellow of the Society of Fire Protection Engineers.

He holds B.S. and M.S. degrees in fire protection engineering from the University of Maryland, a B.A. in mathematics from Western Maryland College, and NICET Level IV certifications in Automatic Sprinkler Systems Layout and Special Hazards Systems Layout. He is a registered professional engineer in Maryland, Virginia, Pennsylvania, and the District of Columbia. Further, he has served as president of the Howard County Chapter of the National Society of Professional Engineers, president of the Maryland Society of Professional Engineers, president of the University of Maryland Engineering Alumni Association Board of Directors, president of the Baltimore Alumni Chapter of the Tau Beta Pi Engineering Honor Society, president of the Chesapeake Chapter of the Society of Fire Protection Engineers, and as a member of the National Fire Sprinkler Association and the American Fire Sprinkler Association. He also has served on the Northern Virginia Community College fire science advisory committee.

The author is Chair of the NFPA Committee on Foam-Water Sprinkler and Foam-Water Spray Systems (NFPA 16), is a member (special expert category) of the NFPA committee on Automatic Sprinkler Systems (NFPA 13), and is a member of the Technical Correlating Committee on Automatic Sprinkler Systems. He is a principal member (special expert category) and secretary of the NFPA Committee on Water Spray Fixed Systems (NFPA 15), the NFPA Committee on Water Tanks (NFPA 22), and the NFPA Committee on Private Water Supply Piping Systems (NFPA 24 and NFPA 291).

He also is a principal member (special expert category) of the NFPA Committee on Water Cooling Towers (NFPA 214). He authored a chapter in the NFPA *Fire Protection Handbook,* entitled "Ultra High-Speed Suppression Systems for Explosive Hazards," and is the author of numerous journal and scholarly articles. Books he has authored include *Design of Water-Based Fire Protection Systems* (Delmar Cengage Learning), *A Designer's Guide to Automatic Sprinkler Systems* (NFPA), *A Designer's Guide to Fire Alarm Systems* (NFPA), the *Engineering Student's Guide for Professional Development,* and the *Department of Fire Protection Engineering History Book* (40th and 50th Anniversary Editions).

ACKNOWLEDGMENTS

A person's outlook on the world and the manner in which that outlook is expressed is attributable to a large extent to one's education. For this reason, I cumulatively thank all of the teachers who helped me to develop and express my opinions.

Dr. Clyde Spicer and Dr. James Lightner in the mathematics department of Western Maryland College (now McDaniel College) challenged and inspired me, and I always will be guided by the high standards they set.

During my undergraduate education at the University of Maryland, Department of Fire Protection Engineering, Dr. John L. Bryan, Dr. James A. Milke, and Dr. Frederick W. Mowrer provided wisdom and insight. For my graduate education and research in the Department of Fire Protection Engineering, Dr. James G. Quintiere, Dr. Vincent M. Brannigan, and Dr. Steven M. Spivak provided my educational foundation. All of these men are outstanding teachers and role models.

Dr. Richard H. McCuen of the Department of Civil Engineering, University of Maryland, was my honors professor in the engineering honors program at the University of Maryland—and was the inspiration for Chapter 2 of this book. Dr. McCuen, an internationally renowned expert on engineering ethics, was kind enough to edit and add his vast experience to this indispensable chapter.

The professors and staff at the Department of Fire Protection Engineering, University of Maryland, continue to be of enormous support, assistance, and inspiration in my current duties as lecturer with the Department. Those who read this book are encouraged to further their education at this truly exceptional and unique school.

Dr. John L. Bryan deserves special recognition, not only for founding the Department of Fire Protection Engineering, University of Maryland, in 1956, but also for helping it to grow and prosper. After Dr. Bryan's retirement, Dr. Steven M. Spivak ably guided the department into an exciting period of discovery and growth. Dr. Marino DiMarzo, the third chair in the Department's 50-year history, continues to enthusiastically expand the Department in its research, faculty, and facilities. On behalf of all graduates of the Department of Fire Protection Engineering, I offer the highest accolades for their leadership and for their friendship.

So much of a person's expertise is obtained on the job. For a significant portion of my fire protection career, "Automatic" Sprinkler Corporation of America provided excellent mentors, especially Terry Victor and Art O'Neil, in an atmosphere that encouraged me to expand my experience and knowledge, and allowed me to arrange my work schedule to return to school and study fire protection engineering. For this, I am deeply grateful. To all who have helped me perform my duties with Gagnon Engineering, my undying thanks.

Among the many contributors who added new photos, illustrations, and tables to this second edition are Jim Cox of Ansul; Terry Victor, Chris Woodcock, and Bruce Fraser of Simplex-Grinnell; Dennis Berry of the NFPA; Morgan Hurley of SFPE; and Tyler Mosman. To the National Fire Protection Association I owe my deep gratitude, for their enormous contributions to fire safety in the world and also their supplying figures and tables for this book. It should be emphasized that the reprinted material is not the complete and official position of the National Fire Protection Association on the referenced subject, which is represented only by the standard in its entirety. I am proud to be a member of NFPA, and I heartily recommend that all readers of this book become active members of this vital organization.

Delmar Cengage Learning and I acknowledge the following reviewers, who offered valuable advice that improved the quality of this book:

Al Cozby
Simplex-Grinnell
Strongsville, OH

Bruce Fraser
Simplex-Grinnell
Milford, MA

Morgan Hurley, PE
Society of Fire Protection Engineers
Bethesda, MD

Jack Nicholas
Northeast Wisconsin Technical College
Marinette, WI

Niles Ottesen
Milwaukee Area Technical College
Oak Creek, WI

Robert Prytula
Chattanooga State Technical College
Chattanooga, TN

David Saunders
Delgado Community College
Jefferson Parish Fire Department
Harahan, LA

Jerry Watts
Accent Fire Engineering International
Santa Fe, NM

John White
Shasta College
Redding, CA

The family is the rock upon which all accomplishments are based, and the fountain from which all inspiration flows. To my late parents Robert and Martha, my wife Martha, my daughter, Rebecca, her husband John, my four grandchildren Alana, Christian, Sabrina, and Jensen, my sister Katherine and her children Joshua and Jennifer—I bestow my most sincere thanks for your love and encouragement.

From the beginning, everyone at Delmar Cengage Learning has been friendly, professional, helpful, and encouraging in the preparation of this second edition. Delmar fervently believes in its value, and their support inspired me to meet and exceed every deadline. They made the difficult experience of writing a book most enjoyable, and my thanks go to all of you.

Robert M. Gagnon, PE, SET, FSFPE

Unit

1

SPECIAL HAZARD AND FIRE ALARM FUNDAMENTALS

Chapter

1

FUNDAMENTAL CONCEPTS FOR DESIGN OF SPECIAL HAZARD AND FIRE ALARM SYSTEMS

Learning Objectives

Upon completion of this chapter, you should be able to:

- List the items that comprise a set of contract documents for the design of a special hazard or fire alarm system.
- List the categories of drawings that comprise a contract drawing package.
- Explain the differences among the categories of contract drawings.
- Evaluate a set of contract drawings to determine the value of each drawing relative to the accurate development of a special hazard or fire alarm system design.
- Explain the problems associated with the designer of a special hazard or fire alarm system failing to reference drawings relevant to fire protection systems in a contract drawing package.
- List the divisions of the contract specifications.
- Explain the relationship between a set of contract drawings and the contract specifications.
- Determine the divisions of the specifications that are of most value to the development of a fire protection system design.
- Identify the problems that could develop if the contract specifications conflict with the contract drawings.

- List the items that should be found in every designer's survey kit.
- Perform a survey of a building to be used for the design of a fire protection system.
- Survey and accurately determine dimensions of a reflected ceiling plan.
- Survey a building using the structural elements as the primary points of reference.
- Field-check a drawing of a fire protection system.
- Perform metric conversions where appropriate.
- Evaluate design objectives for special hazard suppression systems.
- Know the characteristics that define a special hazard.
- Evaluate branches of the Fire Safety Concepts Tree.
- Know the difference between prescriptive and performance-based design.

This chapter provides a basis for understanding fundamental principles and concerns for the design of special hazard and fire alarm systems. The principles presented here provide the information necessary to understand the characteristics that define a special hazard, the types of hazards covered in this book, the design objectives of a special hazard suppression and detection system, and the design concepts and methodologies involved.

This chapter introduces the reader to the use and interpretation of contract drawings, specifications, and surveys. Fire protection system designers often receive less information than they need to do a complete, competent job of designing a fire protection system. This chapter details the value of each component of the contract documents—the primary reference for the design of a fire protection system. The inherent danger of failing to use these documents to full advantage is that the fire protection design may omit critical information, resulting in installation problems or failure to meet applicable standards, which could compromise or negatively impact the effectiveness of the fire protection system.

The contract document package, or simply contract documents, consists of **contract drawings**, a set of plans that describes a project in pictorial form, and **specifications**, a written description of project requirements. Understanding each component of the contract drawings and each section of the specifications is essential to the successful design of a fire protection system.

A fire protection system designer should keep in mind that the public expects and demands that a fire protection system installed in a building meet its primary **performance objective** of properly activating in a timely manner, providing a minimum acceptable level of life and property protection when called upon to function. A performance objective is an engineering basis for a predetermined end result of a given design.

contract drawings
a set of plans that describe a project in pictorial form

specifications
a written description of project requirements

performance objective
engineering basis for a predetermined end result of a given design

Fire protection system design is a highly respected profession that carries enormous responsibility. This responsibility entails the complete investigation of all pertinent information that may affect the performance of the fire protection system.

FIRE PROTECTION AS A PROFESSION

Professionals who dedicate their lives to the specification, design, installation, approval, and maintenance of fire protection systems and to the operation of fire departments are unique. The dedication and commitment of fire protection professionals is almost uniformly exceptional, and this level of commitment distinguishes the fire protection profession from many other occupations.

The dedication displayed in our work is a reflection of our belief that we are making a difference in improving the safety and welfare of the world in which we live. There can be no more rewarding profession than to dedicate one's life to the welfare of others. This dedication characterizes our lives. It sustains us. By designing fire suppression and detection systems, we save the lives of people whose names we may never know.

Along with the rewards of the profession come responsibilities. As fire protection professionals, we must realize that if we make a serious error or knowingly provide less than the minimum level of fire protection, people may die. The responsibility is enormous, and all aspiring fire protection professionals must accept this responsibility freely and fully. More than any other factor, accepting this responsibility is what differentiates ours from most other occupations.

A fire protection system is so much more than just a collection of assorted hardware. If we are truly committed to the spirit and intent of the goals of our profession, we can never accept being associated with the design and installation of fire protection systems that are anything less than compliant with the applicable codes and standards. The vast majority of this book is intended to help the reader achieve this goal.

The application of technical knowledge to our profession is fundamental, but it does not comprise our total responsibility as fire protection professionals. Chapter 2 covers a topic rarely covered in a technical book: Ethical behavior among fire protection professionals is as important as the application of technical knowledge. Knowingly failing to apply technical knowledge in an ethical manner is the most fundamental error that a fire protection professional can make. All readers of this book are advised to become familiar with the code of ethics that applies to the practice of engineering and technology in your jurisdiction, and to follow it to the letter.

This book assists fire protection professionals in becoming technically proficient with respect to several of the most commonly installed types of fire protection systems and helps to familiarize the reader with the fundamentals of ethical practice. These two essential tools will assist the reader in achieving a successful and rewarding career as a fire protection professional.

FIRE PROTECTION ENGINEERS AND TECHNICIANS

The Society of Fire Protection Engineers (SFPE) issued an updated position statement titled, "The Engineer and the Technician; Designing Fire Protection Systems" in October of 2005. This document describes the roles of engineers and technicians in fire protection system design.

Fire Protection Engineer

fire protection engineer
a licensed professional engineer who demonstrates sound knowledge and judgment in the application of science and engineering to protect the health, safety, and welfare of the public from the impacts of fire

SFPE defines a **fire protection engineer** as "a licensed professional engineer who demonstrates sound knowledge and judgment in the application of science and engineering to protect the health, safety, and welfare of the public from the impacts of fire. This includes the ability to apply and incorporate a thorough understanding of fundamental systems and practices as they pertain to life safety and to fire protection, detection, alarm, control and extinguishment."

The position statement provides a description of each of the following functions performed by a fire protection engineer (see www.SFPE.org for more information):

- fire protection analysis
- fire protection management
- fire science and human behavior
- fire protection systems
- passive building systems

SFPE's position statement further describes the following functions that a fire protection engineer is qualified to perform:

- evaluation of the broad range of hazards and protection schemes required to develop a workable, integrated solution to a fire safety problem
- preparation of design documents
- layout of fire protection systems
- affixing a professional seal to documents
- review of fire protection installation shop drawings for compliance with the engineer's design
- monitoring of the installation of fire protection systems
- responsibility for designing and maintaining competency through continued education.

Fire Protection Technician

fire protection technician
an individual who has achieved NICET Level III of IV certification in the appropriate subfield and who has the knowledge, experience and skills necessary to lay out fire protection systems

SFPE defines a **fire protection technician** as "an individual who has achieved NICET Level III of IV certification in the appropriate subfield and who has the knowledge, experience and skills necessary to lay out fire protection systems."

Based on engineering design documents, which could include the system(s) design drawings, specifications, and nationally recognized codes and standards—such as those published by the National Fire Protection Association or the ICC-International Code Council—the technician is qualified to prepare the following:

- the system layout in accordance with the engineer's design
- shop drawings in accordance with the engineer's design or as otherwise permitted by state regulations
- supplemental calculations based on the engineer's design

Technicians are responsible for their work and must maintain competence through continued education.

The fundamental difference between engineers and technicians in fire protection is that engineers perform design and technicians perform layout. To avoid repeatedly using the term "design and/or layout," this book primarily uses "design" for simplicity.

Although this book is based on reference to and concurrence with recognized national standards, engineers or engineering students may use it as a basis for their engineering design career. Engineers are expected to base their design decisions on engineering principles as well as familiarity with requirements of the applicable national standards as a point of reference in their design.

AUTHORITY HAVING JURISDICTION (AHJ)

authority having jurisdiction (AHJ)
the individual or agency placed in responsible charge of reviewing and approving drawings and completed installations

The SFPE position paper further defines the role of the **authority having jurisdiction (AHJ)**, who interrelates with fire protection engineers and fire protection technicians, as follows:

The Authority Having Jurisdiction, also commonly referred to as the AHJ, is the individual or agency that has the responsibility for reviewing and accepting the design provided.

Examples of AHJ are

- municipal permitting organization
- fire prevention officer of the municipality
- insurance company
- governmental organization
- code official
- university fire marshal

METRIC CONVERSIONS, SIGNIFICANT FIGURES, AND ROUNDING

Metric conversions can be made using the appropriate tables in Appendix A. The U.S. government has made a commitment to convert to the metric system, and some agencies have done so already.

Relevant Legislation

Metric Conversion Act of 1975

legislation that created a requirement for conversion to the metric system for all federal projects by 1992

The position of the U.S. government was established by the **Metric Conversion Act of 1975**, which created a requirement for conversion to the metric system for all federal projects by 1992. President George H.W. Bush issued Executive Order 12770 on July 25, 1991, which provided impetus for transition to the metric system. Numerous federal departments, such as the U.S. Army Corps of Engineers, converted to the metric system immediately for all construction projects.

Although the original timetable for metric conversion has not been met, federal construction will be uniformly converted to the metric system in the future. The smoothness of the transition by contractors performing federal construction has surprised those who predicted chaos.

The Metric Conversion Act of 1975 was amended by the Omnibus Trade and Competitiveness Act of 1988, the Savings in Construction Act of 1996, and the Department of Energy High-End Computing Revitalization Act of 2004. The amendments primarily provided explanation, clarification, and justification of the 1975 Act. Although it is interesting to note that the metric system has been the official U.S. standard measure since the 1875 Treaty of the Meter, governmental regulation has done little to alter the commonplace usage of inch/foot units in the United States.

The reason for the conversion is that the United States is one of the few countries in the world that does not base its system of weights and measures on the metric system, which is the international standard. The ability of the United States to compete on an international scale depends on our nation's ability to design and manufacture commodities that can be sold and used overseas, and to design federal projects that can be installed with foreign tools and labor.

Using the metric system can create problems when renovating existing facilities that were not designed using the metric system. For example, a building undergoing renovation may use "soft" metric conversions, which consist of direct conversion of the existing feet-and-inch units to a fractional metric unit. Working with these decimals can be cumbersome, and it may be simpler to use "hard" metric, which involves rounding to the nearest whole metric unit for convenience.

Metric Units

For fire protection system designs, the following metric units are used:

Quantity	Unit	Symbol
Length	Meter	m
Mass	Gram	g
Time	Second	s
Temperature	Kelvin	K

When very large or very small quantities are encountered, these units are broken down for ease of use by using decimal prefixes. Prefixes most commonly used in construction are:

Prefix	Symbol	Multiplier	Examples
kilo	k	1000	km, kg
milli	m	0.001	mm, mg

To provide uniformity in construction units, the *Metric Guide for Federal Construction* advises against use of the centimeter in construction. The meter and millimeter are to be used for fire protection system design, and the meter and kilometer are to be used for fire protection underground piping design.

Common conversions for the evaluation of U.S. and metric scales are listed below:

Inch-Foot Scale	Ratio	Closest Metric Scale
1/8" = 1'-0"	1:96	1:100
1/4" = 1'-0"	1:48	1:50
1/2" = 1'-0"	1:24	1:25

Ceiling tile measurement conversions are:

Inch-Foot Scale	Metric
2' × 2'	600 mm × 600 mm
2' × 4'	600 mm × 1200 mm

Pipe sizes:

Nominal Pipe Size	Metric	Nominal Pipe Size	Metric
1/8"	6 mm	3/4"	20 mm
3/16"	7 mm	1"	25 mm
1/4"	8 mm	1-1/4"	32 mm
3/8"	10 mm	1-1/2"	40 mm
1/2"	15 mm	2"	50 mm
5/8"	18 mm	2-1/2"	65 mm
3"	80 mm	6"	150 mm

Nominal Pipe Size	Metric	Nominal Pipe Size	Metric
3-1/2"	90 mm	8"	200 mm
4"	100 mm	10"	250 mm
5"	125 mm	12"	300 mm

For larger pipe, use 1 inch equals 25 mm. Other metric conversions can be found in Appendix A.

Significant Figures and Rounding

Most fire protection calculations are performed with no more than two significant figures to the right of the decimal point. A significant figure is any digit of a number beginning with the leftmost non-zero digit and ending with the rightmost non-zero digit. As an example, the number 20.205 has five significant figures, and the number 0.000514 has three significant figures. Where greater levels of precision are appropriate, the recommended number of significant figures will be displayed in examples and sample problems.

If the value to be rounded is 5 or more, it is rounded up. If the value to be rounded is 4 or less, it is rounded down.

THE NATIONAL FIRE PROTECTION ASSOCIATION

codes
mandatory requirement suitable for adoption into law

standards
mandatory NFPA requirements that may be used by authorities to approve a fire protection system

recommended practices
NFPA documents that provide nonmandatory advice

guides
informative but nonbinding NFPA documents

The National Fire Protection Association (NFPA) publishes more than 290 **codes**, **standards**, **recommended practices**, and **guides** that apply to fire safety and the design of fire protection systems. For each standard published by the NFPA, a committee of volunteers, appointed by the NFPA, creates, modifies, and revises the standard. The committee membership is balanced to include categories such as special expert, system user, insurance, research, installer, manufacturer, laborer, consumer, and enforcer. Each committee is assigned a full-time employee of the NFPA to serve as a nonvoting liaison to the committee. A chairperson and secretary are selected for the committee, and voting members, or principal members are permitted to have alternate members serve in their place, provided that an application as an alternate member is submitted and accepted.

The NFPA creates codes, standards, recommended practices, and guides for use by fire protection professionals. A code sets forth mandatory requirements that are suitable for adoption into law. A standard consists of mandatory requirements that a designated authority may use to approve a fire protection system. A recommended practice is a document that provides nonmandatory advice. Guides are informative, but nonbinding, NFPA documents.

Each NFPA document follows a standardized format. The document begins with an overview of the publication history of the document, a list of committee members, and a table of contents. The body of the document consists of statements of scope and purpose for the document, a list of definitions, and detailed technical requirements.

The NFPA system of democratic committee activity and document publication is a marvel for its efficiency and its ability to produce consensus documents that are capable of being adopted readily by local governments and approving authorities. Readers are strongly encouraged to join the NFPA and become a part of the standards-making process.

ORAL, WRITTEN, AND GRAPHIC COMMUNICATION

Students taking courses in a fire protection curriculum may wonder why they may be asked or required to take courses in writing, speech making, and literature. Success as a fire protection professional depends to a large extent on the ability to clearly communicate complex ideas to a wide audience. Communications skills include written communication, oral communication, and graphic communication, each of which is equally vital and important.

For a fire protection professional to achieve effective oral, written, and graphic communication, he or she must receive specific training. All fire protection professionals are urged to seek specific instruction on technical writing and effective oral communication and practice these skills at every opportunity. Further, we recommend that those reading this book complete as many of the drawing exercises as possible and seek advanced instruction on fire protection system design.

GRAPHIC COMMUNICATION: DRAWING FIRE PROTECTION SYSTEMS

Graphic communication is conveyed through drawings. Although manual drafting is still performed, it is becoming obsolete, in favor of computer-aided design (CAD). Graphic communication also involves scale and the drawing medium, plus a cover sheet, discussed below.

The Blank Page and the Design Process

drawing
graphic representation of a designer's ideas

A **drawing** is a graphical representation of a designer's ideas. Drawings are needed because attempts to verbally explain a mental idea can become cumbersome, and for some of the more complex designs, impossible.

Fire protection professionals who lack the ability to represent their ideas graphically may be in for a frustrating career with limited options. This book introduces graphic concepts to help fire protection professionals become more well-rounded and effective through the ability to represent ideas in a clearly understood graphic form and to be able to clearly understand the drawings of others.

Technical drawing is a unique and fascinating form of communication, partly an art form and partly a technical media. The skillful combination of these two seemingly opposite concepts is fundamental to the success of a technician, an engineer, or any fire protection professional. For those who have mastered the art and science of special hazard and fire alarm design and layout, success is ensured—not to mention pure enjoyment in this form of communication.

For the novice designer, few concepts may be more intimidating than starting a drawing with a blank page. With the ideas in your head, you are facing a large piece of blank drawing medium, an enormous book of specifications, and a huge roll of drawings.

Before starting a shop drawing, all contract drawings and specifications have to be reviewed. This chapter will help you to extract the necessary information from these references and obtain the needed information from a fire protection survey. The subsequent technical chapters will give you a method for system design and calculation. If possible, seek out a mentor, an instructor, or a person experienced in fire protection system design who can help you organize your thoughts and your reference material. To begin, a simple drawing is suggested, as a prelude for more complex designs.

Regardless of your experience, the technical information on the drawing is of little value if it cannot be read. Therefore, an essential piece of advice is to be as neat and as organized as possible on your first drawings. Your presentation style will improve with subsequent drawings.

Even though starting a first design may be somewhat intimidating, you will be elated after completing the design. The challenge of filling a blank sheet with your ideas and representing them in a clear, usable fashion is one of the most cherished rewards for a fire protection professional. Completing your first drawing will generate the confidence to do more difficult designs, so hang in there, and you will succeed!

Computer-Aided Design (CAD)

computer-aided design (CAD)

a computerized method of preparing drawings

Computer-aided design (CAD), a computerized method of preparing drawings, has become the dominant methodology in special hazard and fire alarm drawing. The fire protection and detection system design process features CAD at every juncture. This process, shown in **Figure 1-1**, develops a design from a conception phase, through a design phase, to a shop drawing phase, to an as-built phase. As you go along, you may discover an error or a better or more efficient way of presenting your ideas. Do not be afraid of changing your drawing. Revisions in CAD are much easier than in manually drafted designs.

Because most fire protection firms, architectural offices, and engineering offices have converted from manual drafting to computer-aided design, and specifications commonly require CAD for fire protection system design. All fire protection professionals must become acquainted with CAD before graduation, or as early in their careers as possible. Most community colleges offer courses in CAD, and it

CONCEPTION PHASE (CAD)	DESIGN PHASE (CAD)	SHOP DRAWING LAYOUT PHASE (CAD)	AS-BUILT PHASE (CAD)
• PRELIMINARY DESIGN BY FIRE PROTECTION ENGINEER • OWNER APPROVAL OF CONCEPT	• COMPLETED CONTRACT DRAWING BY FIRE PROTECTION ENGINEER • OWNER APPROVAL OF DESIGN • DISTRIBUTION OF DRAWINGS AND SPECS TO CONTRACTORS FOR BID	• COMPLETED SHOP DRAWING BY TECHNICIAN • APPROVAL BY AHJ • APPROVAL BY FIRE PROTECTION ENGINEER • INSTALLATION OF SYSTEMS • APPROVAL OF INSTALLATION BY AHJ	• FIELD CHANGES RECORDED ON SHOP DRAWINGS • FIRE PROTECTION ENGINEER APPROVES INSTALLATION • OWNER RECEIVES AS-BUILT DRAWINGS FOR PERMANENT RECORDS

Figure 1-1
Computer-aided design process for fire protection drawings.

Figure 1-2   *A fire protection engineer using CAD to design an alarm and detection system. (Photo courtesy of Tyler Mosman, PE.)*

is recommended that all fire protection professionals register for this instruction. CAD programs are available for use on the personal computer (see **Figure 1-2**).

CAD originally was developed in the 1960s, and expanded into commercial use as a two-dimensional drafting medium in the 1970s. In subsequent iterations it evolved into a three-dimensional multifaceted presentation. The use of CAD requires specialized software and compatible hardware. The software may be selected from a variety of commercially available computer programs and is loaded onto the hard disk drive in the computer. Some CAD software is created especially for fire protection system design and contains the appropriate symbols and design tools needed, including hydraulic calculation software on certain CAD programs.

The hardware required consists of a computer with sufficient speed, memory, and mathematics capability, a terminal screen, and a keyboard. Input of information may be facilitated with a digitizer and puck, which serves as a drawing tablet. Another handy tool for data input is a scanner, which transfers a print from a drawing into the computer. A drawing that has already been completed on CAD can be loaded onto a hard drive to serve as a background for fire protection system design.

Output of a completed design is performed with a CAD plotter, shown on **Figure 1-3**, which draws what you have designed onto a reproducible medium. If you don't have a plotter, you can take your CAD disk to a copying shop for plotting.

Figure 1-3   *A fire protection engineer plots out a completed suppression and detection system design on a CAD plotter. (Photo courtesy of Tyler Mosman, PE.)*

CAD has many advantages over manual drafting. Although some manual designs are quite attractive, CAD is neater than most manual designs, and because most CAD software adds plan dimensions, it potentially is more accurate than manual drafting. Making revisions and corrections is considerably easier with CAD, and disk storage is more efficient and space-saving than the plan file storage rooms that once housed manual designs.

People who hire fire protection professionals expect a solid background in fire protection system design. With increasing regularity, they are looking for those who have experience with CAD.

Scales

Fire protection system drawings are drawn to scale. For this reason, at least one scale should be purchased. An architect's scale, shown in **Figure 1-4**, is

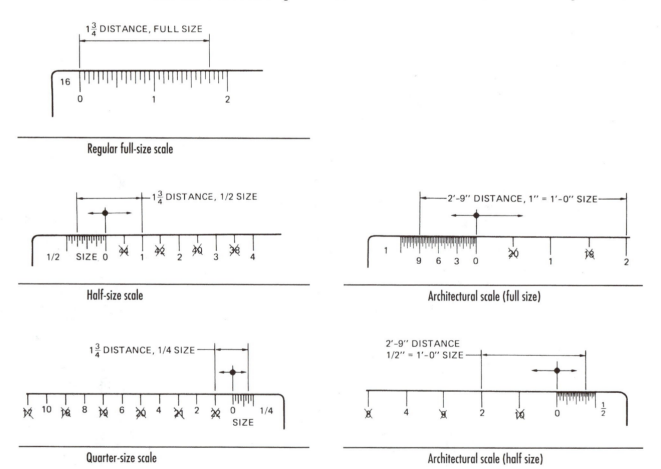

Figure 1-4   *An architect's scale is needed for design of fire protection systems.*

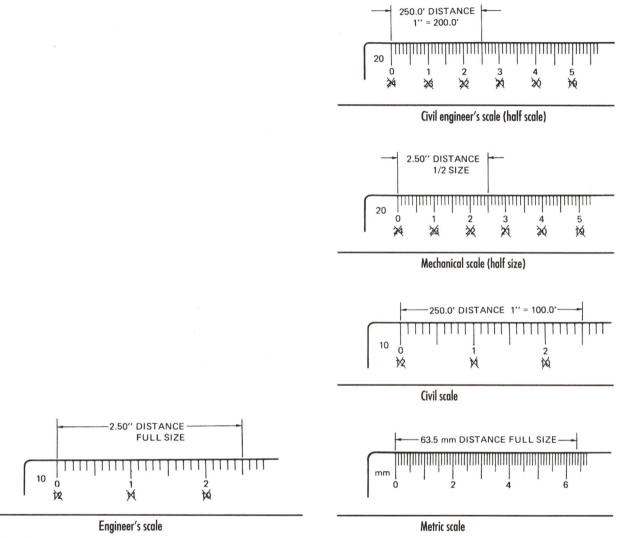

Figure 1-5   *An engineer's scale is required for underground piping drawings.*

needed for the design of special hazards and fire alarm systems, and an engineer's scale, shown in **Figure 1-5**, is needed for underground fire protection piping design, which is necessary when designing water supplies for low- and high-expansion foam systems, foam-water systems, and ultra-high-speed deluge explosion suppression systems. These scales can be purchased for well under ten dollars each, and plastic scales are much less expensive than wooden scales.

Print Media

The drawing medium is the reproducible surface on which a drawing exists. A large piece of plain white paper may suffice, but a blueprint of this medium cannot be made effectively because it is not translucent. Some copying shops have the equipment to make large photocopies, but this process usually is four or more times as expensive as a blueprint.

Translucent media allow light to shine through—the process used in making a blueprint. Translucent media include Mylar, vellum, and onion skin, with Mylar being the thickest and onion skin being the thinnest. CAD media are electronic and can be reproduced by e-mailing or sending a CD-ROM to an office supply or design reproduction firm.

Drawing Arrangement

The size of the print medium you select must be larger than the size of the drawing you are printing, to allow room for notes and a title block. A common drawing size is 30" × 42" (70.2 cm × 106.68 cm), but it is a good idea to coordinate with the owner or general contractor to determine whether a specific drawing size is specified for a project.

The building or object is oriented to the upper left-hand corner of the drawing, allowing sufficient space for dimensions, which are drawn outside of the building walls. Additional floors or areas, if space permits, may be drawn below the floor already drawn, as shown in **Figure 1-6**. If the building is too large to fit on your drawing, break the building into sections or use a smaller scale. Drawings utilizing a scale smaller than 1/8" = 1'-0" should not be used for fire protection.

A title block is added to the lower right-hand corner of the drawing. Using a format similar to **Figure 1-7** should suffice for most design projects. Feel free to use your creativity to develop a title block with which you are comfortable.

Contract Drawings Cover Sheet

The cover sheet of a contract drawing contains a wealth of information for the fire protection designer. The location map—a large-scale drawing showing roads and intersections—helps you get to the job site. Also, it can, if scanned onto a special hazard or fire alarm drawing, speed plans through the approval and permit process by assisting the authority having jurisdiction in referencing the building during the permit process.

NFPA

NFPA 170, Standard for Fire Safety Symbols

A symbol and abbreviation index serves as a reference in your interpretation and use of the drawings contained within the contract drawing set. NFPA 170, *Standard for Fire Safety Symbols,* should be used as the basis for symbols used in the design of fire protection systems.

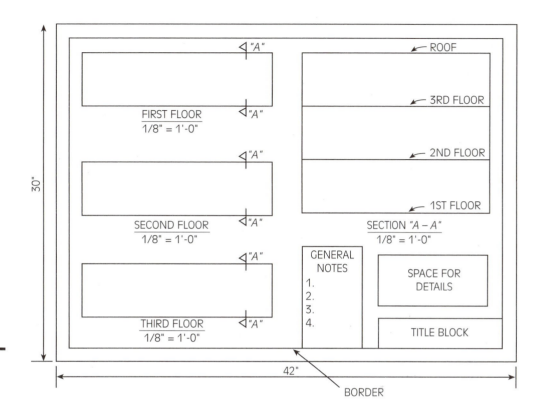

Figure 1-6 *General drawing arrangement.*

Figure 1-7 *Sample title block.*

civil drawings
scaled drawings that coordinate underground utilities entering and leaving a building or group of buildings

contour lines
lines indicating the elevation of the finished exterior grade

plan reference elevation
sea level, or an elevation chosen specifically for a building or group of buildings

reference grid
a system of parallel reference lines aligned either to magnetic north or to plant north

depth of cover
lineal distance from the top of an underground pipe to the finished grade

profile plan
a plan that specifically shows reference elevations with respect to finished grade

invert elevation
an elevation referencing the bottom of a pipe with respect to the reference elevation

boring table
compilation of data obtained by drilling cores of earth at several strategic locations

SITE (UNDERGROUND) DRAWINGS

Water-based special hazard systems, such as low-expansion foam systems, high expansion foam systems, and foam-water systems, require a water supply. This often is supplied by an underground piping system.

Site plans, often called **civil drawings**, sometimes designated as drawing C-1, C-2, C-3, etc., are scaled drawings used to coordinate underground utilities entering and leaving a building or group of buildings. Shown on this plan are existing underground and above ground utilities, usually shown as dotted lines, and proposed new utilities, shown with solid lines.

Civil drawings are uniquely characterized by a series of lines, called **contour lines**, which indicate the elevation of the finished exterior grade with respect to a **plan reference elevation**.

The elevation of reference could be sea level, or a reference specifically chosen for a building or group of buildings (elevation 100'-0" is commonly used as a reference elevation).

Site plans also can be referenced to a **reference grid**. Reference grids provide a system of parallel reference lines, aligned either to magnetic north or to plant north. The use of reference grids is an issue of convenience for the architect or engineer, allowing for buildings to be oriented uniformly throughout the contract drawing set.

Site plans are used to graphically show the **depth of cover** for a pipe, the depth of an underground pipe below finished grade. A plan that specifically shows reference elevations with respect to finished grade is called a **profile plan**. A profile plan of a fire protection underground pipe, shown in **Figure 1-8**, shows the entire pipe, from source to building entry, in a sectional view. A profile plan resembles a graph, with the Y axis representing elevation (in feet) and the X axis representing horizontal travel of pipe (in feet). An underground pipe elevation is identified as an **invert elevation**, an elevation referencing the bottom of the pipe with respect to the reference elevation.

Another feature of site plans is the **boring table**. These tables display data obtained by drilling cores of earth at several strategic locations. A fire protection designer refers to these tables to determine the composition of the earth through which an underground fire protection pipe must travel. A bore that yields solid rock indicates an area to be avoided with respect to placement of an underground fire protection pipe. A bore that yields very loose sand may indicate an area that is unsuitable for placement of underground piping without special backfill.

ARCHITECTURAL PLANS

Architectural drawings are identified by sequential drawing numbers preceded by an A, such as A-1 and A-2, and show dimensions of walls, floors, ceilings, and other building features. These drawings can vary widely in the level of detail.

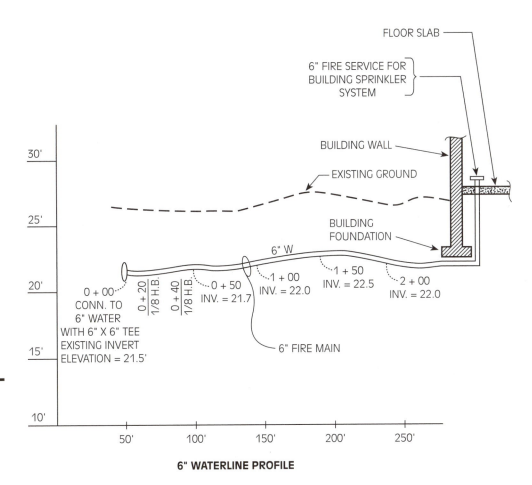

Figure 1-8 *Profile plan for a water supply to a low- or high-expansion foam system or a foam-water system.*

6" WATERLINE PROFILE

architectural drawings
drawings that show dimensions of walls, floors, ceilings, and other building features

plan job
design performed using new architectural plans as the basis for design

survey job
a project involving an existing building for which plans cannot be obtained

Architectural drawings ordinarily are fully dimensioned plans that are usable for special hazard and fire alarm design and involve little or no survey work. When a fire protection designer has a complete set of well-dimensioned architectural drawings, the design can be done entirely using the plans for reference. In the fire protection industry this procedure commonly is called a **plan job**. A project involving an existing building for which a complete set of fully dimensioned plans cannot be obtained is called a **survey job**.

Cutaway views through a building included within the architectural set are called **building sections**. A building section, chosen thoughtfully to represent an area involving unusual conditions, most likely can be traced or scanned directly from an architectural elevation detail. Fire protection piping then can be added to the section. A sectional view is required for each unusual condition, to aid the plan reviewer in understanding the proposed design.

Plan views of a ceiling, usually found in the architectural section of a contract drawing package when suspended ceilings are installed, are called

building sections
cutaway views through a building

reflected ceiling plans
plan views of suspended ceilings

finish schedule
a contract drawing that lists all rooms in a building and provides details of several room features

structural drawings
drawings that provide details related to the floors, roof, and structural elements of a building

foundation plans
plans that show floor and wall structural details and sectional views

I-beam
a solid steel member that looks like an "I" from its end

bar joist
a webbed member supported by I-beams

framing plans
plans that show beam and joist size and elevation

reflected ceiling plans. An architectural drawing that lists all rooms in a building by room number and provides details of room features is called a **finish schedule**. Design details related to the floors, walls, and ceilings are included for each room.

STRUCTURAL DRAWINGS

For buildings without suspended ceilings, **structural drawings**, which provide details related to the floors, roof, and structural elements, should be the central reference for fire protection design. The structural set consists of foundation drawings and framing plans, identified with an "S" preceding the drawing number (as S-1).

Foundation plans show finished floor structural details with which one must coordinate when designing underground entries into the building. For multistory buildings, the foundation plans detail the floor slab thicknesses and show building elevations that may contain details not shown on architectural elevations.

An **I-beam** is a solid steel structural member that looks like an "I" from its end. A **bar joist** is a webbed member supported by the I-beams, as shown in **Figure 1-9**. **Framing plans** show I-beam and bar joist size and elevation. The tops of I-beams are referenced to a baseline elevation, such as finished floor. Sloping of I-beams can be deduced by noting change in the elevation of the tops of the beams, noted on the framing plans.

Because bar joists rest on top of the I-beams, a **bar joist bearing dimension**, the depth of the top of a joist, must be ascertained for the elevation of the top of the bar joist to be determined. This information can be found either in the structural general notes or on structural sectional details. The elevation of the top of the bar joist is important to the fire protection designer because it is from this point that piping is predominately supported.

Other information obtained from the structural drawings includes the depth and location of **cross bracing**. This bracing provides structural rigidity between bar joists, to counteract wind load or earthquake seismic load.

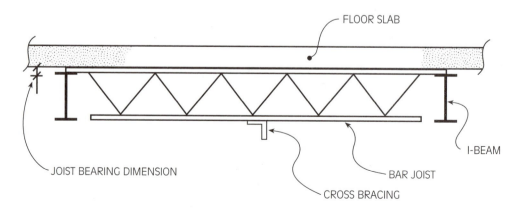

Figure 1-9 *Partial building sectional view showing structural components.*

**bar joist bearing
dimension**
depth of the top of
a joist

Foam-water systems, low-expansion foam systems, and other gaseous, particulate, or water-based special hazard systems may be arranged similarly to the layout shown on **Figure 1-10**.

For projects that contain few suspended ceilings where the piping is to be installed primarily exposed, the designer should trace the framing plan and use it

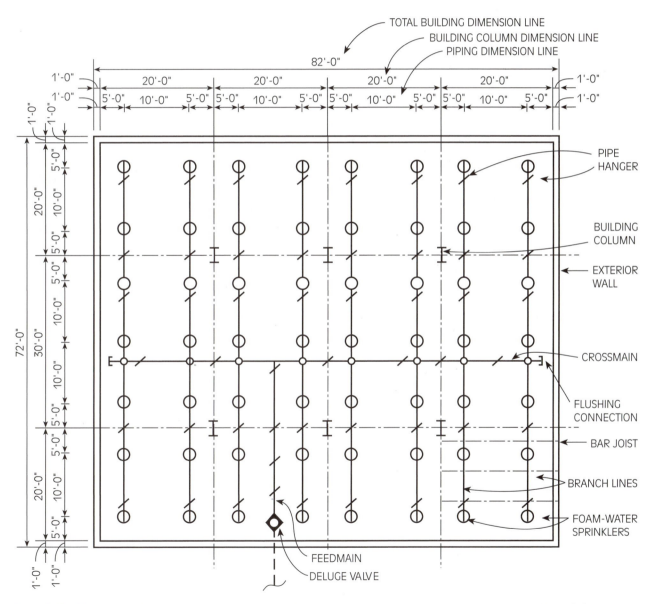

Figure 1-10   *Plan view showing dimensioning guidelines for a foam-water fire protection system; branch lines always run perpendicular to bar joists.*

cross bracing
supports that provide structural rigidity between bar joists

branch lines
sprinkler pipes that have sprinklers or discharge devices installed directly on them

as the basis for the fire protection layout. **Branch lines**, the pipes on which the discharge devices are installed directly, are oriented perpendicular to the bar joists for ease of hanging, as shown in Figure 1-10.

The centers of the building columns should be the points of reference for all fire protection piping dimensions. This advice pertains to both exposed construction and concealed construction with suspended ceilings. A good practice for fire protection designers is to have three lines of running dimensions in each direction, referenced to the structural members, as shown in Figure 1-10.

Before submitting the drawing for approval and again before releasing the drawing for fabrication, the three lines of dimensions are added until all dimensions are in total agreement.

HVAC DRAWINGS

HVAC drawings
heating, ventilating, and air conditioning drawings

There are two types of heating, ventilating, and air conditioning drawings, which are **HVAC drawings** identified with an "M" (for mechanical) drawing number: (1) conceptual HVAC drawings, and (2) shop HVAC drawings.

The most common type of HVAC plan, found in most contract drawing packages, is the conceptual HVAC drawing. These plans ordinarily are not dimensioned to show the exact location of HVAC ductwork with respect to the steel beams and columns but, rather, show the widths, depths and locations of the ducts.

The other type of HVAC plan is the shop drawing, or sheet metal drawing. HVAC designers use conceptual HVAC plans to derive fully dimensioned and elevated duct drawings, in which all ducts are dimensioned to the building columns and all supply and return diffusers are dimensioned. These plans may appear somewhat complicated, because they contain additional information that the conceptual HVAC plans do not show.

supply diffuser
ceiling element used to distribute fresh air to a room

return diffuser
ceiling element used to draw stale air from a room

Another important function of conceptual HVAC drawings is to show the locations of supply and return diffusers. A **supply diffuser**, connected to a system of supply ductwork, distributes fresh air to a room. A **return diffuser** draws stale air from the room.

plenum space
a space above a suspended ceiling that is kept under negative pressure for return air

Return diffusers are connected either to ductwork that draws air out of the rooms of a building and exhausts it outside, or they serve as return air grilles mounted to a ceiling to draw air into a **plenum space** above the ceiling. The plenum space is the volume of air encountered above the ceiling, maintained under negative pressure by a fan, and exhausted through the fan to the outside.

PLUMBING DRAWINGS

Designers should refer to the plumbing drawings, designated in the contract drawing set with a "P" prefix, because in some buildings, plumbing and fire protection piping involve a common underground feed pipe.

A fire protection system piping layout or schematic also may be shown on the plumbing drawings—perhaps in the case of a high-rise building requiring a standpipe system. Water supply and other fire protection-related information might be found in the plumbing general notes.

Preferably, fire protection and detection systems are portrayed in a separate fire protection set. These are described subsequently.

Buried plumbing pipe is installed in the floor slab. This piping most likely consists of floor drains or toilet and sewer piping. **Rain leaders** and **domestic plumbing piping** are exposed below the ceiling slab or concealed between a suspended ceiling and the ceiling slab. Rain leaders are rainwater drainage piping, and domestic plumbing piping is a water supply for drinking water and toilets.

rain leaders
rainwater drainage piping

domestic plumbing piping
piping that supplies water fountains, sinks, and toilets

FIRE PROTECTION CONTRACT DRAWINGS

The fire protection profession has a strong preference for displaying all fire protection and detection requirements on a separate set of contract drawings. Information that can be obtained might be a riser schematic and a floor plan containing the requirements for fire protection piping, as well as all details that a special hazard or fire alarm designer will need to meet the performance objectives of the systems.

ELECTRICAL DRAWINGS

Information on electrical drawings, designated within the contract drawing set with an "E" prefix, are useful to special hazard and fire alarm system designers because they provide the location and power requirements of lighting and other electrical devices.

Flush-mounted lights, mounted flush to the ceiling, could interfere with the fire protection agent distribution below the ceiling. **Recessed fixtures**, whose faces are flat to the ceiling but protrude into the ceiling space, could interfere with fire protection system piping. **Suspended fixtures**, hanging from rods or chains, could be a source of conflict with the piping or agent distribution, depending on the elevation and dimension of the fixture.

flush-mounted lights
lights mounted below and flush to the ceiling

recessed fixtures
lights whose faces are flat to the ceiling and whose bodies protrude into the ceiling space

suspended fixtures
lights hanging from rods or chains

water-flow switches
electronic devices that indicate water movement in a sprinkler pipe

tamper switches
electronic devices that indicate a sprinkler valve closure

solenoids
switches used to open and close electrically actuated fire protection valves

water-level switches
devices that indicate the level of water in a fire protection water storage tank

pressure switches
electrical devices that indicate high or low pressure in a water tank or in a sprinkler system

Locations of other electrical devices also can be shown on electrical drawings. Included might be some equipment associated with a fire alarm system, such as **water-flow switches** that give electrical indications of water movement in a sprinkler pipe, **tamper switches** that give electrical indications of a sprinkler valve closure, **solenoids** that are used to open and close electrically actuated fire protection valves, **water-level switches** that indicate the level of water in a fire protection water storage tank, and **pressure switches** that give electrical indications of high or low pressure in a water tank or a foam concentrate storage tank. Pressure switches also provide an electronic indication of agent flow in a special hazards suppression system.

Coordinating the power availabilities shown on the plans and the power requirements of electrical fire protection devices is extremely important. If a fire pump is required to boost the pressure of a water supply for a water-based special hazard system, it must be ordered in a way that ensures congruence of the power to the controller and motor. Other fire protection devices, such as fire alarm control units, air compressors, flow and tamper switches, and alarm devices, require the same attention to coordination to ensure applicability to a fire protection system in a given building.

Electrical riser diagrams may detail the locations of electrical devices associated with a fire alarm system, especially in high-rise buildings.

SPECIFICATIONS

Architects and engineers develop specifications from either a standardized computer specification database or from a revision to the specifications developed for a previous project of similar scope. The revision process sometimes results in specifications for a project that may conflict with the contract drawings.

Contract specifications are a functional component of the contract documents, no less important than the contract drawings. In most cases the specifications prevail when a conflict arises between the contract drawings and the specifications, even when an extremely detailed set of contract drawings is compared to a broadly general set of specifications. To prevent major changes, the best approach is to discover and correct discrepancies before a system is designed and installed.

A specification writer selects one of two methods to develop specifications for a fire protection system: a performance specification or a detailed specification. A **performance specification** is a general specification that provides the minimum information necessary to estimate, design, and install a fire protection system. A performance specification contains broad requirements that offer few restrictions, allowing for economical or innovative design methods that comply with the applicable codes and standards.

A performance specification addresses all aspects of the design and installation of a suppression or detection system. On the one hand, it can provide requirements

performance specification
a general specification that provides the minimum information necessary to estimate, design, and install a fire protection system

for coordination of systems without mandating the use of a specific manufacturer or design methodology. On the other hand, a simple reference to "design in accordance with a standard" is an insufficient basis for any specification. Engineers developing a specification must, at minimum, ensure that recent water flow test data are provided, must ensure that occupancies are properly selected and identified, must ensure that the appropriate densities or delivery requirements of the special hazards system are stated, and must ensure that new systems will be an effective protection metholodology.

A **detailed specification** provides in-depth requirements for the design of a fire protection system, allowing little latitude for interpretation or implementation of alternative design proposals. A specification writer who selects this method may see a heightened fire scenario or a specific application that a performance specification may not address adequately.

A detailed specification may exclude or prohibit the use of newer or innovative piping or sprinkler technologies. The sprinkler contractor should propose any variations from the fire protection concept outlined in a detailed specification in advance of submission of a cost estimate for the system.

Among architects and engineers there is a growing consensus that the performance-based specification is preferred for new buildings. Many building owners find that performance specifications have the potential to provide a less expensive fire protection system when detailed requirements are removed and the latest technologies permitted.

Specification Format

Most specifications follow a format standardized by the Construction Specifications Institute (CSI), consisting of **specification divisions**, broad categories of building component groupings. Each division consists of **specification sections** that outline the detailed requirements of each division. The divisions provide a road map for building a structure and its internal components to meet the needs of the owner and to comply with the applicable codes and standards. An outline of the CSI 2004 division list is as follows:

detailed specification
in-depth requirements for the design of a fire protection system that allow little latitude for interpretation or alternative design proposals

specification divisions
broad categories of building component groupings standardized by the Construction Specifications Institute

specification sections
detailed requirements for each CSI division

SPECIFICATION DIVISION NUMBERS AND TITLES

Procurement and Contracting Requirements Group
  Division 00   Procurement and Contracting Requirements
Specifications Group
  General Requirements Subgroup
    Division 01   General Requirements

Facility Construction Subgroup

Division 02   Existing Conditions

Division 03   Concrete

Division 04   Masonry

Division 05   Metals

Division 06   Wood, Plastics, and Composites

Division 07   Thermal and Moisture Protection

Division 08   Openings

Division 09   Finishes

Division 10   Specialties

Division 11   Equipment

Division 12   Furnishings

Division 13   Special Construction

Division 14   Conveying Equipment

Division 15   Reserved

Division 16   Reserved

Division 17   Reserved

Division 18   Reserved

Division 19   Reserved

Facility Services Subgroup

Division 20   Reserved

Division 21   Fire Suppression—See detailed section listing for special hazards suppression systems.

Division 22   Plumbing

Division 23   Heating, Ventilating, and Air Conditioning

Division 24   Reserved

Division 25   Integrated Automation

Division 26   Electrical

Division 27   Communications

Division 28   Electronic Safety and Security—See detailed section listing for fire alarm systems.

Division 29   Reserved

Site and Infrastructure Subgroup

Division 30   Reserved

Division 31   Earthwork

Division 32   Exterior Improvements

Division 33   Utilities

Division 34   Transportation

Division 35   Waterway and Marine Construction

Division 36   Reserved

Division 37   Reserved

Division 38   Reserved

Division 39   Reserved

Process Equipment Subgroup

Division 40   Process Integration

Division 41   Material Processing and Handling Equipment

Division 42   Process Heating, Cooling, and Drying Equipment

Division 43   Process Gas and Liquid Handling, Purification, and Storage Equipment

Division 44   Pollution Control Equipment

Division 45   Industry-Specific Manufacturing Equipment

Division 46   Reserved

Division 47   Reserved

Division 48   Electrical Power Generation

Division 49   Reserved

Although all divisions may not apply and may not be included in a specification for a proposed building, most divisions of a specification contain information of interest to the special hazard or fire alarm designer or contractor.

Fire Suppression Specification Sections

The standardized specification format has been revised to accommodate fire protection system and fire alarm system specifications within Division 21, which represents a significant departure from previous practice. Before this change, fire suppression systems were specified in Division 15, Mechanical, and fire alarm systems were detailed in Division 16, Electrical.

By moving these two separate specification provisions into Division 21, section 21 20 00, "Fire Extinguishing Systems," and section 28 31 00, "Electronic Detection and Alarm," better coordination and reduced conflict will result by perhaps having one contractor perform or be responsible for both suppression and detection services, or by having two contractors subcontracted by the same general contractor or mechanical contractor. Under the previous format, coordination of fire protection components was considerably more difficult when the special hazard contractor reported to a mechanical contractor under the old Division 15 and the fire alarm contractor reported to the electrical contractor under the former Division 16.

A complete listing of fire suppression sections that may be of interest to a special hazards designer under the 2004 CSI specifications format follows:

Division 21—Fire Suppression Sections

21 00 00   Fire Suppression

    21 01 00   Operation and Maintenance of Fire Suppression

        21 01 10   Operation and Maintenance of Water-Based Fire-Suppression Systems

        21 01 20   Operation and Maintenance of Fire-Extinguishing Systems

        21 01 30   Operation and Maintenance of Fire-Suppression Equipment

    21 05 00   Common Work Results for Fire Suppression

        21 05 13   Common Motor Requirements for Fire-Suppression Equipment

        21 05 16   Expansion Fittings and Loops for Fire-Suppression Piping

        21 05 19   Meters and Gages for Fire-Suppression Systems

        21 05 23   General-Duty Valves for Water-Based Fire-Suppression Piping

        21 05 29   Hangers and Supports for Fire-Suppression Piping and Equipment

        21 05 33   Heat Tracing for Fire-Suppression Piping

        21 05 48   Vibration and Seismic Controls for Fire-Suppression Piping and Equipment

        21 05 53   Identification for Fire-Suppression Piping and Equipment

    21 06 00   Schedules for Fire Suppression

        21 06 10   Schedules for Water-Based Fire-Suppression Systems

        21 06 20   Schedules for Fire-Extinguishing Systems

        21 06 30   Schedules for Fire-Suppression Equipment

    21 07 00   Fire Suppression Systems Insulation

        21 07 16   Fire-Suppression Equipment Insulation

        21 07 19   Fire-Suppression Piping Insulation

    21 08 00   Commissioning of Fire Suppression

    21 09 00   Instrumentation and Control for Fire-Suppression Systems

21 10 00   Water-Based Fire-Suppression Systems

   21 11 00   Facility Fire-Suppression Water-Service Piping

      21 11 16   Facility Fire Hydrants

      21 11 19   Fire-Department Connections

   21 12 00   Fire-Suppression Standpipes

      21 12 13   Fire-Suppression Hoses and Nozzles

      21 12 16   Fire-Suppression Hose Reels

      21 12 19   Fire-Suppression Hose Racks

      21 12 23   Fire-Suppression Hose Valves

      21 12 26   Fire-Suppression Valve and Hose Cabinets

   21 13 00   Fire-Suppression Sprinkler Systems

      21 13 13   Wet-Pipe Sprinkler Systems

      21 13 16   Dry-Pipe Sprinkler Systems

      21 13 19   Preaction Sprinkler Systems

      21 13 23   Combined Dry-Pipe and Preaction Sprinkler Systems

      21 13 26   Deluge Fire-Suppression Sprinkler Systems

      21 13 29   Water Spray Fixed Systems

      21 13 36   Antifreeze Sprinkler Systems

      21 13 39   Foam-Water Systems

21 20 00   Fire-Extinguishing Systems

   21 21 00   Carbon-Dioxide Fire-Extinguishing Systems

      21 21 13   Carbon-Dioxide Fire-Extinguishing Piping

      21 21 16   Carbon-Dioxide Fire-Extinguishing Equipment

   21 22 00   Clean-Agent Fire-Extinguishing Systems

      21 22 13   Clean-Agent Fire-Extinguishing Piping

      21 22 16   Clean-Agent Fire-Extinguishing Equipment

   21 23 00   Wet-Chemical Fire-Extinguishing Systems

      21 23 13   Wet-Chemical Fire-Extinguishing Piping

      21 23 16   Wet-Chemical Fire-Extinguishing Equipment

   21 24 00   Dry-Chemical Fire-Extinguishing Systems

      21 24 13   Dry-Chemical Fire-Extinguishing Piping

      21 24 16   Dry-Chemical Fire-Extinguishing Equipment

21 30 00   Fire Pumps

   21 31 00   Centrifugal Fire Pumps

      21 31 13   Electric-Drive, Centrifugal Fire Pumps

      21 31 16   Diesel-Drive, Centrifugal Fire Pumps

21 32 00   Vertical-Turbine Fire Pumps

21 32 13   Electric-Drive, Vertical-Turbine Fire Pumps

21 32 16   Diesel-Drive, Vertical-Turbine Fire Pumps

21 33 00   Positive-Displacement Fire Pumps

21 33 13   Electric-Drive, Positive-Displacement Fire Pumps

21 33 16   Diesel-Drive, Positive-Displacement Fire Pumps

21 40 00   Fire-Suppression Water Storage

21 41 00   Storage Tanks for Fire-Suppression Water

21 41 13   Pressurized Storage Tanks for Fire-Suppression Water

21 41 16   Elevated Storage Tanks for Fire-Suppression Water

21 41 19   Roof-Mounted Storage Tanks for Fire-Suppression Water

21 41 23   Ground Suction Storage Tanks for Fire-Suppression Water

21 41 26   Underground Storage Tanks for Fire-Suppression Water

21 41 29   Storage Tanks for Fire-Suppression Water Additives

Fire Alarm Specification Sections

Fire alarm designers use CSI (2004) Section 28 31 00.

28 30 00   Electronic Detection and Alarm

28 31 00   Fire Detection and Alarm

28 31 13   Fire Detection and Alarm Control, GUI, and Logic Systems

28 31 23   Fire Detection and Alarm Annunciation Panels and Fire Stations

28 31 33   Fire Detection and Alarm Interfaces

28 31 33.13   Fire Detection and Alarm Interfaces to Remote Monitoring

28 31 33.16   Fire Detection and Alarm Interfaces to Access Control Hardware

28 31 33.23   Fire Detection and Alarm Interfaces to Access Control System

28 31 33.26   Fire Detection and Alarm Interfaces to Intrusion Detection

28 31 33.33   Fire Detection and Alarm Interfaces to Video Surveillance

28 31 33.43   Fire Detection and Alarm Interfaces to Elevator Control

28 31 43   Fire Detection Sensors

28 31 46   Smoke Detection Sensors

28 31 49   Carbon-Monoxide Detection Sensors

28 31 53   Fire Alarm Initiating Devices
    28 31 53.13   Fire Alarm Pull Stations
    28 31 53.23   Fire Alarm Level Detectors Switches
    28 31 53.33   Fire Alarm Flow Switches
    28 31 53.43   Fire Alarm Pressure Sensors
28 31 63   Fire Alarm Integrated Audio Visual Evacuation Systems
    28 31 63.13   Fire Alarm Horns and Strobes

FIRE PROTECTION SURVEYS

survey
a thorough investigation of a building and its components for the purpose of taking detailed measurements of the building to serve as reference for a fire protection drawing

In the course of our jobs as fire service officials, special hazard and fire alarm layout technicians, and fire protection engineers, we often are called on to conduct a **survey** of an existing facility and to generate an original drawing that clearly depicts the building conditions, fire protection features, and fire protection systems. A survey is a thorough investigation of a building and its components for the purpose of taking detailed measurements of the building that will serve as a reference for a fire protection drawing.

A fire protection designer performing a survey for the first time may need a methodical survey procedure to ensure a successful effort. To take some of the mystery out of an experience that at first can seem overwhelming, we break the big picture into smaller, more manageable parts below.

Get Your Bearings

Before taking measurements, you should consult some basic information that is required on your plans by the applicable NFPA standard. Start by obtaining the official name of the building, the juxtaposition of the building with respect to other buildings on the property and with other adjacent buildings, and the orientation of the building with respect to magnetic north.

Careful compiling of this sort of basic information will help to speed your plans through the permit process. It also is necessary to obtain the name, mailing address, and phone number of the building owner, the building tenant or tenants, and the owners' representative or contact person.

Determine the General Building Layout

point of contact
a person who meets you at the job site, shows you the area to be surveyed, and remains available if any questions arise

Planning before you arrive on the job site will save many frustrating hours of confusion. Be certain to make an appointment and to arrange for a **point of contact**—a person who meets you at the job site, shows you the area to be surveyed, and remains available if any questions arise. Having a point of contact is especially important if security procedures are involved. The point of contact ideally should be a person who is familiar with the provisions of the contract and with the portion of the building affected by the contract.

Preparations should be made for a guided tour of areas that are not in the immediate area of work but that affect your system, such as mechanical rooms containing fire protection control equipment. Additional assistance must be prearranged to open locked doors, lift ceiling tiles, and obtain security clearances.

Whatever time is necessary to gain a general understanding of the layout of the building will make the detailed measurements to follow more meaningful. Walk every area, taking note of the occupancy of each room. Develop a solid understanding of the limits of your contract.

Even if you were assured that no plans of the building are in existence, you still should diplomatically request a complete set of plans for the building. Many buildings maintain files, records, and drawings, which could be there for the asking. Any drawings that you are able to obtain, no matter how outdated or incomplete, will save you a considerable amount of survey time.

Bring Proper Survey Equipment

As illustrated in **Figure 1-11**, equipment found in every technician's or engineer's survey kit should include as a minimum:

- Two 8-foot (2.4 m) folding rules. With practice, two rules can be used to take quick elevations in the 8- to 12-foot (2.4 m to 3.66 m) range. Two rules are especially handy in the event one breaks.

Figure 1-11 *Survey equipment.*

Note: Top row left to right: 50-foot roll tape, two 8-foot folding rules, flashlight, sonic distance measuring device, survey bag. Middle row: 30-foot digital elevation measuring device. Bottom row: manual feet-and-inch adding device, electronic feet-and-inch adding device, safety glasses, pipe-diameter measuring gauge, string, crayon, ear protection, calculator, screwdriver, awl, notebook.

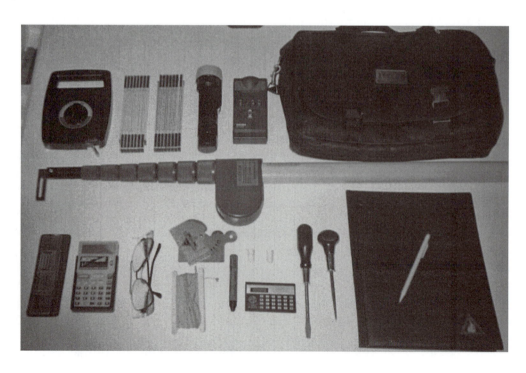

- A 50-foot (15.24 m) tape measure with a hook to attach to columns or square-edged walls. A 100-foot (30.48 m) tape is a highly recommended option. An infrared or ultrasonic measuring device, which sends signals to a specified target and determines the distance traveled by the soundwaves, can be of assistance for quick approximations or in unusually tight quarters.

- Two or more clamps to attach an end of a tape rule to a column or other reliable point of measure. These can turn a two-person survey into a one-person survey.

- A **pipe diameter measurement gauge**, used to determine the diameter of existing pipes. This is an essential tool, especially since 3½-inch (8.89 cm) pipe and 5-inch (12.7 cm) pipe usually are extremely difficult to identify using other measurement methods.

- A **telescoping elevation measurement pole**, capable of measuring elevations in the 30- and 50-foot (9.144 m and 15.24 m) range. Once you have such a device, it quickly becomes inconceivable that a survey could have been completed successfully without it. Operation of the elevation pole is illustrated in **Figure 1-12**.

- A feet-and-inch or metric manual adding device or electronic calculator. Such a device can reduce or eliminate measuring errors and could save you an extra survey trip.

- A flashlight for use in surveying areas above ceilings and other unlit areas.

- A string at least 50-feet (15.24 m) long with a weight on one end to use as a vertical frame of reference when it is suspended from a key point at the ceiling.

- Survey clothing, including personal protective equipment such as a hard hat, raincoat, safety eyeglasses, goggles, and ear protection. Long sleeves are necessary in most industrial plants, because chemicals can drip from pipes at the ceiling. Old or worn clothes are advised if the area surveyed is especially dirty. Steel-toed boots, hard hats, and safety glasses are required in most industrial buildings and buildings under construction.

pipe diameter measurement gauge
a device used to determine the diameter of an existing pipe into which you are tapping a new fire protection main

telescoping elevation measurement pole
a device used to measure elevations in the 30- to 50-foot range

survey strategy
the approach by which a survey begins—the layout of all structural elements, such as building columns and structural beams, as a reference for all subsequent measurements

Survey Building Details

This crucial step requires a sound **survey strategy**. In the approach that yields the most success, the survey begins with the layout of the centerlines of structural elements, such as building columns and structural beams. This establishes a building skeleton from which all other measurements can be taken.

After obtaining a comprehensive structural plan, walls can be measured using the column centerlines as a point of reference.

This phase concludes by recording the location of ceiling obstructions and potential conflicts with other existing mechanical facilities.

Figure 1-12 *A fire protection engineer uses a telescoping elevation measurement pole to determine structural elevations. (Photo courtesy of Tyler Mosman, PE.)*

Develop a System Design Strategy

Determine the most logical system arrangement, and develop a basic understanding of probable special hazard system piping locations or alarm and detection placement. This arrangement may change once layout begins, but a frame of reference has to be established before returning to your office, especially because field conditions may dictate the system arrangement chosen.

● **Caution**

Asbestos is a fibrous, usually white substance which, if inhaled may cause cancer. It usually is found in the form of insulation on old pipes or as a component of older floor tiles or shingles. As the first step, you must provide formal, written notification to the owner of the presence of asbestos before any plans are made to deal with this problem. Do not attempt to handle this material yourself.

survey reference elevation
the lowest, flattest, most reliable elevation that can be found

line-of-sight
a visual path above a ceiling created by removing ceiling tiles at regular intervals along a length or dimension of particular interest, such as a fire protection pipe, with the aid of a flashlight and a telescoping elevation pole

Building Elevations Your first action in the measurement of building elevations should be to determine a **survey reference elevation** from which all other elevations are measured. This point is the lowest, flattest, most reliable elevation that can be found. Finding a reliable reference elevation may not be a simple task in a ramped parking garage or in an extremely old building where sagging and settling has taken a considerable toll.

If all else fails, you may have to use a laser transit to establish usable reference elevations. Stairwells usually offer a reliable means of measuring floor-to-floor elevations in multistory buildings. During this phase, be certain to note potential mechanical or structural conflicts. A good tactic is to take the elevator to the highest floor and measure each floor subsequently, down to the ground floor, being careful to tie your dimensions into the reference elevation you have chosen.

Ceiling Measurements Many regularly spaced ceiling tiles can be surveyed accurately by measuring the end pieces adjacent to the walls and counting the tiles from wall to wall. Ceiling tile dimensions also must be tied into the building skeleton you drew in an earlier phase, to serve as a dimensional check in your survey. Carrying and using a feet-and-inch calculator, or its metric equivalent, during your survey may uncover problems or discrepancies before you return to the office and may make a resurvey unnecessary.

Surveying above a ceiling can be frustrating. A **line-of-sight** can be created by removing tiles at regular intervals along a dimension of particular interest, such as locating a fire protection pipe or a line of smoke detectors. A flashlight and telescoping elevation pole can be used to shoot a visual path along your line of sight. Extra tiles may require removal where potential conflicts are found.

Careful, respectful handling and replacement of the tiles, especially in the vicinity of others working below you, especially in a room where food is served, is essential. Whenever possible, have the owner find someone to handle the ceiling tile removal for you.

Water Supply Information Water-based special hazard suppression systems, such as low-expansion foam systems, high-expansion foam systems, foam-water systems, water mist systems, and ultra-high speed deluge suppression systems, require a reliable water supply.

Accurate surveying of the point of tie-in to an existing fire protection pipe is one of the more important phases of a survey. The measurements you take must be related to your building skeleton and your chosen reference elevation. Precise measurement of the pipe size into which you intend to tap is essential.

When tapping into an underground water main, obtain a drawing that accurately shows the size and location of the existing underground main. A strategy for installing the new underground pipe in the building must be laid out, preferably in coordination with the owner and trenching contractor. Choice of a location for entry into the building must be coordinated with the owner, and location of a fire department connection or foam system manual connection must be in accordance with NFPA standards and local codes.

Draw the Building and Lay Out the System

At this juncture, you are ready to return to your office and begin layout of the building and the fire protection system. Shortly after you begin your layout, you probably will begin to compile a list of items that were missed in your survey. Some things that seemed clear at the job site suddenly may seem mysterious on the CAD screen. This is to be expected, especially on your first survey or with large or complicated surveys. Clearly note areas of conflict or areas in question so they can be clarified on your return trip. Complete the drawing to the fullest extent possible during this phase.

Field-Check the Drawing

field check
a thorough survey
of a proposed fire
protection system
using a completed fire
protection system
design as a basis

Whatever you do, don't fail to field-check your drawing before installing the fire protection system. A **field check** is a thorough survey of a proposed fire protection system using a completed fire protection system design as a basis. Your responsibility at this juncture is to check each pipe on your layout and try to find errors. It is much better that you fail-test your design and discover conflicts and problems than for someone using your drawing to find them. An excellent idea is to perform your field check with the fire protection system installer providing assistance.

Begin your field check from the point of connection to each remote point by measuring piping locations, using your telescoping pole to establish the proposed elevation, then walk the full length of the main to identify conflicts. An engineer performing a field check is shown on Figure 1-12.

After finishing the field check, you are ready to return to your office to complete the layout and perform calculations. On large or complicated surveys, make an appointment for future visits before leaving.

Inspect the System after Installation

Visiting the job site during or after installation can be valuable, especially for the initial projects by a fire protection or detection designer. In doing so, one can learn most effectively from any mistakes. What might have been difficult to visualize on the drawing board should become clear after seeing the finished product.

You also will have a sense of accomplishment when seeing the finished product. During the final field check before installation, a fire protection designer surveys only imaginary pipes, but during the inspection of the completed system, the accomplishments can be fully visualized and appreciated. Seeing the finished product of one's work, armed with the knowledge that the building is considerably safer after installation of the system, is the most rewarding benefit of fire protection system design.

AUTOMATIC SPRINKLER SYSTEMS

The vast majority of protected hazards are protected satisfactorily by automatic sprinkler systems, which are not covered in this book. Automatic sprinkler systems historically have been successful in protecting life and also in protecting hazards that are capable of being protected by water. Automatic sprinkler systems protect hazards that fall into the categories of

- light hazard (churches, hospitals, schools),
- ordinary hazard (parking garages, laundries, shopping centers), and
- extra hazard (saw mills, textile operations, paint dipping).

Hazards protected by automatic sprinkler systems also are characterized by their amenability for standard suppression system response times, usually up to 60 seconds, and water delivery demands correlating to the occupancy of the hazard. Automatic sprinkler systems feature heat-responsive automatic sprinklers that are individually actuated, and most fires involving commonly encountered combustibles are suppressed with only one or two individually activated automatic sprinklers discharging.

In an example related to the number of discharging sprinklers, NFPA 13D, *Standard for the Installation of One- and Two-Family Dwellings and Manufactured Homes,* requires no more than two sprinklers to be simulated to be flowing on a system. A study conducted in Australia and New Zealand, evaluating 82 years of automatic sprinkler experience, found that 82% of fires were controlled by two or fewer automatic sprinklers.

NFPA

NFPA 13D, Standard for the Installation of One- and Two-Family Dwellings and Manufactured Homes

SPECIAL HAZARD SUPPRESSION SYSTEMS

Not all hazards are protected satisfactorily by an automatic sprinkler system. A special hazard suppression system, as detailed in this book, protects hazards that are not amenable for protection by an automatic sprinkler system. Special hazard suppression systems ordinarily protect a variety of hazards displaying unique challenges in which standard sprinkler systems may not be the most efficient choice of suppression methodology, as listed below.

Special Hazard Categories

Special hazards are categorized by one or more of several characteristics:

- Large quantities of flammable liquids, such as in an aircraft hangar, usually mandate the use of foam for the most efficient and effective extinguishing of a flammable liquid-pool fire.
- Facilities containing valuable or irreplaceable commodities, such as an art storage facility, may consider non-water-based special hazards media, such as a clean-agent gas to protect commodities without damaging them.

- Facilities where water may pose a danger, such as those involving the manufacture or storage of water-reacting metals or chemicals, may require a special hazard media congruent with the commodity protected.

- In some facilities automatic sprinkler water-delivery times—up to 60 seconds for a dry-pipe automatic sprinkler system—are not fast enough to achieve efficient suppression. Examples of such facilities are a nitroglycerin manufacturing plant and one of many varieties of munitions or demilitarizing facilities that may require detection and suppression response within milliseconds, must have a rapidly-acting special hazards suppression system, such as an ultra-high speed deluge suppression system or an explosion-suppression system.

- Mobile facilities where the transport of water is unfeasible, such as in the orbiting space station or on commercial aircraft, most likely will require a special hazard-suppression system that requires less space and has less weight than an automatic sprinkler suppression system. An example of such a system is a water-mist suppression system or a clean-agent gas suppression system.

- Facilities where service loss is intrinsically linked to facility loss or water damage, such as in an Internet switching facility or telephone switchgear facility, may require special media, media delivery methods, and/or agent delivery times to enhance the maintenance of service continuity. Another example of service loss is a pre-engineered special hazards suppression system for restaurant cooking areas, industrial paint spray booths, and large vehicles such as buses, logging, and construction vehicles.

- Facilities where high-tech research lost by a fire or water damage could be significantly more serious than the dollar loss, such as in a cleanroom or anechoic chamber, or Department of Defense research laboratory, may require a special hazard media with enhanced delivery time.

DESIGN APPROACHES FOR SPECIAL HAZARD DESIGN

Special hazard design requires evaluation of design objectives and design concepts, and implementation of a design methodology.

Design Objectives

Special hazard fire suppression design is intended to fulfill one or more of the following objectives:

- property protection and preservation
- life safety and preservation
- business continuity
- protection of the environment

Property protection of an object or a building is the predominant objective of most special hazards suppression systems. A special hazards system designed for this objective must suppress an expected fire involving the object or building, and the suppressing agent further should be of such composition as not to damage or destroy the object in the act of suppression.

An example of such a hazard is an irreplaceable object, such as a priceless painting, in which an improper choice of suppressant could control or extinguish an expected fire but render significant damage to the painting. In this case, selection of a suppression system could either involve a non-water-based agent or a water-based agent in combination with other protective measures, such as a water-proof enclosure for the painting, or safeguards on the suppression system to ensure accurate detection and efficient suppression.

Life safety should be a primary consideration in special hazards design, even where the stated contractual role of the suppression system emphasizes protection or preservation of a specific object or structure. Although protection of an object—whether it be a solid munitions cutting or an extruding machine or an exhibit in an art gallery—is important, it also is vitally important that operators, observers, or other personnel adjacent to the object of protection are likewise protected.

Continuity of business operations is a fundamental requirement of a business owner, and also is a fundamental objective of a special hazard suppression system. Assuming that the special hazard suppression system has met its objectives of protecting property and life safety, the system also should be amenable for recommencement of business in a timely fashion. For example, a clean-agent suppression system of a data processing facility is expected to operate in a manner in which

- suppression is efficient
- life safety notification and preservation are expected
- post-fire clean-up is simplified by removing and repairing or replacing the item that caught fire, exhausting the expelled clean agent
- resetting the suppression and detection system is easy, and includes continued protection with a reserve supply of agent

Environmental Protection Agency (EPA)
U.S. Governmental agency responsible for protecting air and water supplies

Increasingly in recent years, the **Environmental Protection Agency (EPA)**, a United States governmental agency responsible for protecting U.S. air and water supplies, has enforced regulations that have a direct effect on the design of special hazards suppression systems. For example, the EPA regulations relative to the discharge of high- and low-expansion foam requires that the foam and a hazardous material solution, such as a spilled flammable liquid, be contained and properly disposed to prevent such materials from entering sewers or seeping into the groundwater system. Containment and disposal strategies increasingly are required in conjunction with the design of most special hazards suppression systems.

Design Concepts

The NFPA uses the Fire Safety Concepts Tree, as shown in **Figure 1-13**, to evaluate a hazard, and make decisions relative to protection of the hazard.

As can be seen, a significant focus of the Fire Safety Concepts Tree involves

- preventing fire ignition (i.e., handling combustibles safely, enforcing a no-smoking policy)
- protecting persons exposed to an expected fire (i.e., providing notification and instructions for egress)
- controlling the fire by construction (i.e., isolating the hazard with fire rated walls) or by control of the combustion process (i.e., automatically turning off the heat to a container of flammable liquid)

Use of the "or" gates (represented by a circle containing a plus sign) in the figure means that one of the branches shown below the symbol must be achieved

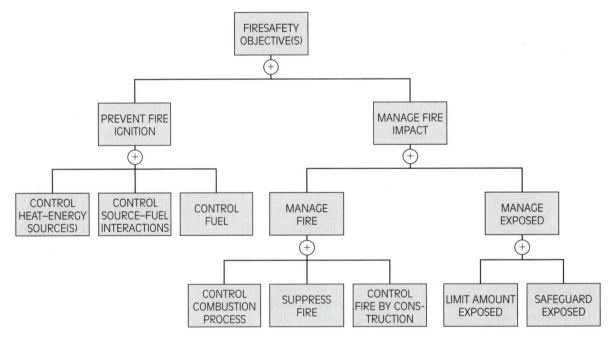

Figure 1-13   *Fire Safety Concepts Tree: Designers use this figure to make fundamental decisions on special hazards and detection design.*

Source: NFPA 550, Fig. 4.3 (2002). Reprinted with permission from NFPA 550, *Fire Safety Concepts Tree,* Copyright © 2007, National Fire Protection Association, Quincy, MA 02169. This reprinted material is not the complete and official position of the National Fire Protection Association on the referenced subject, which is represented only by the standard in its entirety.

for the objective above the symbol to be achieved. Elsewhere in NFPA 550 are diagrams displaying "and" gates shown (represented by a circle containing a dot), which means that all of the branches shown below the symbol must be achieved for the objective above the symbol to be achieved.

The "Prevent Fire Ignition" branch of the Fire Safety Concepts Tree, as outlined in NFPA 550, is further divided as follows:

- Control heat-energy sources
  - Eliminate heat-energy sources, or
  - Control rate of heat-energy release
- Control source-fuel interactions
  - Control heat-energy source transport
    - Provide barrier, or
    - Provide separation
  - Control heat-energy transfer processes
    - Control conduction
    - Control convection
    - Control radiation
  - Control fuel transport
    - Provide barrier, or
    - Provide separation

- Control fuel
  - Eliminate fuel, or
  - Control fuel ignitibility
    - Control fuel properties, or
    - Control the environment

The "Manage Exposed" branch of the Fire Safety Concepts Tree includes:

- Limit amount of persons exposed
- Safeguard exposed
  - Defend exposed in place
    - Restrict movement of exposed
    - Defend the place
      - Defend against fire products
      - Provide structural stability
    - Maintain essential movement
  - Move exposed

- Cause movement of exposed
  - Detect need
  - Signal need
  - Provide instructions
- Provide movement means
  - Provide capacity
  - Provide route completeness
  - Provide protected path
  - Provide route access
- Provide safe destination

The "Manage Fire" branch of the Fire Safety Concepts Tree includes:

- Control combustion process
  - Control fuel
    - Control fuel properties
    - Limit the quantity
    - Control fuel distribution
  - Control the environment
    - Control physical properties of the environment
    - Control chemical composition of the environment
- Control fire by construction
  - Control movement of fire
    - Vent fire
    - Confine/contain fire
  - Provide structural stability
- Suppress fire
  - Manually suppress fire (fire department or fire brigade)
    - Detect fire
    - Communicate signal
    - Decide action
    - Respond to site
    - Apply sufficient suppressant
  - Automatically suppress fire (automatic suppression and detection system)
    - Detect fire
    - Apply sufficient suppressant

This book concentrates on the "Automatically Suppress Fire" branch of the Fire Safety Concepts Tree. The first several chapters deal with the "Apply sufficient suppressant" branch of the Fire Safety Concepts Tree, and the fire alarm and detection chapters deal with the "Detect fire" branch, which includes notification of persons exposed to a fire, as well as timely actuation of special hazards suppression systems.

Design Methodologies

Special hazards designers use one of two methods to perform design of special hazards suppression and alarm and detection systems: prescriptive and performance-based design.

Prescriptive design is the direct use of national, local, or manufacturer's standards to design every aspect of a suppression or detection system, including pipe sizing and dimensions of piping and nozzles. Examples of prescriptive-based design for suppression systems are the sole use of a national standard such as NFPA 11, NFPA 12, NFPA 12A, NFPA 13, NFPA 15, or NFPA 16 to determine placement of all discharge devices and size all piping. The vast majority of automatic suppression systems use prescriptive-based design.

Performance-based design, as defined by the Society of Fire Protection Engineers in the essential book *SFPE Engineering Guide to Performance-Based Fire Protection,* is

> An engineering approach to fire protection design based on (1) established fire safety goals and objectives; (2) deterministic and probabilistic analysis of fire scenarios; and (3) quantitative assessment of design alternatives against the fire safety goals and objectives using accepted engineering tools, methodologies, and performance criteria.

Figure 1-14 shows the steps in the performance-based design process.

Fire protection engineers use the performance-based design process to evaluate and design fire suppression and detection systems for facilities where prescriptive-based design does not directly apply, or where an extra level of safety is warranted. An example might be a historic structure where stairways, construction materials, and egress routes would not meet today's prescriptive codes, where the placement of suppression equipment and apparatus could have a negative aesthetic effect on centuries-old architectural features, or where protection and preservation of the structure, its contents, and its occupants is of significant national interest. Such an analysis may use engineering fire models to study the growth and effect of a fire at various points of ignition in the structure, and may be a basis for developing computerized simulations or models to study the ancillary effects on detection, suppression, and occupant egress.

Prescriptive design
the direct use of national, local, or manufacturer's standards to design every aspect of a suppression or detection system

Performance-based design
an engineering approach to fire protection design

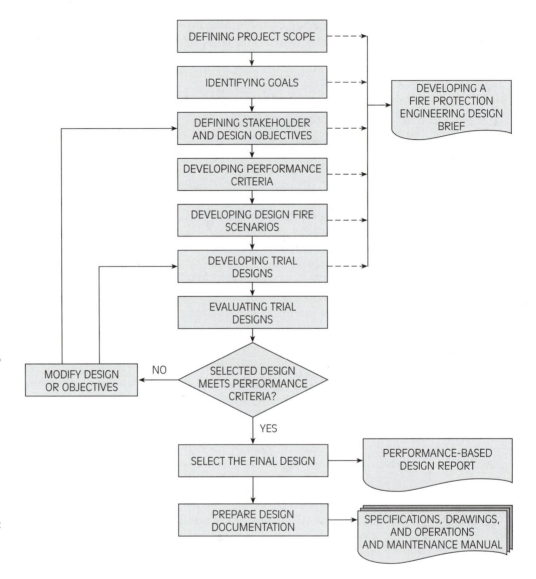

Figure 1-14 *Steps in the Performance-Based Analysis and the Conceptual Design Procedure for Fire Protection Design.*

Source: Figure 3-2 from SFPE Engineering Guide to Performance-Based Fire Protection. Copyright © Society of Fire Protection Engineers. Used with permission.

Results of the analysis of a variety of fire scenarios and proposed solutions are carefully reviewed with the authority having jurisdiction, to determine the design approach that would be most effective for the design scenarios presented. The chosen design scenario and system configuration may be significantly different from one determined by a prescriptive-based design process.

SUMMARY

An understanding of the differences between a fire protection engineer and a fire protection technician is fundamental. Also essential is knowledge of oral and written communication and media, particularly computer-aided design (CAD).

The contract documents consist of contract drawings and specifications. Contract drawings include the cover sheet and civil, architectural, structural, HVAC, plumbing, fire protection, and electrical drawings. All components of the contract drawing package are used to minimize conflicts or omissions that could affect the performance of a fire protection system negatively.

Conflicts arising between contract drawings and specifications usually are resolved with the specifications taking precedence over the contract drawings. A specification is divided into specification divisions, each consisting of several specification sections that contain detailed requirements for each division. Specifications for fire suppression systems are located in Division 21, and fire alarm systems are located in Division 28 of the CSI standardized format.

An organized and well-planned strategy for conducting a survey, using the building skeleton and appropriate tools, assures a successful result. Field checking and viewing the completed system will further understanding of fire protection system design.

The fundamental approaches, objectives, concepts, and methodologies related to special hazard and fire alarm design must be understood and evaluated before commencing design.

REVIEW QUESTIONS

1. What items are contained within the contract documents?

2. List and explain the value of each of the categories of drawings that comprise the contract drawings.

3. Give examples of information that would be missed if only the plumbing drawings were made available to the fire protection system designer.

4. What is the difference between a reference elevation and a reference grid? A plan job and a survey job? A return diffuser and a supply diffuser? A finish schedule and a reflected ceiling plan?

5. What NFPA standard should be used as reference for fire protection symbols?

6. Which specification divisions are most useful when designing a fire protection system?

7. What are the differences between a performance specification and a detailed specification, and the advantages and disadvantages of each?

8. What section does a designer reference to find information relative to fire protection systems under the new specification format? Why has the specification format changed?

9. List the tools used on a fire protection survey and explain their use.

10. What is the best way to arrange for a survey appointment?

11. Describe the procedure for surveying above an existing suspended ceiling.

12. What cautions would you suggest to a person performing a fire protection survey?

DISCUSSION QUESTIONS

1. What could happen if a fire protection system designer neglected to take full advantage of the wealth of information available in a complete set of contract drawings?

2. What problems could ensue if an incomplete set of specifications were used in the design of a fire protection system?

3. How should a surveyor develop a survey strategy for a building?

ACTIVITIES

1. Interview one or more experienced fire protection designers to examine their experience with contract drawings. What contract drawings do they ordinarily receive when they are assigned a project? What methods do they use to obtain a complete set of contract drawings? What difficulties have they had in obtaining a complete set of contract drawings? What experiences have they had when using an incomplete set of contract drawings?

2. Obtain or borrow a complete set of contract drawings from a fire protection contractor, builder, architect, or engineer. Examine and compare the drawings contained within the set. Are any drawings discussed within this chapter not present? Would the absence of any of the drawings discussed affect the proper performance of a fire protection design negatively?

3. Contact a fire protection contractor and determine the reasons that a contractor might be given an incomplete set of contract drawings by a builder or architect.

4. Obtain a reflected ceiling plan. Does the plan show dimensioned locations for the ceiling tiles? If not, is the location of the tiles centered in rooms or centered with respect to column lines?

5. Obtain a complete book of specifications from a fire protection contractor, architect, or engineer, and review each division for its relevance to the design and installation of a fire protection system.

6. Locate a set of contract documents, and compare the specification to the contract drawings to discover any inconsistencies. Discuss the problems that may ensue for a fire protection designer who is presented with such conflicts.

7. Interview an architect or engineer to determine his or her criteria for using a performance specification versus a detailed specification. Determine whether a performance specification or a detailed specification supersedes the contract drawings in a case of conflict.

8. Review a specification that contains the revised provisions for fire protection systems and detection systems. Compare the new format to the older format that specified fire protection systems in Division 15, and detection systems in Division 16, and evaluate their similarities and differences.

9. Contact an experienced fire protection surveyor and arrange to accompany that person on a fire protection survey of a building. Take note of survey equipment used, strategies employed, and use of the point of contact.

10. Accompany an experienced person on a field check of a drawing that has been developed from survey measurements. Employ the fail-test method for checking fire protection piping locations.

11. Survey a completed fire protection system, and compare the system to the drawing from which it was installed.

Chapter 2

ETHICS AND PROFESSIONAL DEVELOPMENT

Learning Objectives

Upon completion of this chapter, you should be able to:

- Discuss the code of ethics applicable to your professional qualifications.
- Discuss the ethical responsibilities of students, technicians, and engineers.
- Compare and contrast the respective ethical responsibilities to the individual, the company, society, the client, and the fire protection profession.
- Cite the differences between the NICET and NSPE codes of ethical conduct.
- Describe a methodology for determining whether a fire protection professional should become involved in whistle-blowing.
- Evaluate ethics case studies and apply the appropriate professional standards of conduct described in this chapter to these cases.

The proper application of technical knowledge to the design of a fire protection system is essential to your success as a fire protection professional. Most of this text is dedicated to helping you acquire or sharpen technical skills. In addition to the minimum requisite technical skills, ethical standards of professional conduct—not covered in most technical books—may contribute as much or more to the success of a fire protection professional. This chapter continues our introduction to fire protection system design with a discussion of the procedures and codes that guide professional development and ethical practice. Application of the fundamental principles of professional conduct, in combination with the proper application of technical knowledge, will assist fire protection professionals in having a successful and rewarding career.

Ethical practice of fire protection system design ensures that the designs for which you are responsible serve the best interests of the public. Active participation in professional organizations and continuing education and study will advance your knowledge and your career, with the ultimate result of benefiting the society we serve.

ETHICS IN ENGINEERING

Engineering is one of the most widely respected professions in the world. In some countries an engineer is honored with the title Engineer Smith and is greeted with the highest respect. The public willingly extends blind trust to engineers to protect the public from harm through application of specialized technical knowledge—not unlike the trust the public confers on pilots of commercial airliners.

The early railroad era represented a significant benchmark in the development of the engineering profession. As our nation expanded and rail service followed, railroad bridge failures became commonplace. Society demanded an analytical and methodological approach to railroad trestle design. The profession instituted a requirement that the safety of railroad bridges be determined mathematically in advance of construction to demonstrate their ability to perform properly. Tau Beta Pi, the National Engineering Honor Society, uses a railroad trestle as its symbol to commemorate this significant development in the profession. Today, complex computer models demonstrate the adequacy of each structural member and each connection on a railroad or bridge structure, and calculates resultant stresses to determine the integrity of the entire assembly.

Although it is commonly thought that engineers got their name from locomotive engineers or that engineers were once primarily designers or operators of machinery, the term *engineer* dates a thousand or more years before the colorful locomotive era. The term is based loosely on the word *ingenuity*. The responsibility of engineers goes far beyond the design of machinery. It is a philosophical pact between a design professional and the public to use all available proven methodologies to solve engineering problems and ensure the safety of the public.

ENCOUNTERING ETHICAL SITUATIONS

Some writers, philosophers, and psychologists theorize that, to some extent, bad behavior or the potential for bad behavior resides within each individual. Visualize this complex dichotomy simply as the "angel and devil syndrome," with the forces of good and evil perched on our opposite shoulders, competing for attention. Our ethical dilemma in fire protection design would be to overcome the inclination to exhibit bad behavior so the angel always defeats the devil.

This analogy is overly simplistic in light of the vast number and range of ethical dilemmas that engineering professionals are encountering today. With the current complexity of technological dilemmas, the application of expert knowledge carries responsibility, and the public trusts fire protection professionals to act in good faith. A system of ethical guidelines has been established to assist fire protection professionals to meet the societal demand for proper performance of their responsibilities.

The Ethical Dilemma

ethical dilemmas
situations in which pressure is placed upon an individual to do something that is not ethically proper, or when it is not immediately obvious which option would be ethically proper

In every professional's life, situations arise when they are tempted to stray from accepted ethical behavior. These situations pose **ethical dilemmas**. For fire protection professionals, the following situations may occur:

- situations in which a company or other outside entity places significant pressure on an individual to do something other than what the profession clearly considers ethically proper
- situations in which it is not immediately obvious which option would be ethically proper

In our attempts to deal with these situations, we encounter ethical dilemmas that may be difficult to solve. Knowledge of what the profession expects of its members can be helpful in properly resolving the dilemma.

RESPONSIBILITIES OF A FIRE PROTECTION PROFESSIONAL

When we perform an engineering design or layout, we have a responsibility to conduct ourselves in accordance with several complex, and sometimes intermingled or contradictory, values:

- our personal values
- our company's values
- society's values
- values of the client
- values of the fire protection profession

These values reflect the diverse loyalties that a professional must respect.

Personal Values

We receive our system of **personal values** primarily from our parents in our earliest years of life. These early impressions are of lasting and influential value throughout life.

As we get older and more sophisticated, our value capacity increases. We begin to recognize the importance of other values, and the weight we give to differing personal values changes. Many forces influence our value system. Our personal values expand to include the values of our peers and mentors, the values we obtain from television, film, and the print media, the values of the religious or social community in which we live, and the values of the schools from which we obtain our technical and literary knowledge.

Our personal value system, therefore, becomes a fascinating patchwork quilt, fabricated from the totality of our varied personal experiences. Our complex personal value system is tested and sometimes modified with each encounter with an ethical dilemma.

Values often reflect responsibilities. In professional life you are not required to neglect these personal responsibilities. In resolving ethical dilemmas, these personal responsibilities must be balanced with the responsibilities of a professional toward the company, the client, the profession, and society.

Company Values

Your employer may expect or require that, as a representative of the company, you adopt the **company values** and behave in a way that meets the goals of the company and brings credit to the company. For example, a company would not want you to divulge **proprietary information**—unique data that are the sole property of the company—to a competitor or to behave in a manner that would bring dishonor to the company.

Company values rarely are written down for an individual to read; they are assumed to be congruous with the personal values of each employee. In a few cases they are printed in a **company manual**, a written set of company regulations that requires the signature of employees. Most company manuals require that all inventions and written material become the property of the company. Some companies prohibit an employee from interacting with a client of the company for a certain period of time after an employee has ceased employment with the company, or require a new employee to submit to a drug test before the employee is hired, or periodically after employment.

Even when a company has a written policy, it has many other values as well, which a young professional may not recognize. Therefore, every entry-level professional should seek a mentor to provide guidance in the many unwritten values a professional is expected to respect and honor.

The company can exert considerable influence on an employee in either a positive or a negative manner. A company that permits fire protection professionals to resolve ethical concerns in accordance with their personal societal, and professional values will benefit from an enhanced reputation for ethical behavior.

By contrast, companies that pressure fire protection professionals to betray personal, societal, or professional values could create an uncomfortable atmosphere for employees. A company's improper actions possibly could place the lives of people in jeopardy and, in time, place the welfare of the company itself in danger. An employee must decide whether working for such a company is in his or her best interest.

An extreme example of negative company influence is a company's convincing a fire protection system installer to glue phony clean-agent discharge nozzles and phony or inoperable detection devices to a ceiling with no connection to a source of suppressant or electric power. Among the several eventualities that could ensue from such a crime are the following:

- If the company is not caught, people may die.
- If the company is caught, all individuals involved in the plot are in jeopardy of being convicted of a serious crime and going to jail.
- The company may have to be dissolved to pay damage claims and legal fees.
- Personal property may be confiscated to pay for damage claims related to personal responsibility.
- The professional reputation of company employees who were not directly involved in the unprofessional conduct may be damaged by such conduct.
- The crime could seriously damage the reputation of all fire protection professionals, even those who do not work for the company.

Societal Values

When the public enters a facility protected by a special hazards fire protection and detection system, it has every right to expect that the system has been designed and installed in a manner that provides adequate fire protection, that it meets all expected performance objectives, and that the system provides adequate opportunity for the safe egress of all occupants. In addition, the public reasonably expects that the system will not malfunction, resulting in injury or property damage.

societal values

values related to the public expectation that all fire protection professionals competently discharge their responsibilities and thereby ensure the safety of all who enter a building that is protected by a fire protection system

Societal values are related to the public expectation that all fire protection professionals will discharge their responsibilities competently to ensure the safety of all who enter a building that is protected by a fire protection and detection system. Failing to do so can incur the wrath of society and unleash a whirlwind of bad publicity, indictments, financial implications, and perhaps jail. Fire protection design involves an element of public trust that must not be betrayed.

The public is often fickle when it comes to awareness and appreciation for fire protection and detection systems. Sometimes fire protection systems are treated with benign neglect by the general public when problems with the systems do not exist. People rarely even know if a building they enter is protected by a suppression system, but the public can become righteously indignant and even vengeful when a suppression system is not installed or fails to perform as expected. News reports are increasingly citing the presence or absence of a fire suppression system when a tragedy strikes.

To ensure the protection of the public, society demands a system of checks and balances consisting of a competent review of fire protection plans, and the inspection, testing, and maintenance of all fire protection systems. For example, a suppression system, no matter how competently designed, fails in its primary performance objective if the main control valve is turned off. Society has little patience with such failure and goes to great lengths to ensure that the individuals responsible are punished.

The National Fire Protection Association (NFPA) is an effective standards-making institution that enforces the societal demand by providing minimum standards for public safety and welfare. This complex, layered system of standards, review, and inspection requirements seeks to establish a basis for comparison, to correct errors, and to prevent unethical behavior. These requirements work well when all professionals discharge their responsibilities competently and ethically.

Values of the Client

client's value system
values of the receiver of the fire protection system that demand a code-compliant fire protection system at the lowest possible price

The client may be the building owner or a firm that represents the owner of the building. In the case of a government-owned building, the client is the public. The **client's value system** demands

- a code-compliant fire protection and detection system that meets or exceeds the minimum accepted standard of care

- a fire protection and detection system obtained at the lowest possible price

These two goals of the client may be at cross-purposes. For example, a fire protection design and construction company designated as the lowest bidder may discover, subsequent to acceptance of the contract, that supplying a code-compliant system would render a cost for the system that exceeds the contracted amount, perhaps because an estimator based the system cost on less than the correct number of floors in a multistory building. The manner in which the company reacts to such a situation cuts to the heart of the ethical behavior of a company. Two options could ensue:

1. The company installs a code-compliant fire protection system and absorbs the loss.

2. The company attempts to install a system that it knows to be substandard in an effort to make a profit.

Although a client may expect to buy a limousine for the price of a motor scooter, the client can no more afford the societal cost of a major fire protection system failure than can the installing and design firm. A client might react to the situation described in two ways:

1. Force the company to install the system as specified at a loss.

2. Negotiate terms for an extra to the contract to have the job done correctly.

It should be noted that the client may not be contractually obligated to issue authorization for extra money to the contractor but may opt to do so for fear of a major catastrophe resulting from improper installation or fear of problems that may ensue if the company declares bankruptcy before construction is completed.

Values of the Fire Protection Profession

values of the fire protection profession
values summarized in the Code of Ethics for NICET-Certified Engineering Technicians and Technologists and the National Society of Professional Engineers (NSPE) *Code of Ethics for Engineers*

Professionals engaged in fire protection design are obliged to constantly exhibit the **values of the fire protection profession**. These values are summarized in the Code of Ethics for NICET-Certified Engineering Technicians and Technologists, **Figure 2-1**, and the NSPE *Code of Ethics for Engineers,* **Figure 2-2**.

These codes are not a list of do's and don'ts. Instead, they are a series of value statements that reflect what the profession believes is important. The codes may be easy to read, but they are much more difficult to interpret and apply when faced with a specific ethical dilemma. Thus, professionals have to read these codes carefully, identify the values, and reflect on how they must be applied in professional practice.

PROFESSIONAL STANDARDS OF CONDUCT

The professional standards of conduct that guide fire protection professionals include those of the National Institute for Certification in Engineering Technologies (NICET) and the National Society of Professional Engineers (NSPE).

Code of Ethics for NICET-Certified Engineering Technicians and Technologists

NICET
abbreviation for the National Institute for Certification in Engineering Technologies

Practitioners who work as technicians for fire protection, fire alarm, and architectural or engineering firms are strongly encouraged to seek certification with **NICET**, the National Institute for Certification in Engineering Technologies. NICET provides evaluations and examinations for fire protection technicians in the field of fire protection engineering technology by sponsoring examinations in the subfields of automatic sprinkler system layout, special hazards systems layout, fire alarm systems, and fire protection system inspections. These subfields comprise the principal areas in which the application of technical knowledge must be performed in an ethical manner.

Practitioners of fire protection technology can become certified in one of four grades, depending on their experience and their ability to pass a series of examinations. Many authorities having jurisdiction (AHJ) are requiring a minimum level of certification for an individual performing a layout, or certification for an

CODE OF RESPONSIBILITY FOR NICET-CERTIFIED ENGINEERING TECHNICIANS AND TECHNOLOGISTS

NICET-certified engineering technicians and technologists recognize that the services they render have a significant impact on the quality of life for everyone. As they perform their duties and responsibilities on behalf of the public, employers, and clients, they should demonstrate personal integrity and competence. Accordingly, certificants should:

1. Have due regard for the physical environment and for public safety, health, and wellbeing. If their judgment is overruled under circumstances where the safety, health, property, or welfare of the public may be endangered, they should notify their employer, client, and/or such other authority as may be appropriate.

2. Undertake only those assignments for which they are competent by way of their education, training, and experience.

3. Discharge their duties to their employer or client in an efficient and competent manner with complete fidelity and honesty.

4. Conduct themselves in a dignified manner. They will admit and accept their own errors when proven wrong and refrain from distorting or altering the facts in an attempt to justify their decisions.

5. Disclose to their employers or clients, in writing, any conflict of interest or other interest which might create the appearance of a conflict of interest.

6. Not receive, either directly or indirectly, any gratuity, commission, or other financial benefit in connection with any work which they are performing unless such benefit has been authorized by their employer or client.

7. Neither personally nor through any other agency or persons improperly seek to obtain work or, by way of commission or otherwise, make or offer to make payment to clients or prospective clients for obtaining such work.

8. Strive to maintain their proficiency by updating their technical knowledge and skills in engineering technology.

9. Not misrepresent or permit misrepresentation of their own or their associate's academic or professional qualifications nor exaggerate their degree of responsibility for any work.

10. Not reveal facts, data, or information obtained in connection with services rendered without the prior consent of the client or employer except as authorized or required by law.

Figure 2-1 *Code of Ethics for NICET-Certified Engineering Technicians and Technologists. (Courtesy of National Institute for Certification in Engineering Technologies)*

individual performing a quality assurance review of a drawing in advance of submission to the AHJ for approval. Such requirements are an effort to conform to societal values.

The NICET Code of Ethics has a broad similarity to many other codes of ethics that have been devised. Among the salient points are the following.

1. Public safety is paramount and takes precedence over pressures that may be brought to bear on a fire protection technician. If a technician is ordered or

National Society of Professional Engineers

Code Of Ethics for Engineers

Preamble

Engineering is an important and learned profession. The members of the profession recognize that their work has a direct and vital impact on the quality of life for all people. Accordingly, the services provided by engineers require honesty, impartiality, fairness and equity, and must be dedicated to the protection of the public health, safety and welfare. In the practice of their profession, engineers must perform under a standard of professional behavior which requires adherence to the highest principles of ethical conduct on behalf of the public, clients, employers and the profession.

I. Fundamental Canons

Engineers, in the fulfillment of their professional duties, shall:
1. Hold paramount the safety, health and welfare of the public in the performance of their professional duties.
2. Perform services only in areas of their competence.
3. Issue public statements only in an objective and truthful manner.
4. Act in professional matters for each employer or client as faithful agents or trustees.
5. Avoid deceptive acts in the solicitation of professional employment.

II. Rules of Practice

1. Engineers shall hold paramount the safety, health and welfare of the public in the performance of their professional duties.
   a. Engineers shall at all times recognize that their primary obligation is to protect the safety, health, property and welfare of the public. If their professional judgment is overruled under circumstances where the safety, health, property or welfare of the public are endangered, they shall notify their employer or client and such other authority as may be appropriate.
   b. Engineers shall approve only those engineering documents which are safe for public health, property and welfare in conformity with accepted standards.
   c. Engineers shall not reveal facts, data or information obtained in a professional capacity without the prior consent of the client or employer except as authorized or required by law or this Code.
   d. Engineers shall not permit the use of their name or firm name nor associate in business ventures with any person or firm which they have reason to believe is engaging in fraudulent or dishonest business or professional practices.
   e. Engineers having knowledge of any alleged violation of this Code shall cooperate with the proper authorities in furnishing such information or assistance as may be required.
2. Engineers shall perform services only in the areas of their competence.
   a. Engineers shall undertake assignments only when qualified by education or experience in the specific technical fields involved.
   b. Engineers shall not affix their signatures to any plans or documents dealing with subject matter in which they lack competence, nor to any plan or document not prepared under their direction and control.
   c. Engineers may accept assignments and assume responsibility for coordination of an entire project and sign and seal the engineering documents for the entire project, provided that each technical segment is signed and sealed only by the qualified engineers who prepared the segment.
3. Engineers shall issue public statements only in an objective and truthful manner.
   a. Engineers shall be objective and truthful in professional reports, statements or testimony. They shall include all relevant and pertinent information in such reports, statements or testimony.
   b. Engineers may express publicly a professional opinion on technical subjects only when that opinion is founded upon adequate knowledge of the facts and competence in the subject matter.
   c. Engineers shall issue no statements, criticisms or arguments on technical matters which are inspired or paid for by interested parties, unless they have prefaced their comments by explicitly identifying the interested parties on whose behalf they are speaking, and by revealing the existence of any interest the engineers may have in the matters.
4. Engineers shall act in professional matters for each employer or client as faithful agents or trustees.
   a. Engineers shall disclose all known or potential conflicts of interest to their employers or clients by promptly informing them of any business association, interest, or other circumstances which could influence or appear to influence their judgment or the quality of their services.
   b. Engineers shall not accept compensation, financial or otherwise, from more than one party for services on the same project, or for services pertaining to the same project, unless the circumstances are fully disclosed to, and agreed to by, all interested parties.

c. Engineers shall not solicit or accept financial or other valuable consideration directly or indirectly, from contractors, their agents, or other parties in connection with work for employers or clients for which they are responsible.
   d. Engineers in public service as members, advisors or employees of a governmental or quasi-governmental body or department shall not participate in decisions with respect to professional services solicited or provided by them or their organizations in private or public engineering practice.
   e. Engineers shall not solicit or accept a professional contract from a governmental body on which a principal or officer of their organization serves as a member.
5. Engineers shall avoid deceptive acts in the solicitation of professional employment.
   a. Engineers shall not falsify or permit misrepresentation of their, or their associates', academic or professional qualifications. They shall not misrepresent or exaggerate their degree of responsibility in or for the subject matter of prior assignments. Brochures or other presentations incident to the solicitation of employment shall not misrepresent pertinent facts concerning employers, employees, associates, joint venturers or past accomplishments with the intent and purpose of enhancing their qualifications and their work.
   b. Engineers shall not offer, give, solicit or receive, either directly or indirectly, any political contribution in an amount intended to influence the award of a contract by public authority, or which may be reasonably construed by the public of having the effect or intent to influence the award of a contract. They shall not offer any gift, or other valuable consideration in order to secure work. They shall not pay a commission, percentage or brokerage fee in order to secure work except to a bona fide employee or bona fide established commercial or marketing agencies retained by them.

III. Professional Obligations

1. Engineers shall be guided in all their professional relations by the highest standards of integrity.
   a. Engineers shall admit and accept their own errors when proven wrong and refrain from distorting or altering the facts in an attempt to justify their decisions.
   b. Engineers shall advise their clients or employers when they believe a project will not be successful.
   c. Engineers shall not accept outside employment to the detriment of their regular work or interest. Before accepting any outside employment they will notify their employers.
   d. Engineers shall not attempt to attract an engineer from another employer by false or misleading pretenses.
   e. Engineers shall not actively participate in strikes, picket lines, or other collective coercive action.
   f. Engineers shall avoid any act tending to promote their own interest at the expense of the dignity and integrity of the profession.
2. Engineers shall at all times strive to serve the public interest.
   a. Engineers shall seek opportunities to be of constructive service in civic affairs and work for the advancement of the safety, health and well-being of their community.
   b. Engineers shall not complete, sign or seal plans and/or specifications that are not of a design safe to the public health and welfare and in conformity with accepted engineering standards. If the client or employer insists on such unprofessional conduct, they shall notify the proper authorities and withdraw from further service on the project.
   c. Engineers shall endeavor to extend public knowledge and appreciation of engineering and its achievements and to protect the engineering profession from misrepresentation and misunderstanding.
3. Engineers shall avoid all conduct or practice which is likely to discredit the profession or deceive the public.
   a. Engineers shall avoid the use of statements containing a material misrepresentation of fact or omitting a material fact necessary to keep statements from being misleading or intended or likely to create an unjustified expectation, or statements containing prediction of future success.
   b. Consistent with the foregoing, Engineers may advertise for recruitment of personnel.
   c. Consistent with the foregoing, Engineers may prepare articles for the lay or technical press, but such articles shall not imply credit to the author for work performed by others.
4. Engineers shall not disclose confidential information concerning the business affairs or technical processes of any present or former client or employer without his consent.

Figure 2-2   *NSPE Code of Ethics for Engineers. (Courtesy of National Society of Professional Engineers)*

a. Engineers in the employ of others shall not without the consent of all interested parties enter promotional efforts or negotiations for work or make arrangements for other employment as a principal or to practice in connection with a specific project for which the Engineer has gained particular and specialized knowledge.

b. Engineers shall not, without the consent of all interested parties, participate in or represent an adversary interest in connection with a specific project or proceeding in which the Engineer has gained particular specialized knowledge on behalf of a former client or employer.

5. Engineers shall not be influenced in their professional duties by conflicting interests.

a. Engineers shall not accept financial or other considerations, including free engineering designs, from material or equipment suppliers for specifying their product.

b. Engineers shall not accept commissions or allowances, directly or indirectly, from contractors or other parties dealing with clients or employers of the Engineer in connection with work for which the Engineer is responsible.

6. Engineers shall uphold the principle of appropriate and adequate compensation for those engaged in engineering work.

a. Engineers shall not accept remuneration from either an employee or employment agency for giving employment.

b. Engineers, when employing other engineers, shall offer a salary according to professional qualifications.

7. Engineers shall not attempt to obtain employment or advancement or professional engagements by untruthfully criticizing other engineers, or by other improper or questionable methods.

a. Engineers shall not request, propose, or accept a professional commission on a contingent basis under circumstances in which their professional judgment may be compromised.

b. Engineers in salaried positions shall accept part-time engineering work only to the extent consistent with policies of the employer and in accordance with ethical considerations.

c. Engineers shall not use equipment, supplies, laboratory, or office facilities of an employer to carry on outside private practice without consent.

8. Engineers shall not attempt to injure, maliciously or falsely, directly or indirectly, the professional reputation, prospects, practice or employment of other engineers, nor untruthfully criticize other engineers' work. Engineers who believe others are guilty of unethical or illegal practice shall present such information to the proper authority for action.

a. Engineers in private practice shall not review the work of another engineer for the same client, except with the knowledge of such engineer, or unless the connection of such engineer with the work has been terminated.

b. Engineers in governmental, industrial or educational employ are entitled to review and evaluate the work of other engineers when so required by their employment duties.

c. Engineers in sales or industrial employ are entitled to make engineering comparisons of represented products with products of other suppliers.

9. Engineers shall accept personal responsibility for their professional activities; provided, however, that Engineers may seek indemnification for professional services arising out of their practice for other than gross negligence, where the Engineer's interests cannot otherwise be protected.

a. Engineers shall conform with state registration laws in the practice of engineering.

b. Engineers shall not use association with a nonengineer, a corporation, or partnership as a "cloak" for unethical acts, but must accept personal responsibility for all professional acts.

10. Engineers shall give credit for engineering work to those to whom credit is due, and will recognize the proprietary interests of others.

a. Engineers shall, whenever possible, name the person or persons who may be individually responsible for designs, inventions, writings, or other accomplishments.

b. Engineers using designs supplied by a client recognize that the designs remain the property of the client and may not be duplicated by the Engineer for others without express permission.

c. Engineers, before undertaking work for others in connection with which the Engineer may make improvements, plans, designs, inventions, or other records which may justify copyrights or patents, should enter into a positive agreement regarding ownership.

d. Engineers' designs, data, records, and notes referring exclusively to an employer's work are the employer's property.

11. Engineers shall cooperate in extending the effectiveness of the profession by interchanging information and experience with other engineers and students, and will endeavor to provide opportunity for the professional development and advancement of engineers under their supervision.

a. Engineers shall encourage engineering employees' efforts to improve their education.

b. Engineers shall encourage engineering employees to attend and present papers at professional and technical society meetings.

c. Engineers shall urge engineering employees to become registered at the earliest possible date.

d. Engineers shall assign a professional engineer duties of a nature to utilize full training and experience, insofar as possible, and delegate lesser functions to subprofessionals or to technicians.

e. Engineers shall provide a prospective engineering employee with complete information on working conditions and proposed status of employment, and after employment will keep employees informed of any changes.

"By order of the United States District Court for the District of Columbia, former Section 11(c) of the NSPE Code of Ethics prohibiting competitive bidding, and all policy statements, opinions, rulings or other guidelines interpreting its scope, have been rescinded as unlawfully interfering with the legal right of engineers, protected under the antitrust laws, to provide price information to prospective clients; accordingly, nothing contained in the NSPE Code of Ethics, policy statements, opinions, rulings or other guidelines prohibits the submission of price quotations or competitive bids for engineering services at any time or in any amount."

Statement by NSPE Executive Committee

In order to correct misunderstandings which have been indicated in some instances since the issuance of the Supreme Court decision and the entry of the Final Judgment, it is noted that in its decision of April 25, 1978, the Supreme Court of the United States declared: "The Sherman Act does not require competitive bidding."

It is further noted that as made clear in the Supreme Court decision:

1. Engineers and firms may individually refuse to bid for engineering services.
2. Clients are not required to seek bids for engineering services.
3. Federal, state, and local laws governing procedures to procure engineering services are not affected, and remain in full force and effect.
4. State societies and local chapters are free to actively and aggressively seek legislation for professional selection and negotiation procedures by public agencies.
5. State registration board rules of professional conduct, including rules prohibiting competitive bidding for engineering services, are not affected and remain in full force and effect. State registration boards with authority to adopt rules of professional conduct may adopt rules governing procedures to obtain engineering services.
6. As noted by the Supreme Court, "nothing in the judgment prevents NSPE and its members from attempting to influence governmental action . . . ."

Note:

In regard to the question of application of the Code to corporations vis-a-vis real persons, business form or type should not negate nor influence conformance of individuals to the Code. The Code deals with professional services, which services must be performed by real persons. Real persons in turn establish and implement policies within business structures. The Code is clearly written to apply to the Engineer and it is incumbent on a member of NSPE to endeavor to live up to its provisions. This applies to all pertinent sections of the Code.

National Society of Professional Engineers
1420 King Street
Alexandria, Virginia 22314-2794
703/684-2800   FAX: 703/836-4875
Publication date as revised: July 1993 • Publication #1102

Figure 2-2   *(continued)*

advised to do something that violates the Code of Ethics, the technician must provide timely notification and be prepared to take appropriate action.

2. Professionals must work only within the subfield of their expertise. If you are certified in special hazard layout or fire alarm system layout, your practice of fire protection engineering technology should be limited exclusively to that specific field. You should obtain certification for all subfields applicable to your education and experience.

3. Loyalty, honesty, dignity, and lack of conflict of interest are an employee's responsibilities to an employer.

4. All gifts received as part of one's job must be reported. Accepting gifts gives the appearance that a professional's judgment has been influenced by factors other than the important technical details.

5. Clients should not be used for personal gain by an employee. Undercutting an employer with respect to a client reflects a conflict of interest and poor professional judgment. An example of such a violation is establishing a private design practice using an employer's clients.

6. The professional must stay current with fire protection technology, through education, training, reading, and vigilance on the job. Failing to maintain competency implies that you are not providing the client or society with the best engineering judgment. NICET has instituted a requirement for verification of continuing professional education and competency, and several states have begun to require the same of licensed professional engineers. Employees must be motivated to seek continuing education, and employers must assist employees in so doing.

7. The professional must represent professional qualifications accurately, and maintain the confidentiality of the employer's information.

The NICET Code of Ethics are intended to express the values that are important to the profession and to help professionals resolve ethical dilemmas. NICET staff members may be able to assist a certified technician in resolving specific ethical problems that the Code does not address specifically. Failing to comply with the NICET Code of Ethics may result in loss of certification.

The process of obtaining certification is not easy, and maintaining certification requires vigilance, but certification confers honor to the recipient and reflects a professional's responsibility to the public. All technicians are advised to become certified at the highest level commensurate with their experience.

Code of Ethics for Engineers

Code of Ethics for Engineers
a compilation of rules and obligations published by the National Society of Professional Engineers and used by engineers to resolve ethical conflicts in fire protection system design

Professional engineers are bound by the ***Code of Ethics for Engineers***, published by the National Society of Professional Engineers (NSPE). It expands on the NICET Code of Ethics, and can be used by engineers to resolve ethical conflicts in fire protection system design. Registration as a professional engineer is required of fire protection engineers in private practice and is becoming increasingly important for engineers in government, business, and industry. The Code of Ethics is a legal requirement and differs from the NICET Code of Ethics, which is administered and enforced by the NICET internal board of review. Licensure as a professional engineer is like an insurance policy for your career. It qualifies you for promotion in many companies, and in some companies, licensure is a requirement for employment.

Many states have adopted into law codes of ethics based upon the NSPE *Code of Ethics for Engineers*. As a part of the licensing procedure for engineers, they are legally required to become familiar with and follow the code of ethics prescribed by each state in which they practice. Each state has established a board of ethical review that schedules hearings to evaluate claims of unethical conduct and levies penalties for conduct that is judged by the board to be unethical, which includes fines and revocation of licenses.

The fundamental canons of the Code of Ethics are similar to the primary responsibilities required of fire protection technicians listed in the NICET Code of Ethics. The rules of practice and the professional obligations expand on the fundamental canons and provide detailed advice for resolving several specific situations. Some of the specific instances addressed by the *Code of Ethics* are the following:

1. Engineers must avoid guilt by association and not become involved in business ventures with firms or individuals known to be violators of ethical procedures.

2. Engineers must seal or sign only that portion of the work for which they have expertise and have supplied supervision.

3. Engineers must be careful in dealing with political representatives, especially in the realm of contributions and payments.

4. Engineers must give an honest evaluation of a project in a timely manner and cannot accept being guided into an inaccurate conclusion.

5. Engineers must protect the profession by avoiding criticism of others and any behavior that may discredit the profession.

LEGAL/ETHICAL CONFLICTS

legal/ethical conflicts
disagreements related to the absence of congruence between our system of laws and our system of ethical values

If something is legal, is it necessarily ethical, and vice versa? Does ethical behavior supersede the law? Does the law supersede ethical behavior? **Legal/ethical conflicts** arise from the absence of congruence between our system of laws and our system of ethical values. These disagreements historically have been among the more difficult areas of ethical behavior to interpret.

Two of the many examples of behavior that once was legal but subsequently became illegal, primarily because these practices were judged to be unethical, are racial segregation and the denial of voting rights to women. Other behaviors, such as speeding or running a red light to deliver a severely injured person to the hospital, are clearly illegal but perhaps are morally justified on ethical grounds in certain cases. Other behaviors, such as selling items deemed to be unsafe in the United States to countries whose safety laws are less restrictive, are clearly legal but may be viewed as unethical.

The rapid and seemingly unbridled growth of technological development and innovation increase the difficulty of our laws' remaining precisely congruent with our system of perceived ethical behavior. Conflicts between our laws and our code of ethics are becoming more numerous and more noteworthy, so design professionals have a harder time determining how to behave ethically. Picking up the newspaper on any given day presents the reader with an array of legal/ethical conflicts including international piracy of recordings and computer programs, child labor, trade quotas, stock fraud, munitions sales, and price fixing.

An example of a legal/ethical conflict related to fire protection is that of the fire inspector who inspects a nightclub for occupancy violations during daytime working hours. Some may think that the letter of the law is being met, but the spirit of the law is lacking in the worst case scenario—a fire that breaks out during nighttime hours with over-occupancy. In the final analysis, the public will expect the spirit of the law, not the letter of the law, to be enforced.

RESOLVING ETHICAL DILEMMAS

Ethical dilemmas sometimes are dealt with by rationalizing them, and less commonly by whistle-blowing. The method advocated here is conflict resolution, examples of which are provided.

Rationalization

rationalization

an attempt to justify one's own action that is known to be wrong

In almost every case of unethical behavior, **rationalization** is evident. We have all used this word in our personal lives, usually without a comprehensive understanding of its meaning. Rationalization is an attempt to justify one's own action that is known to be wrong. Individuals use rationalization to avoid modifying their internal value system, therefore allowing for unethical conduct to be repeated in the future.

Some examples of rationalization are:

- I included some inaccurate information on my resume because others are doing it.
- I falsified my expense statement because the company owes me.
- If I hadn't signed the falsified report, someone else would have done so.
- I made an illegal copy of the software program because the company is making too much profit off of it.

Whistle-Blowing

If a company is involved in illegal behavior of which an employee is aware, an employee may be placed in an ethical dilemma that could result in whistle-blowing. **Whistle-blowing** is an external action in which an individual exposes a situation perceived to be unethical. Whistle-blowing may violate the NSPE Code of Ethics relative to an employee's responsibility to an employer and the responsibility of an employee to maintain a company's proprietary information. For this reason, a specific procedure should be followed before an employee undertakes the serious steps that constitute whistle-blowing:

1. Prepare for an internal appeal within a company:
   - Evaluate the problem clinically.
   - Look at the problem from more than one point of view.
   - Collect evidence and deal with factual data only.
   - Develop a synopsis of the problem.
   - Explore and identify alternative courses of action and alternate solutions.
   - Examine the firm's internal appeal process.
   - Evaluate the problem with respect to the code of ethics.
   - Determine the outcome of a negotiated appeal with which you would be satisfied.
2. Commence the internal appeal process:
   - Bring the problem and proposed solutions to an immediate supervisor in an informal manner.
   - Make a formal appeal.
   - Continue the appeal to the highest level within the company.
3. Consider the external appeal options:
   - If possible, seek additional professional opinions in a confidential manner.
   - List available alternatives.
   - Protect yourself: Cover your options and keep detailed records.
   - Seek legal advice: Check local, state, and federal laws pertaining to the problem.
   - Contact a professional society and, if possible, obtain support or an endorsement.
   - If additional evidence or support is obtained, try to revive an internal appeal.
   - Contact the client and reveal the problem.
   - If all else fails, go public with the support of a lawyer as a last resort.

Whistle-blowing can result in severe repercussions for an employee, such as demotion, firing, legal problems, harassment, and financial implications. In the example of exposing a situation in which suppression discharge nozzles and phony smoke detectors were glued to a ceiling, whistle-blowing clearly is ethically justified, because failure to expose such a serious situation could lead to death. An employee who exposes that situation can be rewarded with the knowledge of performing a public service and eventually may triumph over any negative repercussions.

ETHICS CASE STUDIES

An effective way to illustrate the importance of proper ethical behavior, and to demonstrate how the resolution of ethical dilemmas either worked or failed, is to provide some case studies in ethics and to evaluate the effectiveness of ethical conflict resolution. The following case studies illustrate classical ethical dilemmas.

Citicorp Center

Imagine designing a high-profile, high-rise building, and overseeing the successful completion of construction, only to subsequently find a serious design flaw that made the building considerably less structurally sound than originally contemplated. Structural engineer William Le Messurier found himself in this predicament.

The Citicorp Center, a 59-story, 910-foot-tall office tower that was completed in New York City in 1978, became the seventh tallest building in the world at the time of its occupancy. An unusual feature of the design, and an important performance objective for the project, was to construct the tower around and over a precious church building, which involved a measure of innovative structural design and was hailed for its uniqueness.

Engineer Le Messurier discovered, after construction, that an assumption he had made with respect to the calculation of wind loads on the Citicorp tower was incorrect. In addition, he discovered a construction discrepancy when subsequently designing a building of similar design. Upon re-inspection of the Citicorp building, he found that structural joints that he had designed as butt-welded joints had been bolted by the contractor instead, for economic reasons. Le Messurier determined that this discrepancy, combined with the incorrect assumption relative to wind loads, weakened the structure and left it vulnerable to collapse in a severe storm.

Some professionals might be tempted to cover up such a flaw. After all, the construction company could have been set up as the sole culprit, and what were the probabilities of high winds capable of bringing down the building? But Le Messurier immediately informed the architect and the building owner of his findings relative to the discrepancy and made recommendations for correction.

To the credit of the client, steel plates subsequently were fillet-welded to the joints in question.

Despite some critical reporting of the belated discovery of the design flaw, many consider Le Messurier a hero for the manner in which he resolved this complex ethical dilemma. He performed engineering services in accordance with the *Code of Ethics for Engineers*, and in the long run preserved the profession of engineering as one deserving of the public's trust and confidence.

Space Shuttle Challenger

In one of the most prominent engineering failures in history, space shuttle Challenger exploded on liftoff, killing all astronauts aboard. During the post-crash investigation, attention quickly focused on the flexible "O" rings that connect sections of the solid rocket boosters. In a spectacular video, flames were seen shooting through the "O" rings shortly before the flames overtook the entire spacecraft.

Subsequent to the tragedy, it was reported that aerospace engineers issued warnings before launch, cautioning that space shuttle launches at freezing temperatures could compromise the flexibility and reliability of the "O" rings under those conditions. Pre-launch videos showed icicles dripping from the "O" rings.

The ethical dilemma in this case is far from simple to evaluate, and perhaps can be adequately illustrated only by asking a series of questions:

- How certain does an engineer have to be that failure is inevitable, and if not 100% inevitable, what are the probabilistic scenarios for failure? Perhaps in answering this question, one can appreciate Le Messurier's actions with respect to the Citicorp Center.

- How loudly and forcefully should you state your case in a bureaucratic atmosphere in which engineers may be managed by non-engineers who may not fully understand the risks? Does the public consider an after-the-fact "I told you so" as an adequate remedy?

- As an engineering manager, how do you deal with engineers who speak of probabilistic uncertainty, and how do you deal with superiors who show an attitude that schedule is the predominant consideration, that cost is paramount, and that risk is inevitable in the performance of complex ventures? Perhaps when answering this question, one can appreciate the joint actions of the engineer, the architect, and the building owner of the Citicorp Center, who initiated repairs despite the probabilistic risk and spiraling cost.

Hyatt Regency Walkway Collapse

In Kansas City, Missouri, the construction of a new Hyatt Regency hotel featured a walkway that appeared to be suspended by thin air, traversing the entire length of the main lobby. During a music concert, enthused fans were swaying and dancing to the music on the walkway, until suddenly it came down in a mass of metal and bodies. Many people were trapped below the walkway, some of them dead.

Clearly, the rhythmic movement of the dancers created an undulating motion in the walkway, and the intermittent upward motion of the walkway stressed and snapped suspension cables (which are very strong in tension and comparatively weak in compression) that held the walkway in place. A post-incident investigation revealed a discrepancy: Rod connectors, which were not shown on the engineering drawings, were installed in the suspension cables between the walkway and the roof, and some joints failed at the rod connectors, which are not as strong as the cables themselves.

The engineer's position was that the construction company was solely at fault and made an unauthorized engineering revision without the approval or knowledge of the engineer. The construction company's position was that the engineer's design was flawed and unconstructable, and that a field change was necessary to make the design work.

A question to be asked is: What responsibility does an engineer have with respect to inspection and certification of a completed design? In answering this question, one might see a parallel in the failure to catch the Citicorp Center design flaw before occupancy, but also can appreciate the correctness of action after discovering the flaw.

RESPONSIBILITIES OF A STUDENT

Students should join and actively participate in student organizations and societies related to their intended profession. This activity gives students a practical idea of how professional societies operate and also may help them obtain jobs by networking with alumni and practicing professionals. Further, students should seek leadership positions in student organizations to obtain executive skills, experience in public speaking, organizational skills, and the development of self-confidence.

Also, students should seek relevant work opportunities before graduation. The combination of scholastic and professional experience enhances the development of professional behavior.

Finally, students should seek a mentor or role model and pattern themselves after that person's positive qualities. A mentor can positively guide a student's career long after graduation.

RESPONSIBILITIES OF A PROFESSIONAL

Beyond behaving ethically, fire protection professionals should become active in advancing the profession by participating in as many professional organizations as time allows, and by updating their skills by attending seminars and by reading applicable publications.

Professional Development Organizations

Organizations for fire protection professionals provide regular onsite and Internet-based educational programs for their members, and many publish technical literature as well. Although time is a valuable commodity for fire protection professionals, it is strongly recommended that serious consideration be given to active participation in as many of the following applicable organizations associated with fire protection system design as possible.

National Fire Protection Association (NFPA)
1 Batterymarch Park
P.O. Box 9101
Quincy, Massachusetts 02269
(617) 770-3000
www.nfpa.org

NFPA publications: *NFPA Journal; Fire Technology; National Fire Codes; Fire Protection Handbook; Automatic Sprinkler and Standpipe Systems; Automatic Sprinkler Systems Handbook; Building Construction for the Fire Service, A Designer's Guide to Automatic Sprinkler Systems; A Designer's Guide to Fire Alarm Systems*

Society of Fire Protection Engineers (SFPE)
7315 Wisconsin Avenue
Suite 620E
Bethesda, Maryland 20814
(301) 718-2910; Fax (301) 718-2242
www.sfpe.org

SFPE publications: *SFPE Bulletin; SFPE Handbook of Fire Protection Engineering; Journal of Fire Protection Engineering*

National Institute for Certification in Engineering Technologies (NICET)
1420 King Street
Alexandria, Virginia 22314
(888) IS-NICET
www.nicet.org

NICET publications: *NICET News;* NICET certification booklets

National Fire Sprinkler Association (NFSA)
40 Jon Barrett Road
Patterson, New York 12563
(845) 878-4200, ext. 133; Fax: (845) 878-4215
www.nfsa.org

NFSA publications: *Sprinkler Quarterly; Tech Notes; Grass Roots; Regional Report*

American Fire Sprinkler Association (AFSA)
12750 Merit Drive, Suite 350
Dallas, Texas 75251
214 349 5965; Fax: 214 343 8898
www.firesprinkler.org

AFSA publications: Sprinkler Age; Membergram

National Society of Professional Engineers (NSPE)
1420 King Street
Alexandria, Virginia 22314
(703) 684-2800
www.nspe.org

NSPE publications include: *Engineering Times; State News Letter*

The SFPE, NFSA, AFSA, and NSPE have local chapters that offer opportunities to develop leadership skills and provide exposure to valuable seminars, industry tours, and noted speakers. Readers of this book are encouraged to join and participate actively.

Professional Publications

Publications that can be of assistance to the fire protection professional include:

CASA Notes
Canadian Automatic Sprinkler Association
335 Renfrew Drive, Suite 302
Markhem, Ontario, Canada L3R 9S9
(905) 447-2270; Fax (905) 477-3611

California Construction Law
Law Offices of Abdulaziz & Grossbart
P.O. Box 15458
North Hollywood, California 91615-5458
(818) 760-2000, (213) 877-5776; Fax (818) 760-3908
AG@pacificnet.net

Cockshaw's Construction Labor News and Opinion
P.O. Box 427
Newtown Square, Pennsylvania
(610) 353-0123, Fax (610) 353-0111

Codes & Standards
Kelly P. Reynolds Standards & Associates, Inc.
833 West Chicago Avenue, Suite 200
Chicago, Illinois 60622-5406
(800) 950-CODE, (312) 829-6000; Fax (312) 829-8855

Consulting-Specifying Engineer
P.O. Box 7525
Highlands Ranch, Colorado 80126-9325

Contractor
1350 E. Touhy Avenue, P.O. Box 5080
Des Plaines, Illinois 60018
(708) 635-8800

ENR/Engineering News Record
1221 Avenue of the Americas
New York, New York 10020
(212) 512-2000

FPC/Fire Protection Contractor
550 High Street, Suite 220
Auburn, California 95603
(530) 823-0706; Fax 530-823-6937
www.fpcmag.com

Industrial Fire World
P.O. Box 9161
College Station, Texas 77842
(409) 693-7105; Fax (409) 764-0691

Operation Life Safety
4025 Fair Ridge Drive
Fairfax, Virginia 22033-2868
(703) 273-0911 ext. 320; Fax (703)-273-9363

PM Engineer, For Plumbing, Hydronics & Sprinkler Engineers
755 West Big Beaver Road, Suite 1000
Troy, Michigan 48084
(810) 362-3700; Fax (810) 362-0317
PMEeditor2@aol.com

Pipeline (an NFSA affiliate)
Florida Fire Sprinkler Association
200 West College Avenue, Suite 313
Tallahassee, Florida 32301
(904) 222-2070; Fax (904) 222-1752

Subcontractor
American Subcontractors Association
1004 Duke Street
Alexandria, Virginia 22314-3588
(703) 684-3450; Fax (703) 836-3482
ASAOffice@aol.com

GUIDELINES FOR SUCCESS IN FIRE PROTECTION PRACTICE

The following guidelines cover professional obligations, personal development, professional attitude, and personal direction for fire protection professionals.

Professional Obligations

- Cherish your title and your accomplishment as a fire protection professional. Accept the title with utmost seriousness and humility. Fire protection is among the world's most respected professions. Protect and maintain this prestigious reputation.
- Be active in the advancement of fire protection. Join professional societies, and work toward betterment of the profession.
- Don't overreach your knowledge and experience. Practice only within your area of education, experience, and competence, but broaden your education and experience to expand your practice.
- Be true to your word. Follow through on your commitments.
- Know the difference between right and wrong. Even when the prevailing winds against you are strong, properly balance your professional obligations to society, the profession, the client, and your employer.
- Seek licensure or certification. This may be your best insurance policy in our evolving and constantly changing profession.
- Be ethical in all aspects of your practice. Adopt your profession's code of ethics as your standard of professional practice.

Personal Development

- Become a leader. Don't be satisfied with minimum efforts. Strive to be the best in your field. Five years from graduation, most of you will be expected to lead or train other people. Accept this role willingly and enthusiastically.
- Continue your education. Consider your current education to be the beginning of a lifelong commitment to continuing education. Be determined

to grow with the changing and expanding technology. Maintain a position on the crest of the wave.

- Develop leadership and communication skills. Dale Carnegie once said that professional success is attributable to 15% technical skills and 85% interpersonal and communications skills. These skills must be learned and constantly practiced.

- Find a mentor. Pattern your work habits after those of the best fire protection professional you know. As you gain experience, you may become a mentor yourself one day.

- Become a spokesperson for your discipline. Fire protection professionals who share knowledge in well-written journal articles and effective speeches are most likely to become leaders. Obtain additional training if necessary, but get started early and practice often.

Professional Attitude

- Be creative. Bring your intellect, talent, and fresh ideas to the quest for newer and better fire protection solutions.

- Be curious. Take on challenging assignments. Diversity of experience will produce a well-rounded professional and a more effective contributor to society.

- Be a voracious reader of technical publications related to fire protection. Your performance will be affected positively by your expanded knowledge.

- Be a volunteer. Perform community service, including consistent service to the fire protection profession. Help others to achieve what you have achieved.

- Be confident. Confidence comes naturally with mastery of the profession's requisite skills. The manner in which you present your ideas may be as important as the ideas themselves.

Personal Direction

- Take an interest in people. Every fire protection analysis, no matter how technically oriented, must consider the effect of that analysis on people. The ability to work effectively with people is a fundamental criterion for leadership in your profession.

- Follow your interests. Although pursuit of money can be a significant motivator, allowing your career to follow your interests will result in a far more enjoyable and satisfying career.

- Be a goal setter, but enjoy the ride. Goal-oriented behavior is important, and your daily progress toward those goals should be satisfying.

- Be proud of your profession, and encourage others to become fire protection professionals.

SUMMARY

Fire protection design professionals must understand the concepts involved in resolving ethical dilemmas and be vigilant in keeping up with the current technology and using the appropriate code of ethics to determine accepted ethical methods of practice. Understanding the competing values that influence fire protection design is just as important as technical knowledge.

Using the concepts of conflict resolution introduced in this chapter is important to the discharge of professional responsibilities. Fire protection students have responsibilities that they must exercise in school as preparation for professional life. Evaluation of ethics case studies provides a framework for understanding the appropriate use of codes of ethics. Active participation in organizations related to the design of fire protection systems sharpens skills and helps to develop a working set of ethical practice procedures. Reading design-related fire protection publications keeps a designer current with the rapidly expanding technology of fire protection system design. Accepting the guidelines for success introduced in this chapter is key to a rewarding career.

REVIEW QUESTIONS

1. Using the *Code of Ethics for Engineers* as a basis, determine which of the following, when received as a gift from a client or someone with whom one has a professional relationship, must be reported to an employer. Identify items that, if accepted, could create a conflict of interest. In each case, state the justification for your determination:

   a. a pen with promotional advertising printed on it

   b. a promotional videotape

   c. a book of engineering data

   d. an engraved briefcase

   e. a car

   f. a calendar

   g. a sterling silver pen engraved with your name

   h. an encyclopedia

2. Using the *Code of Ethics for Engineers* as a basis, determine which activities violate the Code. Provide the Code reference for each item.

   a. performing outside design work for your employer's client without your employer's knowledge

   b. allowing a nonregistered or noncertified individual to use your registration or certification number

   c. accepting a free computer in exchange for specifying a product from a company

   d. claiming credit for the work of another person

   e. paying a public official to get a design contract

   f. designing a bridge with a degree in fire protection engineering and professional registration in fire protection engineering

   g. performing part-time work for another firm with your employer's permission

3. Obtain a copy of the SFPE *Canon of Ethics* (www.sfpe.org) and compare and contrast with the *NSPE Code of Ethics*. How does each code serve its respective organization?

4. A client wants to hire you to perform research but makes clear that your conclusion must match a predetermined objective. Apply the *NSPE Code of Ethics* to this situation.

5. Write a paper on the following topic: "My conduct as a student does or does not reflect on my performance as a professional."

DISCUSSION QUESTIONS

1. Describe a situation related to fire protection system design not cited in this chapter in which whistle-blowing may be justified.

2. Using the NSPE *Code of Ethics for Engineers*, identify the fundamental values, and explain how these relate to the profession and to society.

3. Discuss why cheating on an examination in a fire protection course as a student can be interpreted as being unprofessional conduct by the profession.

4. List some rationalizations that students give for cheating on a test, and explain why they represent rationalization.

5. Would you turn in a classmate who had cheated on an exam? Would you turn in a colleague who falsified his or her resume on a proposal to a client? If your responses are different, explain.

ACTIVITY

From a recent newspaper, select an article related to ethical or unethical conduct. Write an evaluation of the impact that the conduct has had or will have with respect to an individual, a company, society, and the profession. Apply either the NSPE or the NICET code of ethics to the situation, and evaluate ethical courses of action.

Chapter

3

SPECIAL HAZARD SUPPRESSION AGENTS AND THEIR APPLICATIONS

Learning Objectives

Upon completion of this chapter, you should be able to:

- Recognize the attributes of the fire tetrahedron.
- Discuss the differences between the burning rate of a suppressed fire and an unsuppressed fire, and relate them to the fire signature of a detector.
- Understand the capabilities and limitations of each fire suppression agent covered in this book.
- Describe the differences between the capabilities of aqueous and nonaqueous agents.
- Explain a protocol for selecting a fire protection agent.
- Compare and contrast the differences between a local application system and a total flooding system.
- Expand on the extinguishment capabilities and extinguishing mechanisms involved with each component of the fire tetrahedron.
- Determine applications associated with fire extinguishers for the five classes of fires.

Readers of this book are intimately familiar with fire from many perspectives. Perhaps in scouting you learned the most efficient ways to start a campfire. Perhaps you have witnessed a fire, or perhaps you or your loved ones have been the victim of fire. Perhaps you are in the fire service and fire is an everyday occurrence. Fire often is thought to be a rare event, but when it happens to you or to those you love, fire becomes personal.

As can be determined by reading a major metropolitan newspaper daily, fire is a constant threat, and its victims usually are young children and elderly people—those for whom society feels the most responsibility. To fully exercise our societal responsibility to protect the public, fire protection professionals must understand the fundamental mechanisms of fire dynamics and fire extinguishment.

FUNDAMENTAL FIRE PRINCIPLES

NFPA

NFPA 1, Uniform Fire Code (2006)

combustion (fire)
a chemical process of oxidation that occurs at a rate fast enough to produce heat and usually light in the form of either a glow or flame

pyrolysis
a chemical process in which a compound is converted to one or more products in the presence of heat and oxygen

NFPA

NFPA 921 (2004), Guide for Fire and Explosion Investigations

smoldering
combustion without flame, usually with incandescence and smoke

Historically, fire has been a beneficial force when controlled, but it has the potential to be a destructive force when unexpected and uncontrolled. Sometimes it is an unpredictable force of nature, but it can be prevented by sound fire safety practices such as control of smoking and proper storage and segregation of combustibles, and its effects can be efficiently controlled by well-designed suppression and detection systems. Although the study of fire dynamics warrants a book in itself, a simplified summary of the fundamentals of fire dynamics is needed before proceeding to a thorough discussion of suppression and detection concepts.

Combustion and Fire

Combustion is defined by NFPA 1, *Uniform Fire Code* (2006), paragraph 3.3.55, as "a chemical process of oxidation that occurs at a rate fast enough to produce heat and usually light in the form of either a glow or flame." As an example, a campfire that burns down to glowing coals is absent of flame but is still a fire by this definition. Conversely, a lighted cigarette lodged between two cushions of a sofa may smolder for hours, creating heat and a glow, before eventually bursting into flame. In most fires, combustion oxidation involves oxygen, found in our normal atmosphere, which contains 21% oxygen.

Pyrolysis is defined as a chemical process in which a compound is converted to one or more products in the presence of heat and oxygen.

Smoldering is defined by NFPA 921 (2004), *Guide for Fire and Explosion Investigations*, paragraph 3.3.140, as "combustion without flame, usually with incandescence and smoke." Oxidation reactions that are too slow to qualify for the NFPA definition of combustion or smoldering—including the slow rusting of iron and the yellowing of a newspaper over a long time, which produce insufficient heat for ignition or combustion—are not considered to be hazardous oxidation reactions.

Fire occurs in many forms. Students of chemistry are skilled at the use of Bunsen burners which use premixed gases for beneficial effect. Those who have started a campfire know that a campfire is initiated with the ignition of small twigs, whereupon larger sticks, and eventually large logs, can be added as the heat release of combustibles permits. Those who are experienced with prevention and extinguishment of fires in buildings and structures know that some of the many ways in which a fire can be initiated within a building are improper use of smoking materials, unwise placement of portable unit heaters, children playing with matches, and faulty electric wiring.

Classifications of Fire

Combustibles come in many varieties and configurations, from large logs to sawdust, from a mineshaft lined with coal to fine particles of coal dust, from a small oil can to an open vat of lubricating oil. Fires are classified by NFPA 1, *Uniform Fire Code* (2006), paragraph 3.3.99, as follows:

Class A Fires—Fires in ordinary combustible materials, including cellulosics such as wood, cloth, and paper, as well as rubber and many plastics

Class B Fires—Fires in flammable liquids, combustible liquids, petroleum greases, tars; oils, oil-based paints, solvents, lacquers, alcohols, and flammable gases

Class C Fires—Fires that involve energized electrical equipment

Class D Fires—Fires in combustible metals, such as magnesium, titanium, zirconium, sodium, lithium, and potassium

Class K Fires—Fires in cooking appliances that involve combustible cooking media (vegetable or animal oils and fats)

Ignition

ignition
the process of initiating self-sustained combustion

Ignition is defined by NFPA 921, *Guide for Fire and Explosion Investigations* (2004), paragraph 3.3.92, as "the process of initiating self-sustained combustion." This process is not alike for all combustion. For example, a match may singe a hardwood table but may be insufficient to ignite it, whereas a match would be more than sufficient to ignite a single sheet of paper held in the air and ignited from its tip. A sheet of paper floating in a tank of water, or coated in a flame-proof material is likely not to ignite with a match. Consider also a "flame-proof" mattress that will not ignite if a single match is thrown on its surface but may be capable of bursting into flames under the right conditions, such as an entire pack of lighted matches thrown onto its surface.

autoignition
initiation of combustion by heat but without spark or flame

Autoignition is defined by NFPA 921, *Guide for Fire and Explosion Investigations* (2004), paragraph 3.3.12, as "initiation of combustion by heat but without spark or flame." A flammable or combustible liquid may be ignited by heating alone without application of a pilot flame, such as a pan of grease on a stove.

piloted ignition temperature
minimum temperature a substance should attain in order to ignite under specific test condition

spontaneous ignition
initiation of combustion of a material by an internal chemical or biological reaction that has produced sufficient heat to ignite the material

heat transfer
a process that consists of conduction, convection, and radiation

Piloted ignition temperature is defined by NFPA 921, *Guide for Fire and Explosion Investigations* (2004), paragraph 3.3.94, as the "minimum temperature a substance should attain in order to ignite under specific test condition." It may be possible to heat a flammable liquid such that it will ignite with a pilot flame but will not autoignite. The point at which a flammable liquid will ignite under piloted ignition is called the flash point (see discussion later in the chapter).

Spontaneous ignition is defined by NFPA 921, *Guide for Fire and Explosion Investigations* (2004), paragraph 3.3.146, as "initiation of combustion of a material by an internal chemical or biological reaction that has produced sufficient heat to ignite the material." Rags soaked with a flammable liquid may ignite spontaneously if they are stored improperly, as could baled wet hay.

Heat Transfer

Once a substance ignites, it produces heat, which may transfer to adjacent combustibles. A single sheet of paper held in the air away from other combustibles will produce heat, but not enough heat to ignite adjacent combustibles.

Heat transfer is another engineering topic that deserves a semester or more of in-depth study. NFPA 921, *Guide for Fire and Explosion Investigations* (2004), defines heat transfer according to the following categories:

1. *Conduction*: "heat transfer to another body or within a body by direct contact." An example is two cushions on a couch, where fire spreads from one cushion to the other by direct contact, or by direct flame impingement.

2. *Convection*: "heat transfer by circulation within a gas or a liquid." A convection oven heats food not by direct contact but, rather, by heating and circulating the air around the food.

3. *Radiation*: "heat transfer by way of electromagnetic energy." A sofa that is fully involved in flame may produce sufficient heat to raise the temperature of drapes or furniture to their ignition temperature, even when the two objects are not in direct contact, depending on the heat produced by the flaming object, the properties of the object receiving heat, and the distance between the flaming and nonflaming object. A burning building may ignite adjacent buildings in a similar manner.

flashover
a transition phase in the development of a compartment fire in which surfaces exposed to thermal radiation reach ignition temperature more or less simultaneously and fire spreads rapidly throughout the space, resulting in full room involvement or total involvement of the compartment or enclosed space

Flashover

Combustion in a compartment may spread to a critical point called flashover. **Flashover** is defined by NFPA 921, *Guide for Fire and Explosion Investigations* (2004), paragraph 3.3.72, as "a transition phase in the development of a compartment fire in which surfaces exposed to thermal radiation reach ignition temperature more or less simultaneously and fire spreads rapidly throughout the space, resulting in full room involvement or total involvement of the compartment or enclosed space." Flashover of a compartment is a serious and dangerous phase of a fire and may endanger an entire structure. Flashover usually is identified visually by a sudden and violent burst of flame emanating from the doors or windows of a room.

Explosions

Some commodities, such as solid rocket fuel and nitroglycerine, offer special challenges and may explode under certain conditions. An explosion is defined and described by NFPA 921, *Guide for Fire and Explosion Investigations* (2004), paragraph 3.3.44, as "the sudden conversion of potential energy (chemical or mechanical) into kinetic energy with the production and release of gases under pressure, or the release of gas under pressure. These high-pressure gases then do mechanical work such as moving, changing, or shattering nearby materials." An example of an explosion is a vessel containing explosive dust that ignites, creating pressure on the interior of the vessel. The "mechanical work" to which the definition refers is the physical bursting of the vessel when the pressure exceeds the capacity of the vessel.

NFPA 15 (2007) identifies and defines two categories of rapid reactions, deflagration and detonation:

1. deflagration: "propagation of a combustion zone at a velocity that is less than the speed of sound in the unreacted medium."

2. detonation: "propagation of a combustion zone at a velocity that is greater than the speed of sound in the unreacted medium."

Deflagrations may be suppressed by a properly designed high-speed detection and suppression system. Detonations, however, pose challenges that detection and suppression systems are unlikely to address. More information about explosions can be found in Chapter 7.

THE FIRE TETRAHEDRON

Fire is a phenomenon that is not completely understood. Ever since the days of the cave dwellers, when the harnessing of fire was associated with our early technological development, it continues to be essential to our quest for efficient energy. Failure to control fire results in loss of life and property and continues to be a vexing social and technological problem that requires the expertise of fire protection technicians and engineers who possess the skills to devise measures to control and suppress fire.

The many varieties of combustion include the burning of ordinary combustibles such as paper, the combustion of gases, the combustion of flammable liquids, the combustion of dust particles, and the combustion of flammable metals. Most of these fire scenarios can be described generally as chemical reactions involving heat and a combustible in the presence of oxygen. The pictorial representation of the interdependent factors of heat, fuel, oxygen has historically been pictorialized as a fire triangle shown in **Figure 3-1(a)**.

FLUID
• SMOTHERING BY FLUID
• OXYGEN DEPLETION BY STEAM

SUPPRESSANT
• COOLING BY FLUID
• HEAT ABSORPTION BY STEAM

FUEL
• DILUTION OF LIQUID FUEL
• ABSORPTION BY LIQUID OR SOLID FUEL
• COATING OF SOLID FUEL

Figure 3-1a   *The fire triangle.*

fire tetrahedron
the pictorial representation of the interdependent factors of heat, fuel, oxygen, and uninhibited chemical chain reactions

With the addition of another interdependent contributor to the sustainment of combustion, uninhibited chemical chain reactions, the completed figure has come to be recognized as the **fire tetrahedron**, shown as **Figure 3-1(b)**, which expands the traditional fire triangle.

The fire tetrahedron pictorially describes all four interdependent combustion processes and demonstrates the four modes of extinguishment. A tetrahedron is a four-sided geometric figure, and the fire tetrahedron represents one mode of extinguishment for each surface of the three-dimensional, four-sided figure.

Each of the sides of the fire tetrahedron are discussed as follows:

Fuel

The presence and configuration of fuel is certainly a prerequisite for the commencement and sustainment of combustion. A primary objective of a fire safety inspection is to locate combustibles safely and to minimize sources of possible ignition. We must decrease the probability of ignition of the combustibles by

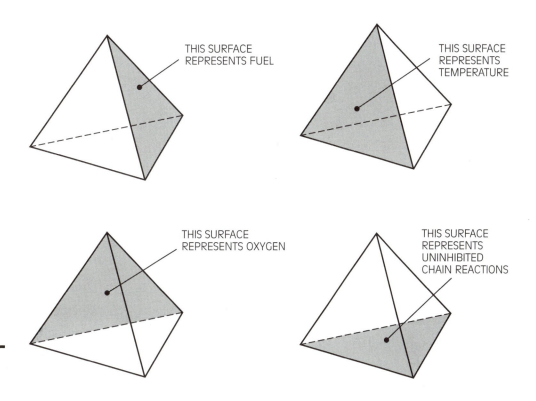

THIS SURFACE
REPRESENTS FUEL

THIS SURFACE
REPRESENTS
TEMPERATURE

THIS SURFACE
REPRESENTS OXYGEN

THIS SURFACE
REPRESENTS
UNINHIBITED
CHAIN REACTIONS

Figure 3-1b *The fire tetrahedron.*

controlling the storage and location of combustibles and by limiting possible ignition scenarios.

We know intuitively that loose, crumpled newspapers are more conducive to ignition by a single match than is a tall stack of tightly bound newspapers. Similarly, we start a campfire with small twigs, not huge logs. The ignition of suspended coal dust or sawdust, in the proper concentration, can result in a more violent combustion phenomenon than the ignition of a lump of coal or a block of wood. In addition, flammable and combustible liquids and gases pose potentially serious dangers, depending on the quantity, arrangement, configuration, and juxtaposition of these items to possible ignition sources.

The broad variety of combustibles, the wide variety of available ignition scenarios, and the numerous phenomena that contribute to sustained combustion make the study of fire a complex subject. Fire protection engineers are continually endeavoring to unravel its mysteries.

Oxygen

Under normal atmospheric conditions, the air we breathe consists of about 79% nitrogen and 21% oxygen, with trace amounts of carbon dioxide and other gases.

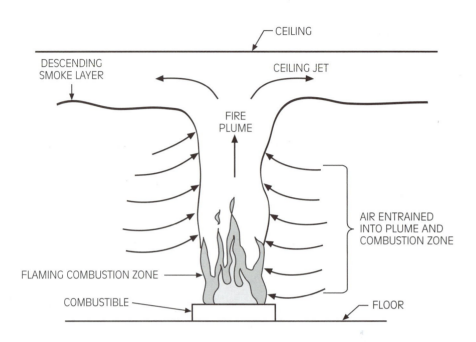

Figure 3-2 *The dynamics of fire.*

Just as our lungs draw oxygen from the air to sustain our bodies, a fire draws oxygen from the air to sustain combustion.

By understanding how oxygen is delivered to a fire, we can begin to understand how combustion is sustained. Interrupting the ready availability of oxygen to a fire can halt the expansion of the combustion process and, depending on the method of oxygen deprivation, may create a scenario for extinguishment.

A fire creates a vertical fire plume of gas and unburnt combustibles that draws air into it, increasing the volume of the descending smoke layer, as shown in **Figure 3-2**. The process of air being drawn into a fire plume is known as **entrainment**. Sustained combustion contributes to an increase in the heat release rate and lowering of the smoke layer in a room to the point at which human survival can be compromised. By limiting the ready access of oxygen to a combustible, we reduce the volume of air available for entrainment and a corresponding reduction in smoke-layer volume.

In a tightly sealed room, a fire may use up all of the oxygen within the room, "starving" the combustion process of a vital resource. A scenario in which the ability of the fire to spread depends on the quantity of oxygen available is called an **air-regulated fire**. This scenario could result in self-extinguishment of the fire without applying any agent. Because sustained combustion in an air-regulated fire depends on the availability of oxygen, the sudden introduction of oxygen to an air-regulated fire, perhaps caused by a firefighter opening a door or window, could result in a **backdraft**, a violent expansion of the combustion process

entrainment
the process of air being drawn into a fire plume

air-regulated fire
a scenario in which the ability of the fire to spread is dependent on the quantity of oxygen available

backdraft
a violent expansion of the combustion process caused by the sudden introduction of air to a fire

caused by the sudden introduction of air to a fire. This phenomenon could endanger occupants or fire service personnel.

A fire that has unlimited access to oxygen depends only on the fuel available and is called a **fuel-regulated fire**. A fuel-regulated fire sustains combustion until all fuel to the fire is gone or until a fire suppression agent is effectively applied. A campfire is an example of a fuel-regulated fire, and a building fire can become fuel-regulated if windows have been broken, doors are open, and other impediments to air flow have been removed.

fuel-regulated fire
a fire that has unlimited access to oxygen and is dependent only on the fuel available

Heat

The first step in starting a campfire is to apply heat with a match to small, dry wooden sticks, called **tinder**, which are easily ignitable. Tinder is used instead of large logs because it can be heated more effectively by a small energy source such as a match. Before combustion can commence, vapor is released from the tinder. When vapor is produced by the heating process in sufficient quantity, the vapor ignites and combustion of the tinder begins. Continued combustion releases increasingly higher levels of heat, which in turn increases the vaporization process and allows for larger sticks and logs to be added to the campfire.

tinder
easily ignitable objects used to start a fire

The key to application of heat to a combustible is an amount of heat sufficient to promote vaporization of a combustible to a level capable of sustained combustion. A forest fire, for example, is capable of providing extremely high, sustained temperatures, which can promote the ignition and combustion of even the largest trees. The forest fire most likely began, however, from a small heat-energy source, such as a match or cigarette that ignited some dry leaves, twigs, or underbrush.

Uninhibited Chemical Chain Reactions

The previous three elements of the fire tetrahedron—fuel, oxygen, and heat—have been recognized as the fire triangle. This was the classical presentation of fire propagation until it became clear that some fires were propagating and were being extinguished in a manner not described by the fire triangle. Sustained combustion can occur when the flame temperature is at such a level as to have a direct effect on increasing the velocity of intermolecular collisions between fuel and oxygen molecules.

As long as the velocity of these intermolecular collisions is permitted to be uninhibited, combustion may continue and potentially spread. Conversely, if an appropriately selected suppression agent is added that inhibits the velocity of these intermolecular collisions, extinguishment may ensue, as discussed later in this chapter.

FIRE SIGNATURES

fire signature

any fire effect (smoke, heat, light, etc.) that can be sensed by a fire detector

A **fire signature** is any fire effect (smoke, heat, light, etc.) that can be sensed by a fire detector. An unsuppressed fire can be represented by a graph that represents the heat released by the fire as a function of time, as shown in **Figure 3-3**. The amount of heat released by a fire varies in accordance with the type of combustible, arrangement of the combustible, availability of oxygen, and numerous other factors. The heat-release graph for a combustible can be the basis for a fire signature because it represents a specialized case of a given combustible burning under specific conditions.

Other fire signatures of use in detection technology include the graph of light emission from a fire, or the graph of the release of specific gaseous combustion byproducts, such as carbon monoxide. A well-selected detector can recognize effects of a fire and perform as specified, including the automatic initiation of a suppressing agent.

Other fire attributes that can be recognized as a fire signature are smoke production, light emmission, carbon monoxide production, and the production of other fire byproducts. Detectors can also be positioned to notice other

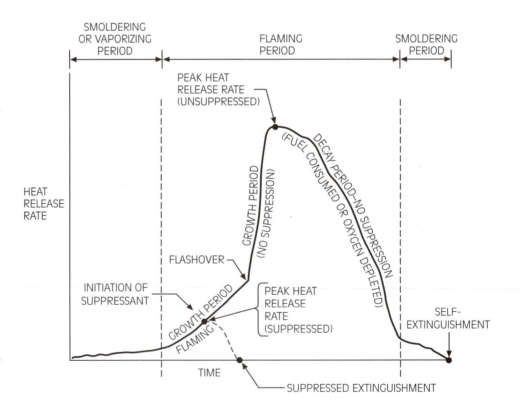

Figure 3-3 *Burning rate of a fire.*

combustion and precombustion changes, such as a rapid decrease in temperature when a liquified natural gas (LNG) leak occurs. Some research in using cameras to view a hazard area has been conducted, in which a camera films a specific area and notifies a control unit if characteristics of flame are in evidence.

The Unsuppressed Fire

A burning rate for an unsuppressed fire consists of three periods, as shown in Figure 3-3:

1. Smoldering or vaporizing period
2. Flaming period (which includes the growth period and the decay period)
3. Post-fire smoldering period

The length of the smoldering or vaporizing period depends on the type of combustible, configuration of the combustible, amount of heat applied, and amount of oxygen available. A fire started by a cigarette wedged between seat cushions could smolder for hours, whereas a dust explosion or piloted ignition of a flammable or combustible liquid could occur in milliseconds.

Once the growth period of an unsuppressed fire starts, it continues to grow until the peak heat release rate is reached. Achievement of the peak release rate is ordinarily associated with the total involvement of all combustibles in the combustion process.

The decay period occurs when the fuel is consumed by the fire in a fuel-regulated fire, or when the oxygen becomes depleted in an air-regulated fire. The decay period proceeds to the post-flaming smoldering period, and continues until the fire is completely out.

The Suppressed Fire

When a fire suppression agent is applied effectively to a fire, the suppressed fire burning rate, shown as a dotted line below the unsuppressed fire burning rate in Figure 3-3, can be altered dramatically from the unsuppressed fire burning rate. The change in the burning rate begins instantaneously upon the effective delivery of a suppression agent to a fire. Effectiveness of the suppression agent depends on the quantity, flow rate, and method of delivery and application of the suppressant to the fire, and can also be affected by such factors as wind, oxygen level, or enclosure temperature. The delivery of suppression agent is initiated by a detector that recognizes the change in burning rate or other changes as a fire signature that warrants activation of the detector.

Figure 3-3 shows that effective delivery of suppressant can dramatically shorten the fire growth period, resulting in a peak release rate significantly lower than the unsuppressed fire. The point on the fire signature related to the commencement of application of suppressing agent is determined by the choice of detection or sensitivity of the releasing mechanism for the suppression system, as discussed in Chapters 11 through 15. The significance of the lowered heat release rate of a suppressed fire signature drives to the heart of the need for automatic suppression systems. A principal reason for installing a suppression system is to prevent flashover, because flashover in a room could occur before the fire service arrives and could threaten all lives in a building and increase the probability that an entire building will be consumed by fire.

SPECIAL HAZARD FIRE SUPPRESSION AGENTS

A fire protection system designer can choose from a variety of available suppression agents. These fall into two major categories: aqueous agents and nonaqueous agents.

Special Hazard Aqueous Agents

aqueous agents
fire suppression agents that are water-based and may involve an additive that enhances the effectiveness of water

Aqueous agents include water and water-based agents mixed with additives that enhance the effectiveness of water. An example is the addition of air and foam concentrate to water, creating a fire-fighting-foam. Foam is significantly lighter than water and forms a blanket over a combustible, making foam more effective at smothering a fire, especially for suppressing flammable liquids in which water is heavier than the liquid being protected. When the flammable or combustible liquid is heated during combustion, a foam blanket forms a separation between the liquid and the vapor produced.

Suppression with foam is discussed in detail in Chapters 4 and 5. For a comprehensive study of water as an extinguishing agent, please refer to *Design of Water-Based Fire Protection Systems,* also published by Delmar Learning.

Special Hazard Nonaqueous Agents

nonaqueous agents
agents in which water is not a component

Nonaqueous agents are agents in which water is not a component. Nonaqueous agents may be selected for a variety of reasons, usually associated with the availability or adjudged effectiveness of water with respect to extinguishing specific commodities. Water may not be an appropriate extinguishing agent in the following instances:

- Water is susceptible to freezing and may not be the best choice in unusual circumstances, such as protection of a racing vehicle during a race in sub-freezing temperatures. In this case, stored water could freeze and be unavailable for delivery when required.

- The freezing and expansion of water may damage the item being protected. Freezing concerns may be overcome by heating or otherwise protecting the source of water supply, or by shielding, reinforcing, or otherwise protecting the object of suppression.

- Water is heavy. When it has to be stored in large quantities, it is impractical for protecting objects where weight is a major concern, such as on the space shuttle and space station. The National Aeronautics and Space Administration (NASA) selected a gaseous agent for these specialized applications.
- Water may react with the item being protected. For example, in a metal-working shop, the combustion of magnesium or other reactive metal shavings could react violently upon the application of water.
- Being conductive, water could cause electrical or electronic equipment to fail.
- The conversion of water to steam could become a hazard.
- Water may not be amenable for the suppression of some flammable-liquid fires.
- Water may leave an unacceptable level of residue or may require extensive clean-up.
- Water may not be available, or the storage of water may be impractical. Examples include protection of a small, remote building on a desert island or on a mountain where a public water supply is not present, and protection of a moving vehicle, which may not allow sufficient space to store water.
- Water may damage the item being protected. Examples include the protection of delicate electronics or a printing press, or irreplaceable items such as valuable paintings.

Limitations to the use of water may be overcome in some instances, such as the following:

- Electrical hazards can be accommodated if the power is shut down by a fire alarm system before water is applied.
- Hazards in which heat may sustain combustion, such as an oven or a vat of heated flammable liquid, can be protected more easily if the power to the item is turned off by the fire alarm system.
- Water damage to delicate commodities can be reduced by careful application of the water to the commodity, in small droplets or a mist.
- Water can be heated or antifreeze can be added to maintain its liquid state in freezing temperatures.
- Water damage can be minimized by shielding, such as installing a waterproof glass enclosure for the Mona Lisa painting.
- Water-mist systems, which require only small amounts of stored water, as discussed in Chapter 6, may successfully address the problems of water unavailability or the impracticality of water storage.

Examples of nonaqueous agents include gaseous agents and particulate agents. Gaseous agents available to the fire protection designer include carbon dioxide, discussed in Chapter 9, and clean agents, discussed in Chapter 8. Halon, a fire protection agent, is no longer available because of its effect on the ozone layer, as discussed in Chapter 8. Particulate agents are powders, such as dry chemical and wet chemical systems. These are covered in Chapter 10.

AGENT SELECTION

A fire protection system designer is responsible not only for the correct system design but also for the proper analysis of the commodity and for the correct selection of agent to protect the commodity. **Table 3-1** gives a comparison of special hazard agents.

Aqueous or water-based agents protect the vast majority of fire protection system applications because of their ready availability and low cost. In specialized cases where water may be unsuitable, or where a special hazard agent may be superior to water, several items must be considered.

- Because special hazard agents are stored in a container of specified size, the quantity must be calculated carefully . A reserve supply of the agent is

Table 3-1   *Comparison of special hazards agents.*

Agent	Extinguishing Mechanism	Predominant Applications	Agent Concentration	Delivery Methodology
CO$_2$	Reduces oxygen. Lowers temperature. Reduces gaseous fuel.	Marine applications flammable liquids electrical equipment fur storage vaults film vaults	34% minimum	Total flooding Local application
Clean agents and halogens	Halogens, cooling or breaking of combustion chain. Other clean agents extinguish by cooling or oxygen deprivation.	Telecommunication equipment computer and control rooms internet facilities	Halocarbons - 5–10% Inert gas - 34% +	Total flooding
Dry chemicals	Breaks combustion chain, reduces gaseous fuel and oxygen.	Flammable liquids and gases	Per manufacturer's guidelines and laboratory listing	Total flooding Local application
Wet chemicals	Smothering, cooling	Kitchen fires	Per Manufacturer's guidelines and Lab. listing.	Local application
Low-expansion foam	Physical separation of air from fuel, suppresses fuel vapors	Flammable liquids	0.1 to 0.16 GPM/sq.ft. (4.1 to 6.5 L/min/m^2)	Local application
High-expansion foam	Smothering	Three-dimensional flammable liquid hazards	Per NFPA 11A formula	Total flooding Local application

desirable and often required to be available. Because the supply of the agent is limited, the system must be designed to extinguish fires in all hazards that could occur simultaneously and should be capable of suppressing flare-ups or reignition after the main supply of agent is depleted.

- The agent must be selected to be compatible with the commodity being protected.
- For the highest probability of complete extinguishment, the special hazard agent system strategy must be compatible with the arrangement and configuration of the commodity. Strategies available to the fire protection designer are the local application method and the total flooding method.

APPLICATION METHODS

local application method
a method of fire suppression in which the agent is applied directly onto the point of hazard

total flooding method
a method of fire suppression involving the complete filling of a room or enclosed volumetric space with a fire protection agent

Some special hazard commodities may be extinguished successfully using agents employing the **local application method**, in which the agent is applied directly onto the point of hazard. An example of the local application method is to shake baking soda on a flaming frying pan to extinguish a grease fire. This method may not be suitable if the commodity is not arranged in such a manner as to be covered by the agent or if the agent is subject to scattering by wind or other factors.

In cases in which commodities are too large for local application or are more amenable to the application and containment of gaseous agents, the **total flooding method** of agent application is most suitable. Total flooding involves completely filling a room or enclosed volumetric space with a fire protection agent. Examples of total flooding are filling a room with an extinguishing gas such as carbon dioxide, or filling an enclosure with high-expansion foam.

EXTINGUISHMENT MECHANISMS

Using the fire tetrahedron as our guide, we can establish a strategy for the use of extinguishing agents to suppress fires.

Cooling—Heat Reduction

Water-based or aqueous agents are highly effective in reducing the temperature of a fire by absorbing heat. A fire grows by heating unburnt combustibles, resulting in vaporization and combustion of an expanding array of burning surfaces. The application of water-based agents coats the surfaces of unburnt combustibles, which inhibits vaporization of the fuel, and also absorbs heat from the combustion zone. Effective and sustained application of aqueous agents can reduce the temperature to a point incapable of sustaining combustion. Carbon dioxide and some clean agents also are capable of absorbing heat from a fire and reducing the flame temperature. Heat can also be reduced by arranging for a detection system to turn off heat-producing equipment, such as ovens, deep-fat fryers, or stoves.

Smothering—Oxygen Deprivation

Aqueous agents can be effective at smothering a fire, either by coating a burning object with a suppression agent, which completely covers a flaming surface with an aqueous or nonaqueous fire-fighting suppressant, or by immersing the object in an aqueous agent. A foam blanket can float on top of a flammable or combustible liquid, preventing oxygen from entering the combustion zone while simultaneously cooling the surface of the liquid.

Carbon dioxide and clean agents can smother a fire by displacing oxygen, using the local application or total flooding method. Dry chemicals can smother a fire when applied so oxygen is prohibited from contributing to the combustion process. Household applications of oxygen deprivation include placing a lid on a flaming frying pan to extinguish a fire or completely covering the contents of a frying pan with baking soda, which can smother a fire.

Fuel Removal

Forest fires can be extinguished by clearing an area of trees around the perimeter of a fire to limit the spread of the fire. A flammable liquid fire can be extinguished by draining the flammable liquid from the container. A campfire can be extinguished by removing the burning logs from the fire and spreading the unburnt fuel in a wider controlled area, where self-extinguishment can be accomplished at a faster rate.

An extinguishing agent can "remove" fuel from the combustion zone by coating the burning object and separating the fuel from adjacent combustibles and from the vapor produced when the object was heated. Coating prevents interaction between the heated fuel and the ignited vapor, starving the fire and promoting extinguishment.

Breaking the Combustion Chain

A method of extinguishment not explained by the fire triangle is not completely understood. Halons and dry chemicals were observed to extinguish fires by absorbing free radicals, which resulted in inhibiting the chain reactions that would have continued the combustion process. For example, the bromine atoms found in halon chemically absorb or attach themselves to hydrogen atoms that foster the spread or continuance of combustion, as shown in the following equation:

$$Br + H \rightarrow HBr$$

Br represents a bromine atom, H represents a hydrogen atom, and HBr represents a compound formed by attachment of the bromine and hydrogen atoms.

The arrow indicates that the reaction on the left side of the chemical equation yields the item(s) on the right side of the equation. Thus, attachment of the hydrogen atom to the bromine atom inhibits the hydrogen atom from fostering the continuance of combustion.

Free radicals such as OH, which foster the spread of the combustion reaction, can combine or attach themselves to bromine atoms, creating H_2O (water):

$$HBr + OH \rightarrow Br + H_2O$$

Similarly, the potassium atoms (K) in some dry chemicals attach themselves to free radicals (OH):

$$K + OH \rightarrow KOH$$

Discovery of this fourth mode of extinguishment—breaking the combustion chain reaction—made the fire triangle obsolete for some nonaqueous agents.

PORTABLE FIRE EXTINGUISHERS

NFPA

NFPA 10, Standard for Portable Fire Extinguishers

incipient fire
a small fire in its early stages of combustion

Portable fire extinguishers can contain a wide variety of extinguishing agents and are specified in accordance with NFPA 10, *Standard for Portable Fire Extinguishers*. NFPA 10 states that portable fire extinguishers enable an individual with minimal training to extinguish an incipient fire. Minimal training assumes little or no training or experience with the use of portable fire extinguishers, and an **incipient fire** is a small fire in its early stages of combustion, usually within 2 minutes of ignition. According to the National Association of Fire Equipment Distributors (NAFED), portable fire extinguishers extinguish 95% of fires on which they are used.

Limitations on Use of Fire Extinguishers

A portable fire extinguisher should not be considered as the sole solution to the fire protection analysis of a building but, rather, only one of many components of a total fire protection plan. Consideration of portable fire extinguishers brings into play three elements, each introducing inherent limitations in the use of fire extinguishers as primary in a fire protection plan.

1. *The person.* An individual may have no idea how to operate an extinguisher or how to operate it effectively. Precious time may be lost in reading directions and in aiming and discharging the extinguisher.

2. *The extinguisher.* Fire extinguishers are selected for specific kinds of fires, with the agent chosen accordingly. Attempts to use an extinguisher on fires other than those for which it was selected may result in ineffective extinguishment if the designer has specified an agent that is incompatible with the fire. Because an extinguisher has a limited amount of agent, with

a limited range and a limited discharge time, individuals with little or no experience who attempt to extinguish a fire instead of seeking safe refuge may be placing themselves in a potentially dangerous situation.

3. *Alarm initiation.* A person using a portable fire extinguisher is doing this instead of initiating an alarm to the fire department, and this delay could enable the fire to grow to dangerous proportions.

When a fire protection professional specifies portable fire extinguishers for a building, its occupants are given a set of unwritten expectations: Someone is expected to use a fire extinguisher to solve a fire problem. It may be unrealistic to expect this of anyone who has no ownership stake in the building. Further, by placing a fire extinguisher in a building, the professional potentially is placing an untrained person at risk. In return for that risk, the design professional should provide the protection of a complete automatic suppression and fire alarm system throughout the building capable of automatically sounding an alarm or of calling the fire department.

Classification of Fires and Portable Fire Extinguishers

Portable fire extinguishers are classified in accordance with the type of fire for which the extinguisher is effective:

Class A—Ordinary combustibles, usually cellulosics such as wood, fabrics, and paper, with limited amounts of rubber or plastic

Class B—Flammable liquids and gases

Class C—Fires involving energized electric components

Class D—Combustible metals or alloys

Class K—Fires in cooking appliances.

Determination of Numerical Ratings

Portable fire extinguishers are further classified by comparative numerical ratings based on test fires with expert operators at a testing laboratory.

- Class A rating: Based on the effectiveness of an extinguisher on a wood crib test fire, and a wood panel (wall) test fire; 1A rating represents the fire fighting equivalent of 1.25 gallons (4.731 L) of water.

- Class B rating: Based on the effectiveness of an extinguisher on a heptane fire in square pans of the appropriate size; a 10B rating represents the extinguishment of a 10-square-foot area of burning heptane by an amateur. To get a 10B rating, an expert in the testing lab must extinguish a 25-square-foot (2.323 m^2) area of 12-inch by 12-inch (30.48 cm $\times$ 30.48 cm) square pans. NFPA 10 establishes that an amateur is capable of extinguishing 40% of the area extinguished by the expert, or 10 square feet of the 25 square foot area—hence the 10B rating.

- Class C, D and K extinguishers: Must pass specific fire tests to obtain listing.

No numerical ratings exist for class C, D, or K extinguishers. The only criterion is that the agent be amenable to extinguishing a fire commensurate with the rating.

EXAMPLE 3-1

An example of a numerical rating for a fire extinguisher is 4A:20B:C. Such a rating would indicate an extinguisher that provides the equivalent of 5 gallons (18.925 L) of water (4A or 4 × 1.25), is effective for the extinguishment of a 20 square foot (1.84 m^2) flammable liquid fire by an amateur (20B or 2 × 10), and also has been tested to be effective for energized electrical components.

Types of Fire Extinguishers

Portable fire extinguishers are listed by Underwriters Laboratories in the varieties shown in **Table 3-2**. The following figures are applicable to UL-listed fire extinguishers:

Figure 3-4: Carbon dioxide extinguisher.

Figure 3-5: Clean agent extinguisher.

Figure 3-6: Dry chemical fire extinguisher—hand portable.

Figure 3-7: Dry chemical fire extinguisher—wheeled.

Table 3-2   *Underwriters Laboratories listed fire extinguishers.*

Water Extinguishers, Expellant Gas-Operated	• Intended for use on Class A fires.
	• Expellant gas water extinguishers contain plain water as the suppressant, and a stored expellant gas.
	• Expellant gas water extinguishers are listed to contain 1¼, 1½, or 2½ gallons of water, and are rated as 1-A or 2-A depending on size.
Water Extinguishers, Pump Tank-Operated	• Intended for use on Class A fires.
	• Expellant gas water extinguishers contain plain water as the suppressant, and a manually-operated pump.
	• Pump tank water extinguishers are listed to contain 1½, or 2½, 4, or 5 gallons of water, and are rates as 1-A to 4-A depending on size.
Water Spray Water Extinguishers	• Operate in temperature ranges of 35° F and plus 120° F.

(Continued)

Table 3-2   *Underwriters Laboratories listed fire extinguishers.* (continued)

	• Intended for use on Class A or C fires.
	• Water spray water extinguishers contain plain water as the suppressant, and are listed as: stored pressure, cartridge, or pressurized cylinder type.
	• Water spray water extinguishers are listed to contain 1.8 or 2½ gallons of water, and are rated as 2-A or 2-A:C.
Antifreeze Solution Extinguishers	• Intended for use on Class A fires
	• Suppression media used in extinguisher is protected and effective at temperatures below minus 40° F.
	• Used where Class A water extinguishers might freeze.
	• 2½ gallon capacity, classification 2-A.
Foam Fire Extinguishers, Hand and Wheeled	• Intended for use on Class A or B fires, and are not to be used on Class C or D fires.
	• Operate in temperature ranges of 35° F and 120° F.
	• Normal ambient temperature for storage is 70° F, and classifications and ratings were obtained at this temperature.
	• Some foam extinguishers are rated for marine use.
	• Foam extinguishers are listed as either the "stored pressure type" or "cylinder activated type."
	• Listed ratings range from 1-A:20-B to 20-A:160-B.
Carbon Dioxide Extinguishers	• Intended for use on Class B and C fires.
	• Operate in temperature ranges of minus 40° F and plus 120° F.
	• Normal ambient temperature for storage is 70° F, and classifications and ratings were obtained at this temperature.
	• Many carbon dioxide extinguishers are rated for use on marine craft.
	• Can be used in closets, small rooms, and confined spaces, but precautions need to be taken, such as use of breathing apparatus when in use.
	• Carbon dioxide extinguishers are listed with ratings ranging from 5-B:C to 20-B:C.

Table 3-2   (continued)

Clean Agent Extinguishers	• Intended for use on Class A, B, and C fires.
	• Operate in temperature ranges of minus 40° F and plus 120° F.
	• Normal ambient temperature for storage is 70° F, and classifications and ratings were obtained at this temperature.
	• Operators are advised to take precautions to avoid breathing clean agents.
	• Many clean agent extinguishers are rated for marine use.
	• Listed ratings range from 1-B:C to 5-B:C, and from 1-A:10-B:C to 10-A:80-B:C.
Dry Chemical Extinguishers, Hand and Wheeled	• Extinguishers charged with ammonium phosphate are intended for use on Class A, B, and C fires.
	• Extinguishers charged with a suppressant other than ammonium phosphate are intended for use on Class B and C fires, and are not amenable for deep-seated Class A fires.
	• Operate in temperature ranges of minus 40° F and plus 120° F.
	• Normal ambient temperature for storage is 70° F, and classifications and ratings were obtained at this temperature.
	• Operators are advised to take precautions to avoid breathing dry chemical.
	• Many dry chemical extinguishers are rated for marine use.
	• Dry chemical extinguishers are listed in the "stored pressure type" and the "cartridge-activated type."
	• Ammonium phosphate extinguishers are listed with ratings that range from 1-A:10-B:C to 40-A:320-B:C.
	• Potassium Bicarbonate extinguishers are listed with ratings that range from 10-B:C to 640-B:C.
	• Sodium bicarbonate extinguishers are listed with ratings that range from 2-B:C to 320-B:C.
Wet Chemical Fire Extinguishers	• Operate in temperature ranges of 35° F and plus 120° F.

(Continued)

Table 3-2 *Underwriters Laboratories listed fire extinguishers.* (continued)

	• Intended for use on Class A, B, C, or K fires. • Wet chemical extinguishers contain wet chemical as the suppressant, and are listed as: stored pressure, cartridge, or pressurized cylinder type. • Water spray water extinguishers are listed to contain 1.8 or 2½ gallons of water, and currently are rated as 2-A:1-B:C:K, 2-A:1-B, 2-A:C:K, 2-A:K, 1-B:C, 2-A:1-B:C, C:K, or K.
Liquified Gas-Type Extinguishers	• Intended for use on Class A, B, and C fires. • Operate in temperature ranges of minus 40° F and plus 120° F. • Normal ambient temperature for storage is 70° F, and classifications and ratings were obtained at this temperature. • Operators are advised to take precautions to avoid breathing gas emitted from these extinguishers. • Many liquefied gas extinguishers are rated for marine use. • Listed ratings range from 2-B:C to 10-B:C, and from 1-A:10-B:C to 30-A:240-B:C. • Liquified gas extinguishers are listed with either Halon 1211, Halon 1301, or Halon 2211.
Combustible Metal Extinguishers, Hand and Wheeled	• Specifically listed for use on Class D fires only. • Operate in temperature ranges of minus 40° F and plus 120° F. • Normal ambient temperature for storage is 70° F, and classifications and ratings were obtained at this temperature. • Extinguishment is possible by use of portable extinguisher or scooped/shoveled powder from storage pails or drums.
Residential Kitchen Cooking Fire Extinguishers	• Intended for use on Class K fires. • Residential kitchen cooking fire extinguishers are to be used on kitchen fires involving combustible vegetable or animal oils and fats.

Figure 3-4 *Carbon dioxide extinguisher. (Courtesy of Ansul, Inc.)*

Figure 3-5 *Clean agent extinguisher. (Courtesy of Ansul, Inc.)*

Figure 3-6 *Dry chemical fire extinguisher— hand portable. (Courtesy of Ansul, Inc.)*

DISTRIBUTION OF PORTABLE FIRE EXTINGUISHERS

Fire extinguishers are placed in a building as a function of the class of fire anticipated, (Class A, B, C, D, or K, as discussed previously), occupancy of the building protected, and travel distance that an occupant must undergo to access and retrieve the extinguisher.

Figure 3-7 *Dry chemical fire extinguisher—wheeled. (Courtesy of Ansul, Inc.)*

Occupancy

The 2007 edition of NFPA 10, Standard for Portable Fire Extinguishers, has added a significant amount of detail relative to the description of occupancies for which fire extinguishers are used—a major improvement over previous editions. Specific descriptions of occupancies are in Annex A on NFPA 10 (2007), not in the body of the standard, so specific occupancy classifications are recommendations, not requirements.

When evaluating occupancy classifications, note the following:

- the difference between classifications of hazards in the main body of NFPA 10, as opposed to associated descriptions in the annex of the standard
- with the exception of the specification of quantities of flammable liquids, quantities and juxtaposition of commodities are not always stated specifically for each occupancy classification
- the increasing severity of hazard when progressing from light to extra hazard occupancy
- the need for specialized training in the use of fire extinguishers becomes more important in areas of higher occupancy

The types of occupancies identified by NFPA 10 for protection by fire extinguishers are:

- *Light (low) hazard occupancy.* Defined by NFPA 10 as a room, space, or enclosure where the quantity and combustibility of Class A combustibles and Class B flammables are considered to be low. NFPA 10 (2007) recommends that buildings or rooms occupied as offices, classrooms, churches, assembly halls, and guest room areas of hotels and motels be classified as a light (low) hazard occupancy. Classification as a light hazard connotes that the majority of items in the room are noncombustible or are arranged so fire spread would not be rapid. Class B flammables may be present in small quantities (less than 1 gallon), in containers, and stored safely, such as might be found in some duplicating machines and art departments. An example of a light hazard occupancy is a classroom containing desks and chairs made of metal, wood, or hard-pressed fiberglass, where heat input to ignite the object would have to be significant, and where storage or placement of other easily ignited materials, such as large stacks of papers or books, is not present or is present in small quantities, or is safely separated from the furniture.

- *Ordinary (moderate) hazard occupancy.* Defined by NFPA 10 as a room, space, or enclosure where the quantity and combustibility of Class A combustibles and Class B flammables (1 to 5 gallons maximum) is considered to be moderate, and where fires of moderate heat release are expected. NFPA 10 (2007) recommends that rooms or buildings should be classified as ordinary (moderate) hazard occupancy when the following are encountered: dining areas, mercantile shops and associated storage, light manufacturing, research operations, auto showrooms, parking garages, and workshop or support service areas (kitchens, storage areas) of light hazard occupancies. Also recommended by NFPA 10 as ordinary hazard occupancy are: warehouses containing Class I commodities (noncombustible products either placed directly on wooden pallets or placed in single-layer corrugated cartons with or without single-thickness cardboard dividers and with or without pallets, or shrink-wrapped or paper-wrapped as a unit load, with or without pallets), and Class II commodities (a noncombustible product in slatted wooden crates, solid wood boxes, multiple-layered corrugated cartons, or equivalent combustible packaging material, with or without pallets). An example of ordinary hazard (moderate) occupancy is a mercantile occupancy such as a shoe store or supermarket, featuring public areas where items are regularly stored and displayed in moderate quantities and contain non-public storage areas with commodities stored on shelves or pallets.

- *Extra (high) hazard occupancy.* Defined by NFPA 10 as a room, space, or enclosure where the combustibility of contents is of the storage, handling, or manufacturing of Class A combustible material in which the quantity of Class A material is high, or where large amounts of Class B flammables (more than 5 gallons) are present, and where rapidly developing fires with high rates of heat release are expected. NFPA 10 recommends that extra hazard occupancies could consist of woodworking, vehicle repair, aircraft and boat servicing, cooking areas, individual product display showrooms, product convention center displays, and storage and manufacturing processes such as painting, dipping, coating, and flammable liquid handling. Also recommended as extra hazard occupancies are warehousing and in-process storage of other than Class I or Class II commodities. An example of an extra hazard occupancy is an auto shop with spray painting facilities, fuel system repair facilities, or automobile bumper stripping and chroming facilities.

- *Mixed occupancies.* Buildings featuring more than one occupancy may be protected on a room or area basis, with extinguishers appropriately placed for the occupancy. An example is a school, which would be expected to be protected with extinguishers rated for Class A hazards and light hazard occupancy, but also may contain a laboratory with a significant quantity of flammable liquid hazards, which would be protected by extinguishers rated for Class B hazards and ordinary hazard occupancy.

- *Specialized occupancies.* Note that NFPA 10 does not specifically list specialized occupancies, such as an aircraft hangar. Such occupancies are likely to contain specialized provisions for manual extinguishment, as would be found in NFPA 408, Standard on Aircraft Hand Portable Fire Extinguishers. NFPA 10 also lists a number of other specialized occupancies, including those covered by NFPA 30A, 32, 58, 86, 96, 120, 122, 241, 302, 303, 385, 407, 410, 418, 430, 498, 1192, and 1194.

Travel Distance

To be able to use the fire extinguisher distribution tables, the concept of travel distance must be understood. If a Class A occupancy has a minimum travel distance of 75 feet (22.875 meters), extinguishers may be spaced 150 feet (45.75 meters) apart, as shown on **Figure 3-8**. As can be seen, an individual directly at the center point between two extinguishers spaced at 150 feet (45.75 meters) apart has a travel distance of 75 feet (22.875 meters) to any fire extinguisher.

travel distance
the actual walking distance from one point to another

Travel distance, the actual walking distance from one point to another, must take into consideration the presence of furniture and other obstructions. Travel distance, therefore, cannot be ascertained by making straight-line determinations from a plan, unless furniture or other compensatory features are present. In Figure 3-8, we evaluate a corridor where travel distance is expected to be unimpeded.

STRAIGHT-LINE ARRANGEMENT

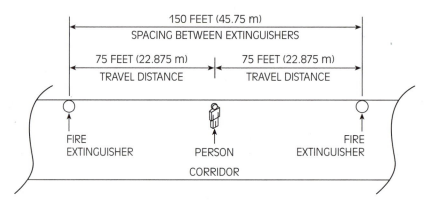

ALTERNATE 90° ARRANGEMENT

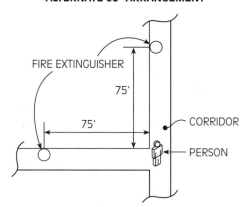

Figure 3-8

Illustration of travel distance. (Note: A Class A hazard is shown, and the occupant must travel 75 feet (22.875m) to reach an extinguisher; extinguishers, therefore, are 150 feet (45.75m) apart.)

Placement of Extinguishers for Class A Hazards

NFPA 10, Standard for Portable Fire Extinguishers (2007), provides a table for placement of fire extinguishers in Class A hazards, as shown in **Table 3-3**. This table shows the minimum size and spacing, and the maximum floor area permitted per extinguisher, for fire extinguishers required for each hazard classification—light, ordinary, and extra. A uniform travel distance of 75 feet (22.875 meters) applies to all Class A hazards.

Up to half of the required complement of fire extinguishers from this table are permitted to be replaced by 1½" water hose stations, uniformly spaced in accordance with NFPA 14, Standard for the Installation of Standpipe, Private Hydrant, and Hose Systems. Hose stations would replace every other fire extinguisher spaced as required by Table 3-3.

Table 3-3   *Fire extinguisher size and placement for Class A hazards.*

Criteria	Light (Low) Hazard Occupancy	Ordinary (Moderate) Hazard Occupancy	Extra (High) Hazard Occupancy
Minimum rated single extinguisher	2-A	2-A	4-A
Maximum floor area per unit of A	3000 ft²	1500 ft²	1000 ft²
Maximum floor area for extinguisher	11,250 ft	11,250 ft	11,250 ft
Maximum travel distance to extinguisher	75 ft	75 ft	75 ft

For SI units, 1 ft = 0.305 m; 1 ft² = 0.0929 m².
Note: For maximum floor area explanations, see E.3.3.

Source: NFPA 10 (2007), Table 6.2.1.1. Reprinted with permission from NFPA 10, *Portable Fire Extinguishers,* Copyright © 2007, National Fire Protection Association, Quincy, MA 02169. This reprinted material is not the complete and official position of the National Fire Protection Association on the referenced subject which is represented only by the standard in its entirety.

Placement of Extinguishers for Class B Hazards

NFPA 10, Standard for Portable Fire Extinguishers (2007) provides a table for placement of fire extinguishers in Class B hazards, as shown on **Table 3-4**.

Table 3-4 shows the minimum size and spacing, and the maximum floor area permitted per extinguisher for Class B hazards, with Class B fire extinguishers required for each hazard classification—light, ordinary, and extra. Travel distances to an extinguisher vary with occupancy, but are less than that for Class A hazards. Some exceptions to Table 3-4 are the following:

- Up to three AFFF or FFFP fire extinguishers of at least 2½ gallons (9.46 liters) shall be permitted to fulfill extra hazard occupancy requirements.

- Two AFFF or FFFP fire extinguishers of at least 1½ gal (6 liters) capacity are permitted to fulfill ordinary hazard requirements.

- Table 3-4 is to be used only for flammable liquid hazards that are not of appreciable depth. Fire extinguishers shall not be used as the sole source of fire protection for Class B hazards whose surface area exceeds 10 ft² (0.93 m²), of appreciable depth, or whose surface area exceeds 20 ft² (1.86 m²), of appreciable depth where personnel are trained in extinguishing fires. The standard does not specify the depth of flammable liquid that qualifies as

Table 3-4   *Fire extinguisher size and placement for Class B hazards.*

Type of Hazard	Basic Minimum Extinguisher Rating	Maximum Travel Distance to Extinguishers	
		ft	m
Light (low)	5-B	30	9.15
	10-B	50	15.25
Ordinary (moderate)	10-B	30	9.15
	20-B	50	15.25
Extra (high)	40-B	30	9.15
	80-B	50	15.25

Notes:

(1) The specified ratings do not imply that fires of the magnitudes indicated by these ratings will occur, but rather they are provided to give the operators more time and agent to handle difficult spill fires that could occur.

(2) For fires involving water-soluble flammable liquids, see 5.5.4.

(3) For specific hazard applications, see Section 5.5.

Source: NFPA 10 (2007), Table 6.3.1.1. Reprinted with permission from NFPA 10, *Portable Fire Extinguishers,* Copyright © 2007, National Fire Protection Association, Quincy, MA 02169. This reprinted material is not the complete and official position of the National Fire Protection Association on the referenced subject which is represented only by the standard in its entirety.

"appreciable," but the annex suggests that an "appreciable depth" be considered as ¼ in. (0.64 cm). A flammable liquid "not of appreciable depth" relates to the probability of most of the fuel being consumed by combustion, reducing the probability of reignition of flammable liquid after the contents of the fire extinguishers have been depleted.

Placement of Extinguishers for Class C Hazards

NFPA 10, Standard for Portable Fire Extinguishers (2007), requires that extinguishers used for Class C fires be rated specifically for Class C fires involving energized electrical equipment, and are permitted to be spaced in accordance with Class A or Class B hazards.

Placement of Extinguishers for Class D Hazards

NFPA 10, Standard for Portable Fire Extinguishers (2007), requires that extinguishers used for Class D fires be rated specifically for Class D fires involving combustible metals, and are permitted to be spaced with a maximum travel distance of 75 feet (23 m). **Table 3-5** shows some capabilities, dangers, and attributes of combustible metals.

Table 3-5   *Combustible metal properties.*

	Aluminum	Lithium	Magnesium	Titanium	Zirconium
Methodology of extinguishment	• Listed class D agents • Dry sand • Dry inert granular material	• Listed class D agents	• Listed class D agents • Tested substitutes	• Listed class D agents • Tested substitutes	• Listed class D agents • Tested substitutes
Water reactivity	• Reactive to water	• Violent reactivity to water	• Magnesium shavings or dust is violently reactive to water	• Reactive to water	• Reactive to water
Reactivity to other extinguishing agents		• Can react to carbon dioxide, halon or foams	• Can react to carbon dioxide, halon, foams, or sand.		• Can react to carbon dioxide, halon, foams, or sand
Reactivity to air	• Reactive to oxygen	• Reacts to moisture in air	• Reactive to oxygen	• Reactive to oxygen	• Reacts to nitrogen, carbon dioxide, and oxygen

Placement of Extinguishers for Class K Hazards

NFPA 10, Standard for Portable Fire Extinguishers (2007), requires that extinguishers used for Class K fires be rated specifically for Class K fires involving cooking equipment and are permitted to be spaced with a maximum travel distance of 30 feet (9.15 m).

Technician Certification

Another addition to NFPA 10 (in 2007) was a paragraph requiring training and certification for technicians who perform annual service, maintenance, and recharging of fire extinguishers after August 17, 2008. The National Institute for Certification in Engineering Technologies (NICET) has a certification program, as well as some equipment manufacturers and suppliers and trade organizations.

The authority having jurisdiction must be consulted to determine the appropriate certification required in a specific jurisdiction. Most certification programs require the applicant to complete a course and/or to pass an examination on the body and annexes of NFPA 10, and on the manufacturer's requirements and recommendations on the care and maintenance of specific types of extinguishers.

Electronic Monitoring

Recognizing that fire extinguishers may be removed improperly from their holders and perhaps misused, as might occur at a fraternity house during spring break, and recognizing that servicing of extinguishers may not always be performed in a timely manner, NFPA 10 contains provisions for electronic monitoring of fire extinguishers.

Electronic monitoring involves an electronic contact that senses when an extinguisher has been removed from its holding bracket and can provide a trouble signal to a fire alarm control unit (FACU) when the extinguisher has been removed. In the event of a fire, fire service personnel can gain valuable information relative to fire location when a control unit indicates that an extinguisher has been removed from a specific location.

Where such devices are installed, NFPA 10 requires that they be inspected and tested annually. Electronic monitoring also can monitor the pressurization of the extinguisher and may be capable of identifying if objects have been placed in front of the extinguisher, inhibiting its use. Designers and authorities having jurisdiction must evaluate the occupancy of the protected facility and make determinations relative to the level of monitoring required.

High-Flow Extinguishers for Specific Fires

Specific applications where flammable liquids or gases are pressurized, potentially resulting in a jet-flow flame; or for three-dimensional fires, high-flow fire extinguishers, as shown on **Figures 3-9** and **3-10**, may be specified. Where specified, large flow dry chemical extinguishers of 10 lb (4.54 kg) or greater, and with a discharge rate of 1 lb/sec (0.45 kg/sec), is mandated. NFPA 10 cautions that extinguishing pressurized gas or liquid fires is hazardous and mandates that fuel supply to such a fire be shut off.

NFPA 10 further mandates that AFFF or FFFP extinguishers not be used for water-soluable flammable liquid fires, such as alcohols, acetone, esters, and ketones. Also, for obstacle fires, an extinguisher containing a vapor-suppressing foam agent, or multiple extinguishers containing non-vapor-suppressing Class B agents, or extinguishers of 10 lb (4.54 kg) or greater and with a minimum discharge of 1 lb/sec (0.45 kg/sec).

In addition, training of personnel is essential. This training is available through many equipment manufacturers and community college fire science programs.

Obsolete Extinguishers

NFPA 10 (2007) expands on the list of fire extinguishers that are required to be removed and replaced with listed fire extinguishers to increase reliability. Included on this list of obsolete extinguishers are soda acid, chemical foam extinguishers, vaporizing liquid extinguishers, cartridge-operated water

Figure 3-9
Extinguishing a flammable liquid under pressure. (Courtesy of Ansul, Inc.)

extinguishers, cartridge-operated loaded-stream extinguishers, copper or brass shell extinguishers joined by soft solder or rivets, carbon dioxide extinguishers with metal horns, solid-charge AFFF extinguishers, pressurized water extinguishers manufactured before 1971, extinguishers required to be inverted to operate, stored-pressure extinguishers manufactured before 1955, extinguishers with 4B, 6B, 8B, 12B or 16B ratings, stored-pressure water extinguishers with fiberglass shells manufactured before 1976, dry chemical stored-pressure extinguishers manufactured prior to 1984, and any fire extinguisher which can no longer be serviced per manufacturer's instructions. In areas where NFPA 10 is accepted, removal and replacement of the previously listed extinguishers is mandatory.

Figure 3-10
Extinguishing an obstacle-spill fire. (Courtesy of Ansul, Inc.)

SUMMARY

The fundamentals of fire propagation, heat transfer, flashover, and extinguishment are important to understanding the suppression system methodologies that follow this chapter. The fire triangle explains extinguishment related to the reduction of oxygen and temperature and the removal of fuel. The fire tetrahedron explains an additional mode of extinguishment related to the inhibition of chain reactions. Special hazard agents are aqueous (water-based) or nonaqueous (non-water-based). Evaluation of a fire-suppressing agent involves a compatibility analysis of the agent with respect to the commodity protected.

A fire signature is a graph of one or more fire characteristics with respect to time for a specific commodity under specific conditions upon which a fire detector's response can be modeled. An unsuppressed fire may grow to the point where flashover occurs and enclosures and combustibles adjacent to the enclosure involved in flashover are threatened.

Underwriters Laboratories provides a resource for selecting a fire extinguisher from appropriate groupings of listed and tested extinguishers. NFPA 10 has added a significant amount of detail in the 2007 edition that assists designers in specifying fire extinguishers.

REVIEW QUESTIONS

1. For each surface of the fire tetrahedron, describe a mode of extinguishment, an example of a fire extinguished by this mode, and an agent that is capable of extinguishing using this mode.

2. Describe the importance of the difference between the peak heat release rates for an unsuppressed fire and a suppressed fire.

3. What are some examples of a fuel-regulated fire and an air-regulated fire? Use these examples to demonstrate possible extinguishment scenarios involving each type of fire.

4. Determine the number and rating of fire extinguishers required to be installed along a 1,000 ft (305 m) corridor in an office building.

DISCUSSION QUESTIONS

1. Would the amount of air entrained into a plume that is 30 feet tall be the same, less, or more than the amount of air entrained in a 10-foot tall plume? Why? If the amount of entrained air differs, compare the relative plume temperature at the top of each plume and explain the reason for the difference.

2. Describe and sketch the relative differences among the following unsuppressed fire signatures, and sketch suppressed fire signatures for each. Which fire scenarios are more likely to result in flashover?

   a. a small piece of scrap paper and a room filled with stacked newspapers.

   b. a large block of wood and a pail of liquid nitroglycerine.

   c. a sofa and a deep vat of flammable liquid.

3. For each fire scenario, determine whether total flooding or local application would be the most appropriate method of extinguishment.

   a. a grease fire in a pan on a stove.

   b. a coal mine filled with coal dust.

   c. a specialized jigsaw that cuts cylinders of solid rocket fuel into discs resembling hockey pucks.

   d. a paint spray booth filled with combustible paint spray and fumes.

ACTIVITIES

1. Visit a facility where fire tests are conducted. Observe or participate in a fire experiment. Determine the object of the experiment, and if data are sufficient for the development of a fire signature. What methodologies would be capable of developing a fire signature for a specific object?

2. Observe the extinguishment of a fire by the fire service. How many of the four modes of extinguishment described by the fire tetrahedron did the fire service employ when attacking a fire? Give examples of each.

3. Interview a fire service official. What aqueous and nonaqueous agents did the fire service employ, and what methodology is used to determine the selection of an agent for a specific fire?

Unit

2

WATER-BASED SPECIAL HAZARD SYSTEMS

Chapter 4

LOW-EXPANSION FOAM SYSTEM DESIGN

Learning Objectives

Upon completion of this chapter, you should be able to:

- Classify high, medium, and low-expansion foams, given their respective expansion ratios.
- Determine whether a liquid is flammable or combustible.
- Know the advantages and disadvantages of the currently available varieties of low-expansion foams.
- Design and calculate a surface or subsurface low-expansion foam system for the exposed fuel surface within a flammable or combustible liquid storage tank.
- Design and calculate a low-expansion foam seal protection system for a floating roof tank.
- Design and calculate a low-expansion foam dike protection system for a tank farm.
- Design and calculate a low-expansion foam system for an aircraft hangar.
- Perform a detailed layout of a low-expansion foam system, designed in accordance with NFPA 11.

Low-expansion foam systems are designed in accordance with NFPA 11 (2005 Edition), *Standard for Low-, Medium-, and High-Expansion Foam*, to protect fires primarily associated with flammable and combustible liquids. A **flammable liquid** is defined by NFPA 11 as a liquid having a flash point below 100°F (37.8°C) and having a vapor pressure not exceeding 40 psi (2068.6 mmHg). A **combustible liquid** is defined as a liquid having a flash point at or above 100°F (37.8°C).

Low-expansion foam systems are used when a blanket of foam is needed to float on the horizontal surface of a flammable or combustible liquid. Limited protection may be provided for the blanketing or coating of a vertical surface.

The specifying engineer or design technician ordinarily specifies a low-expansion foam system when coating and smothering a two-dimensional surface area are determined to be the preferred modes of extinguishment for flammable and combustible liquids. Low-expansion foam, in its role of coating, separates the fuel from the flame, blocks air from entering the combustion process, which effectively smothers the fire, and simultaneously cools the surface of the fuel. Many protection scenarios involving low-expansion foam involve protection of **hydrocarbon** fuels and organic compounds, which contain only carbon and hydrogen (for example, methane, petroleum, coal, etc.).

COMPONENTS OF FOAM

All foams contain three components:

1. Air, contained within foam bubbles
2. Water, delivered at a specified density in gallons per minute per square foot of applied area
3. Foam concentrate, injected into the water stream at a specific predetermined percentage

Foam concentrate is purchased commercially in drums or barrels. When foam concentrate is mixed with water, it creates a foam solution. Foam solution flows from the point of mixing to the hazard through a system of pipes. At the hazard, a system of discharge devices is installed to facilitate mixing of the foam solution with air in the correct expansion ratio. The mixture of the foam solution with air is called **fire-fighting foam**.

The piping and delivery method for low-expansion foam systems varies in accordance with the application and hazard configuration. This chapter covers design and calculation methods for several specific foam applications, including storage tank protection, truck loading racks, dike protection, and aircraft hangar protection.

EXPANSION RATIO

Foams are classified by NFPA 11 according to their ratio of expansion and fall into three major categories:

Classification	NFPA 11 Ratios	Typical Ratio
Low Expansion	up to 20:1	8:1
Medium Expansion	20:1 to 200:1	100:1
High Expansion	200: to 1000:1	500:1

The **expansion ratio** of foam is computed by measuring the volume of the foam produced after air is added to the foam solution and comparing that volume to the original volume of foam solution prior to air addition. A **low-expansion foam system**, therefore, is defined as a system designed to deliver a foam solution possessing an expansion ratio of up to 20:1 to a hazard.

TYPES OF FOAM

Available foams include protein-based, fluoroprotein based, aqueous film-forming foam, alcohol-resistant concentrates, and chemical foams. Fire-fighting foams can have a negative effect on the environment if they are not recovered and removed and disposed of properly after use. Proper drainage to retention tanks and proper retainage is recommended by NFPA 11 Annex F.

- **Protein foam** has an expansion ratio of 8:1 to 10:1 and contains protein-based animal additives, such as hooves, bones, and feathers, which provide protein for the solution. Such solutions can be effective for hydrocarbon fires but do not form a film on the fuel surface and are known to absorb some fuel into the foam, possibly resulting in failure of the foam to maintain a floating posture above the fuel. Protein foams may also have a shorter shelf life than other foams, requiring more frequent replacement.

- **Fluoroprotein foam** is a protein foam that contains fluorochemical additives that make the foam flow more easily with increased resistance to absorption of fuel.

- **Film-forming fluoroprotein foam (FFFP)** is a concentrate that uses fluorinated surfactants to produce a fluid aqueous film for suppressing hydrocarbon fuel vapors. FFFP can be more effective than fluoroprotein foams in applications where a film barrier between the foam and fuel is desirable.

- **Aqueous film-forming foam (AFFF)** is a synthetic foam recommended for the coating of most flammable liquids in storage tanks because of its ability to provide a thin, aqueous film that separates the foam from the fuel, enhancing the ability of the foam to smother, as shown in **Figure 4-1**.

expansion ratio
computed by measuring the volume of the foam produced after air is added to the foam solution and comparing that volume to the original volume of foam solution prior to air addition

low-expansion foam system
a system designed to deliver a foam solution, possessing an expansion ratio of up to 20:1, to a hazard

NFPA

NFPA 11 Annex F

protein foam
a foam that contains protein-based animal additives

fluoroprotein foam
a foam that contains fluorochemical additives that make it flow more easily

film-forming fluoroprotein foam
a concentrate that uses fluorinated surfactants to produce a fluid aqueous film for suppressing hydrocarbon fuel vapors

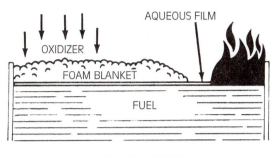

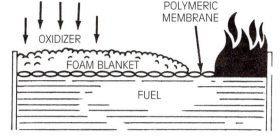

Figure 4-1   *Low-expansion foam. (Courtesy of Ansul, Inc.)*

aqueous film-forming foam (AFFF)
a synthetic foam that forms a thin aqueous film that separates the foam from the fuel

alcohol-resistant foam
a foam used for the protection of alcohol-based fires

chemical foams
a type of foam that depends on the initiation of a chemical reaction within the foam solution to create the air bubbles in the foam; became obsolete with the introduction of AFFF and fluoroprotein foams

foam proportioner
a manufactured product designed to ensure delivery of the precise ratio of foam concentrate to a foam solution

AFFF is the foam of choice for most aircraft fire protection, is readily available, and is commonly employed in many flammable and combustible liquid applications.

- **Alcohol-resistant foam** is used for the protection of alcohol-based flammable liquid fires. Most alcohol flammable liquids collapse the bubbles of standard foam agents because they absorb some of the water from the foam solution. Alcohol-resistant concentrates (ARC) form a polymeric membrane between the foam and the fuel that preserves the air bubbles and their fire-fighting capabilities, as shown in Figure 4-1.
- **Chemical foams** depend upon the initiation of a chemical reaction within the foam solution to create the air bubbles in the foam. These foams have become obsolete with the introduction of AFFF and fluoroprotein foams.
- Synthetic detergent foams are non-fluorochemical hydrocarbon-type surfactants, mixed with solvents and water. No films or membranes are formed, but foam bubbles lay on the surface of the fuel, most commonly, distributed by high expansion foam generators.

PROPORTIONING METHODS

To deliver a foam with the proper expansion ratio and mixed in the proper proportions, a foam concentrate must be mixed with water by a **foam proportioner**. Foam proportioners must be matched exactly to the foam concentrate used, to ensure delivery of the precise ratio of foam concentrate to a foam solution. Proportioners are manufactured specifically to mix concentrates in 1%, 3%, and 6% ratios, and must be matched precisely to the respective concentrate percentage. For example, a 6% concentrate is designed to be used only with a 6% proportioner, and is proportioned at a ratio of 6% foam concentrate to 94% water.

Commonly encountered proportioning methods are the venturi proportioner, the pressure proportioner, and the balanced pressure proportioner. These are described as follows.

Venturi Proportioner

The **venturi proportioner**, also called an in-line proportioner, takes advantage of the laws of physics to perform its function. When water flows through a pipe past an open orifice, a negative pressure is created at that orifice, not unlike the pressure created when using a straw to consume a soft drink. In the case of the venturi proportioner, the negative pressure draws foam concentrate into the water stream, as shown in **Figure 4-2**. By precisely sizing and metering the orifice, the proper amount of concentrate is drawn into the water stream in proportion to the rate of water flow past the orifice. Venturi proportioners are manufactured to mix foam at a specific ratio—1%, 3%, or 6%.

Pressure Proportioner

A **pressure proportioner** draws a portion of the incoming water stream and uses it to pressurize the tank holding the foam concentrate. A **bladder tank** is a tank in which the concentrate is stored within a collapsible bladder, as **Figure 4-3** shows. When water is delivered to the bladder tank, it surrounds and presses on the bladder, squeezing concentrate from the bladder and forcing the concentrate toward the proportioner, as shown in **Figure 4-4**.

venturi proportioner
a device that utilizes the negative pressure created by water flowing past an open orifice to draw foam concentrate into the water stream

pressure proportioner
a device that draws a portion of the incoming water stream and uses it to pressurize the tank holding the foam concentrate

bladder tank
a tank in which the concentrate is stored within a collapsible bladder

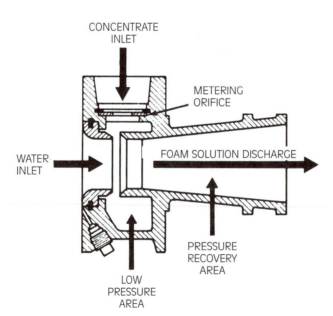

Figure 4-2 *Venturi proportioner. (Courtesy of Ansul, Inc.)*

Balanced Pressure Proportioner

balanced pressure proportioner
a device that uses an atmospheric foam concentrate tank, a pump to pressurize the concentrate and force it toward the proportioner, and a proportioner that balances the pumped concentrate pressure to the water supply pressure

A **balanced pressure proportioner** uses an atmospheric foam concentrate tank, a pump to pressurize the concentrate and force it toward the proportioner, and a proportioner that balances the pumped concentrate pressure to the water supply pressure, mixing the two at the correct ratio, as **Figure 4-5** illustrates.

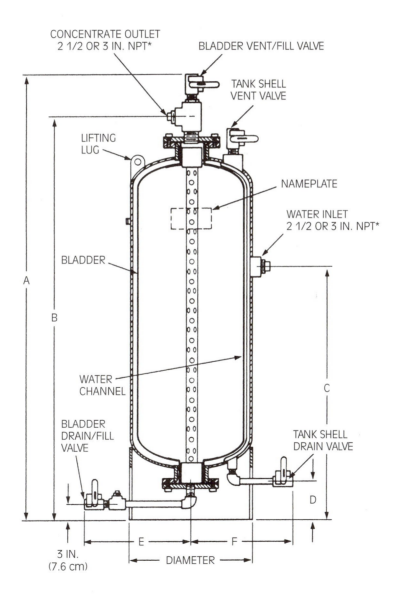

Figure 4-3 *Bladder tank. (Courtesy of Ansul, Inc.)*

NOTES:

1. SPRINKLER VALVE MAY BE ALARM CHECK, DRY PIPE, PRE-ACTION, OR DELUGE TYPES AS REQUIRED BY SYSTEM DESIGN.

2. DISCHARGE DEVICE MAY BE SPRINKLER HEADS (AS SHOWN) OR OTHER TYPE DEVICE SUCH AS MONITOR NOZZLES, HANDLINE NOZZLES, OR FOAM CHAMBERS AS REQUIRED BY SYSTEM DESIGN.

3. ARROWS INDICATE DIRECTION OF FLOW.

4. RECOMMENDED INTERCONNECTING PIPE, FITTINGS, AND VALVE (SEE CHART, NUMBERS 6 THRU 9) SIZES TO BLADDER TANK ARE GIVEN CORRESPONDING TO PROPORTIONER SIZE.

5. PIPE, VALVES, AND FITTINGS MAY HAVE TO BE UPSIZED TO ENSURE NEAR 0 PSI FRICTION LOSS TO MAINTAIN BALANCED PRESSURE OF WATER AND CONCENTRATE AT THE PROPORTIONER.

6. THE HYDRAULIC CONCENTRATE VALVE (VALVE #8) MAY BE ELIMINATED ON AN AUTOMATIC SYSTEM HAVING ONE PROPORTIONER PROVIDED THE PROPORTIONER IS LOCATED AT AN ELEVATION AT OR ABOVE THE BLADDER TANK FOAM CONCENTRATE OUTLET CONNECTION LOCATED AT THE TOP OF THE TANK.

VALVE NO.	VALVE DESCRIPTION	NORMAL POSITION	
	DESCRIPTION	MANUAL SYSTEM	AUTO SYSTEM
1	BLADDER VENT/FILL - 1"	CLOSED	CLOSED
2	TANK SHELL VENT - 1"	CLOSED	CLOSED
3	TANK SHELL DRAIN - 1"	CLOSED	CLOSED
4	BLADDER DRAIN/FILL - 1"	CLOSED	CLOSED
5	SIGHT GAUGE (OPTIONAL) - 1/2"	CLOSED	CLOSED
6	WATER INLET	OPEN	OPEN
7	CONCENTRATE SUPPLY	CLOSED	OPEN
8	HYDRAULIC CONCENTRATE	—	CLOSED
9	SWING CHECK	—	—
10	SPRINKLER ACTUATION/ALARM	—	CLOSED OR OPEN
11	OS&Y	CLOSED	OPEN

PROPORTIONER SIZE	RECOMMENDED PIPE FITTINGS AND VALVE SIZES
2"	1"
2 1/2"	1"
3"	1 1/4"
4"	1 1/2"
6"	2"
8"	2 1/2"

TYPICAL BLADDER TANK SYSTEM
PIPING REQUIREMENTS

ANSUL.	ANSUL FIRE PROTECTION MARINETTE WI 54143-2542		
		SCALE NONE	REV. 0
DATE 12-23-92	CHKD. J. BEHNKE	APPD. K. OLSON	DRAWING NUMBER AE-60059
DWN. T. CAVER			SHEET 1 OF 1

PROPORTIONER

SEE NOTE NUMBER (2)

LOCAL CODES MAY REQUIRE BACKFLOW PREVENTER IN WATER FEED LINE.

BLADDER TANK

SIGHT GAUGE (OPTIONAL)

ANSUL

Figure 4-4   *Bladder tank arrangement. (Courtesy of Ansul, Inc.)*

IN-LINE BALANCED PRESSURE PROPORTIONER

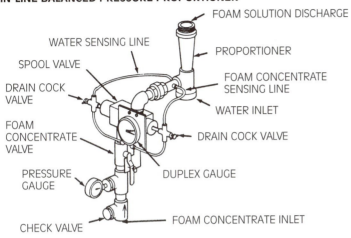

FOAM SOLUTION DISCHARGE

WATER SENSING LINE

SPOOL VALVE

DRAIN COCK VALVE

PROPORTIONER

FOAM CONCENTRATE SENSING LINE

WATER INLET

FOAM CONCENTRATE VALVE

DRAIN COCK VALVE

PRESSURE GAUGE

DUPLEX GAUGE

CHECK VALVE

FOAM CONCENTRATE INLET

TYPICAL IN-LINE BALANCED PRESSURE PROPORTIONING SYSTEM

FOAM CONCENTRATE STORAGE TANK (ATMOSPHERIC TYPE)

PRESSURE/VACUUM VENT VALVE

FOAM RETURN LINE

PRESSURE CONTROL VALVE

FOAM SOLUTION DISCHARGE

SPOOL VALVE

IN-LINE BALANCED PRESSURE PROPORTIONER

WATER SUPPLY

AUTOMATIC CONCENTRATE CONTROL VALVE

FOAM SUCTION LINE

COMPOUND PRESSURE GAUGE

FOAM PUMP AND DRIVER ASSEMBLY

FOAM CONCENTRATE SUPPLY

NOTE: Flush connections are only required with protein based foam concentrates.

LEGEND:

⋈ SHUT-OFF VALVE		⌀ PRESSURE GAUGE		⋈ PROPORTIONER		☐ SPOOL VALVE
Ͷ CHECK VALVE		⊗ DUPLEX PRESSURE GAUGE		⊶⋈ FLUSH CONNECTIONER		→ DIRECTION OF FLOW
PRESSURE RELIEF VALVE		⌀ COMPOUND PRESSURE GAUGE		⋈ STRAINER		

Figure 4-5   *Balanced pressure proportioner. (Courtesy of Ansul, Inc.)*

TYPES OF FOAM SYSTEMS

Types of foam systems include mobile and portable apparatus, semifixed foam systems, and fixed foam systems.

Mobile and Portable Apparatus

Many fire departments have foam distribution equipment that is carried to the fire location by hand (portable) or is mounted onto a truck or wheeled in on a movable platform (mobile). The selection of foam and foam equipment for the mobile or portable application is a function of the expected fire and the flammable or combustible liquid involved.

Semifixed Systems

Semifixed systems have permanent foam makers and outlets spaced as needed or required and are piped to a connection located a safe distance away from the hazard. A mobile operation has to be conducted, whereby the mobile foam equipment is connected to the semifixed piping system. The mobile apparatus would be capable of serving numerous semifixed foam installations. For such arrangements, a detection system, with central station service constantly attended, and an onsite fire brigade is strongly recommended.

Automatic Fixed Foam Systems

This chapter concentrates on automatic fixed foam systems, which are automatically initiated by a detection system, totally self-reliant, and require no manual intervention. Specific design techniques, such as fixed storage tank foam fire protection systems and fixed aircraft hangar fire protection systems, are detailed next.

To deliver low-expansion foam to a fixed storage tank, the following application methodologies are used:

- subsurface injection
- surface application
- seal protection for floating roof tanks
- dike protection

SUBSURFACE INJECTION LOW-EXPANSION FOAM SYSTEMS

subsurface injection
a system in which foam is discharged below the surface of a flammable or combustible liquid

A storage tank that contains a flammable or combustible liquid and is covered by a fixed, permanent roof to prevent the collection of rain water either above or below the fuel surface may be amenable to the application of foam through **subsurface injection**, in which foam is applied below the surface of the liquid, and the foam floats to the surface.

A low-expansion foam is used for this application, and the supply pipe for the injection system may be either tapped into an existing process line at the bottom of the tank, or tapped directly into the tank and dedicated solely to foam injection, with piping and nozzles spaced at the floor of the tank, as shown in **Figure 4-6**.

TYPICAL SCHEMATIC—DEDICATED FOAM LINE

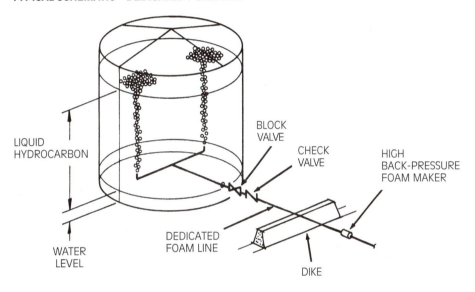

TYPICAL SCHEMATIC—INJECTION INTO PRODUCT LINE

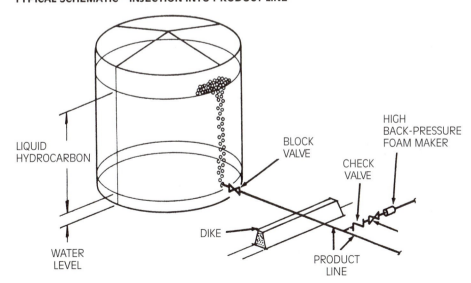

Figure 4-6

Schematic of a cone roof flammable liquid storage tank with subsurface low-expansion foam injection system. (Courtesy of Ansul, Inc.)

MODELS HBPFM-100 THROUGH 300 (FIXED)
AND HBPFM-P-100 THROUGH 300 (PORTABLE)

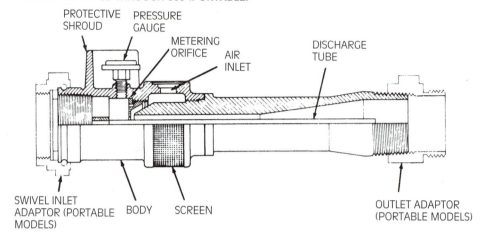

MODELS HBPFM-350 THROUGH 550 (FIXED)

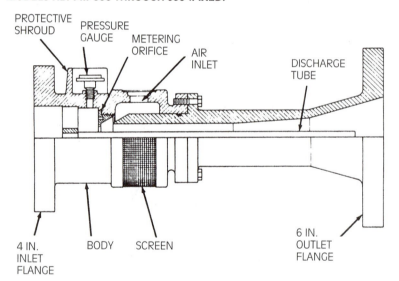

Figure 4-7 *High back pressure foam maker for subsurface injection. (Courtesy of Ansul, Inc.)*

Subsurface injection using a dedicated foam line (as shown on Figure 4-6) requires high back pressure foam makers, as shown on **Figure 4-7**. The spacing of subsurface low-expansion foam nozzles provides for a foam distribution on the surface that may be more gently applied and more uniform than the surface application method. Installing a subsurface injection system on an existing tank may be difficult because it requires that the tank be emptied, that workers enter a tank where potentially lethal fumes could pose a hazard, and that one or more holes be cut into the bottom of the existing tank to allow the new subsurface injection piping to enter. The designer can mitigate these hazards by injecting foam into the product line (as shown on Figure 4-6), which is the preferred methodology for protection of existing vessels.

Design Method for Subsurface Injection

Subsurface injection low-expansion foam systems may be designed successfully by using the following methodology.

1. *Calculate fuel surface area.* The fuel surface area is the circular area of exposed fuel at the upper level of the tank.

$$A = (3.1416) \times (r)^2$$

where A = fuel surface area
r = tank radius

2. *Determine application rate and discharge time.* **Table 4-1** lists application rates and a range of discharge times that vary with respect to the nature of the fuel for subsurface injection. This application rate is distributed over the surface area calculated in step 1 for the duration specified.

Table 4-1 *Minimum discharge times and application rates for subsurface application of fixed-roof storage tanks.*

Hydrocarbon Type	Minimum Discharge Time (min)	Minimum Application Rate	
		L/min · m²	gpm/ft²
Flash point between 37.8°C and 60°C (100°F and 140°F)	30	4.1	0.1
Flash point below 37.8°C (100°F) or liquids heated above their flash points	55	4.1	0.1
Crude petroleum	55	4.1	0.1

Notes:

(1) The maximum application rate shall be 8.1 L/min · m² (0.20 gpm/ft²).
(2) For high-viscosity liquids heated above 93.3°C (200°F), lower initial rates of application might be desirable to minimize frothing and expulsion of the stored liquid. Good judgment should be used in applying foams to tanks containing hot oils, burning asphalts, or burning liquids that are heated above the boiling point of water. Although the comparatively low water content of foams can beneficially cool such liquids at a slow rate, it can also cause violent frothing and "slop-over" of the tank's contents.

Source: NFPA 11 (2005), Table 5.2.6.5.1. Reprinted with permission from NFPA 11, *Low-, Medium-, and High-Expansion Foam*, Copyright © 2005, National Fire Protection Association, Quincy, MA 02169. This reprinted material is not the complete and official position of the National Fire Protection Association on the referenced subject which is represented only by the standard in its entirety.

3. *Calculate minimum foam discharge rate and foam concentrate quantity.*
   Calculate the minimum foam discharge rate

$$D = (A) \times (R)$$

where D = foam discharge rate gpm (L/min)
   A = tank surface area (see step 1) in Ft.^{2}(m^2)
   R = application rate (see Table 4-1)

Calculate the foam concentrate quantity

$$Q = (A) \times (R) \times (T) \times (\%)$$

where Q = primary foam concentrate quantity (gallons)
   A = tank surface area (see step 1)
   R = application rate (see Table 4-1)
   T = discharge time (see Table 4-1)
   $\%$ = concentrate percentage for foam selected—1%, 3%, or 6%
       (percentage is represented as a decimal; i.e., 0.01 for 1%, 0.03
       for 3%, and 0.06 for 6% concentrate)

4. *Determine the number of subsurface foam application outlets.* **Table 4-2** is
   a convenient table that relates number of outlets to tank diameter and
   flash point of the fuel.

5. *Determine supplementary protection requirements.* **Table 4-3** and **Table 4-4**
   allow the designer to determine the foam hose allowance that must be added
   to the minimum subsurface injection quantities calculated previously.

6. *Calculate supplementary foam quantity.*

$$D_s = (N) \times (50 \text{ gpm})$$
$$Q_s = (N) \times (50 \text{ gpm}) \times (T_s) \times (\%)$$

where D_s = supplementary discharge rate (gpm)
   Q_s = supplementary foam concentrate quantity (gals)
   N = number of hose lines (see Table 4-3)
   T_s = hose discharge time (see Table 4-4)
   $\%$ = concentrate percentage, expressed as a decimal
   50 gpm (189.27 LPM) = minimum NFPA 11 flow requirement per
   hose line

7. *Hydraulically calculate the system.* A hydraulic calculation of the system
   must be performed now to ensure that the available water supply is capable of supplying the hazard adequately. For further study on the methodology used to perform hydraulic calculation, please refer to the *Design of
   Water-Based Fire Protection Systems*, also published by Delmar Learning.
   This book provides a thorough treatment of hydraulic calculation.

Table 4-2 *Minimum number of subsurface foam discharge outlets for fixed-roof tanks containing hydrocarbons.*

Tank Diameter		Minimum Number of Discharge Outlets	
m	**ft**	**Flash Point Below 37.8°C (100°F)**	**Flash Point 37.8°C (100°F) or Higher**
Up to 24	Up to 80	1	1
Over 24 to 36	Over 80 to 120	2	1
Over 36 to 42	Over 120 to 140	3	2
Over 42 to 48	Over 140 to 160	4	2
Over 48 to 54	Over 160 to 180	5	2
Over 54 to 60	Over 180 to 200	6	3
Over 60	Over 200	6	3
		Plus 1 outlet for each additional 465 m² (5000 ft²)	Plus 1 outlet for each additional 697 m² (7500 ft²)

Notes:

(1) Liquids with flash points below 22.8°C (73°F), combined with boiling points below 37.8°C (100°F), require special consideration.
(2) Table 5.2.6.2.8 is based on extrapolation of fire test data on 7.5 m (25 ft), 27.9 m (93 ft), and 34.5 m (115 ft) diameter tanks containing gasoline, crude oil, and hexane, respectively.
(3) The most viscous fuel that has been extinguished by subsurface injection where stored at ambient conditions [15.6°C (60°F)] had a viscosity of 2000 ssu (440 centistokes) and a pour point of –9.4°C (15°F). Subsurface injection of foam generally is not recommended for fuels that have a viscosity greater than 440 centistokes (2000 ssu) at their minimum anticipated storage temperature.
(4) In addition to the control provided by the smothering effect of the foam and the cooling effect of the water in the foam that reaches the surface, fire control and extinguishment can be enhanced further by the rolling of cool product to the surface.

Table 4-3 *Supplemental foam hose stream requirements.*

Diameter of Largest Tank	Minimum Number of Hose Streams Required
Up to 19.5 m (65 ft)	1
19.5 to 36 m (65 to 120 ft)	2
Over 36 m (120 ft)	3

Table 4-4 *Hose stream operating times, supplementing tank foam installations.*

Diameter of Largest Tank	Minimum Operating Time*
Up to 10.5 m (35 ft.)	10 min
10.5 to 28.5 m (35 to 95 ft.)	20 min
Over 28.5 m (95 ft.)	30 min

*Based on simultaneous operation of the required minimum number of hose streams discharging at a rate of 189 L/min (50 gpm).

Source: NFPA 11 (2005), Table 5.9.2.4. Reprinted with permission from NFPA 11, *Low-, Medium-, and High-Expansion Foam,* Copyright © 2005, National Fire Protection Association, Quincy, MA 02169. This reprinted material is not the complete and official position of the National Fire Protection Association on the referenced subject which is represented only by the standard in its entirety.

NFPA 11 requires that foam solution velocities in the pipes not exceed 10 feet per second for class 1B flammable liquids and 20 feet per second for other flammable or combustible liquids. This requirement is to minimize the possibility of the destruction of the foam resulting from the entrainment of fuel into the solution at excessive velocities.

EXAMPLE 4-1

Subsurface Injection

An enclosed, fixed-cone roof, flammable liquids storage tank, with a diameter of 100 feet (30.48 m) and a height of 60 feet (18.28 m), containing 100 octane gasoline, requires a low-expansion foam system. Determine the minimum amount of 3% foam concentrate required to protect this hazard.

Solution

1. *Select foam agent.* In conjunction with the foam manufacturer, an agent must be selected that is amenable and appropriate for use with the fuel involved. For 100 octane gasoline, which possesses a flash point of −36°F, the manufacturer recommends 3% AFFF.

2. *Calculate fuel surface area.*

$$A = (3.1416) \times (r)^2$$
$$= (3.1416) \times (50 \text{ ft.})^2$$
$$= 7854 \text{ square feet}$$

3. *Determine application rate and discharge times* per Table 4-1.

$$R = 0.10 \text{ gpm/square foot}$$
$$T = 55 \text{ minutes (because fuel flash point is } -36°F)$$

4. *Calculate foam discharge rate to protect tank surface area.*

$$D = (A) \times (R)$$
$$= (7854) \times (0.10) = 786 \text{ gpm}$$

5. *Calculate foam concentrate quantity to protect tank surface area.*

$$Q = (A) \times (R) \times (T) \times (\%)$$
$$= (7854) \times (0.10) \times (55) \times (0.03)$$
$$= 1296 \text{ gallons of 3\% AFFF foam concentrate required}$$

6. *Determine the number of subsurface foam application outlets* per Table 4-2. Because the tank diameter is 100 feet and the fuel flash point is less than 100°F, two subsurface application outlets are required per Table 4-2.

7. *Determine supplementary hose requirements* per Table 4-3 and Table 4-4. Because the tank diameter is 100 feet, two hose lines must be calculated per Table 4-3, at 50 gpm each. The duration of hose discharge is 30 minutes, per Figure Table 4-4.

8. *Calculate supplementary foam quantity.*

$$D_s = (N) \times (50)$$
$$= (2) \times (50)$$
$$= 100 \text{ gpm}$$

$$Q_s = (N) \times (50) \times (T_s) \times (\%)$$
$$= (2) \times (50) \times (30) \times (0.03)$$
$$= 90 \text{ gallons of 3\% AFFF concentrate required.}$$

9. *Determine total discharge rate.*

$$D_{\text{Total}} = D + D_s$$
$$= 786 + 100$$
$$= 886 \text{ gpm}$$

10. *Determine total foam concentrate quantity.*

$$Q_{\text{Total}} = Q + Q_s$$
$$= 1296 + 90$$
$$= 1386 \text{ gallons of 3\% AFFF concentrate}$$

SURFACE APPLICATION LOW-EXPANSION FOAM SYSTEMS

surface application
a system designed to roll a thin blanket of foam over the surface area of the fuel

Surface application discharge devices are designed to roll a thin blanket of foam over the surface area of the fuel, using the arrangement shown in **Figure 4-8**, with fixed discharge outlets permanently located above the fuel surface, as shown in **Figure 4-9** and **Figure 4-10**.

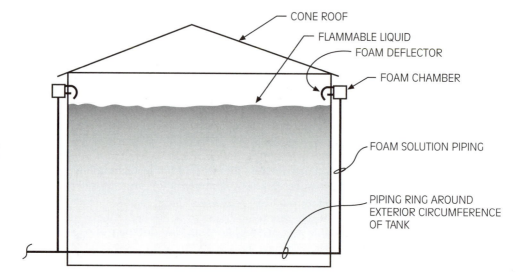

Figure 4-8 *Sectional view of a flammable liquid storage tank with a cone roof and a surface application low-expansion foam system.*

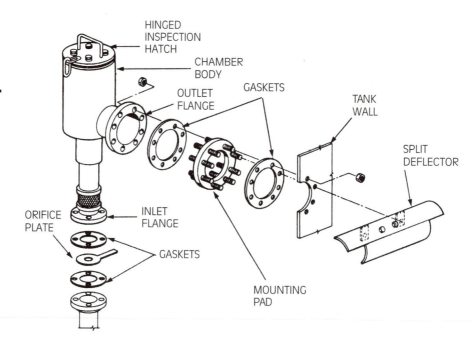

Figure 4-9 *The deflectors and flanges often are provided by the foam system contractor to the tank manufacturer or assembler to permit installation of the chamber by the foam system contractor without need for cutting holes in tanks or entering a vessel during construction.*

Design Method for Surface Application

Surface application low-expansion foam systems are designed in accordance with the following methodology.

Figure 4-10 *Cone roof flammable liquid storage tanks with surface application low-expansion foam systems. Note the pipes with foam chambers installed on the left side of the two tanks in the center of the picture. (Courtesy of National Foam, Inc.)*

1. *Calculate surface area.*

$$A = (3.1416) \times (r)^2$$

where A = tank surface area, ft.^{2}(m^2)
r = tank radius, ft.(m)

2. *Determine application rate and discharge time.* **Table 4-5** gives minimum application rates and discharge times for surface application. Application rates may vary with specific concentrates and manufacturer's recommendations. Note that Table 4-5 refers to two different types of discharge outlets. A **Type I discharge outlet** is designed to deliver foam onto the liquid surface in a gentle fashion, such that the liquid surface will not be agitated and such that foam will not be allowed to submerge below the liquid. A **Type II discharge outlet** applies foam less gently than a Type I outlet, but submergence and agitation are kept to a minimum. Foam chambers, shown in Figure 4-8 and Figure 4-9, are Type II outlets.

3. *Calculate minimum foam discharge rate and foam concentrate quantity.*
   Calculate the minimum foam discharge rate

$$D = (A) \times (R)$$

Type I discharge outlet
a discharge device designed to deliver foam onto the liquid surface in a gentle fashion

Type II discharge outlet
a discharge device designed to deliver foam less gently than a type I outlet while keeping submergence and agitation to a minimum

Calculate the foam concentrate quantity

$$Q = (A) \times (R) \times (T) \times (\%)$$

where Q = primary foam concentrate quantity (gallons) (liters)
           D = foam discharge rate (gpm) (l/min)
           A = tank surface area (see step 1)
           R = application rate (see step 2)
           T = discharge time (see step 2)
           % = concentrate percentage for foam selected, either 1%, 3%, or 6% (percentage is represented as a decimal, i.e., 0.01 for 1%, 0.03 for 3%, and 0.06 for 6%).

4. *Determine number of fixed foam chamber distribution devices* per Table 4-7.

5. *Determine supplementary hose line foam quantity* per Table 4-3 and Table 4-4.

Table 4-5   *Minimum discharge times and application rates for Type I and Type II fixed-foam discharge outlets on fixed-roof (cone) storage tanks containing hydrocarbons.*

Hydrocarbon Type	Minimum Application Rate		Minimum Discharge Time (min)	
	L/min · m²	gpm/ft²	Type I Foam Discharge Outlet	Type II Foam Discharge Outlet
Flash point between 37.8°C and 60°C (100°F and 140°F)	4.1	0.10	20	30
Flash point below 37.8°C (100°F) or liquids heated above their flash points	4.1	0.10	30	55
Crude petroleum	4.1	0.10	30	55

Notes:

(1) Included in this table are gasohols and unleaded gasolines containing no more than 10 percent oxygenated additives by volume. Where oxygenated additives content exceeds 10 percent by volume, protection is normally in accordance with 5.2.5.3. Certain nonalcohol-resistant foams might be suitable for use with fuels containing oxygenated additives of more than 10 percent by volume. The manufacturer shall be consulted for specific listings or approvals.

(2) Flammable liquids having a boiling point of less than 37.8°C (100°F) might require higher rates of application. Suitable rates of application should be determined by test.

(3) For high-viscosity liquids heated above 93.3°C (200°F), lower initial rates of application might be desirable to minimize frothing and expulsion of the stored liquid. Good judgment should be used in applying foams to tanks containing hot oils, burning asphalts, or burning liquids that have boiling points above the boiling point of water. Although the comparatively low water content of foams can beneficially cool such liquids at a slow rate, it can also cause violent frothing and "slop-over" of the tank's contents.

6. *Calculate supplementary foam quantity.*

$$D_s = (N) \times (50 \text{ gpm})$$
$$Q_s = (N) \times (50 \text{ gpm}) \times (T_s) \times (\%)$$

where D_s = supplementary discharge rate (gpm) (l/min)
   Q_s = supplementary foam concentrate quantity (gallons) (liters)
   N = number of hose lines (see step 5)
   T_s = hose discharge time (see step 5)
   % = concentration percentage, expressed as a decimal.
50 gpm (189.27 l) = minimum NFPA 11 flow requirement per
   hose line

7. *Hydraulically calculate the system,* as described in the subsurface injection scenario.

Note that the result obtained in Example 4-2 is same for the subsurface injection method as was determined for the surface application method shown in Example 4-1. This makes sense, because the objective of both methods is to achieve the identical result, which is a thin blanket of foam applied over the surface area of a tank of given diameter, given a flammable liquid of known properties.

EXAMPLE 4-2

Surface Application

An enclosed, fixed cone roof flammable liquids storage tank, with a diameter of 100 feet (30.48 m) and a height of 60 feet (18.28 m), containing 100 octane gasoline, requires a low-expansion foam system. Calculate the minimum amount of 3% foam concentrate required to adequately protect this hazard, assuming foam chambers are used as the discharge devices.

Solution

1. *Select foam agent.* In a manner similar to the subsurface injection example, 3% AFFF is selected.

2. *Calculate fuel surface area.*

$$A = (3.1416) \times (50)^2$$
$$= 7854 \text{ square feet}$$

3. *Determine application rate and discharge time* per Table 4-5, using Type II outlets, foam chambers, as specified in the problem statement. If alcohol-resistant foam is used, refer to **Table 4-6**.

$$R = 0.10 \text{ gpm/square foot}$$
$$T = 55 \text{ minutes}$$

4. *Calculate foam discharge rate and concentrate quantity to protect tank surface area.*

Calculate the minimum foam discharge rate

$$D = (A) \times (R)$$
$$= (7854) \times (0.10)$$
$$= 786 \text{ gpm}$$

Calculate the foam concentrate quantity

$$Q = (A) \times (R) \times (T) \times (\%)$$
$$= (7854) \times (0.10) \times (55) \times (0.03)$$
$$= 1296 \text{ gallons of 3\% AFFF foam concentrate required.}$$

5. *Determine the number of surface foam application outlets* per **Table 4-7**. Because the tank diameter is 100 feet, two foam chambers are required.

6. *Determine supplementary hose requirements* per Table 4-3 and Table 4-4. Because the tank diameter is 100 feet, two hose lines must flow, at 50 gpm each, for 30 minutes.

7. *Calculate supplementary foam quantity.*

$$D = (N) \times (50)$$
$$= 100 \text{ gpm}$$

$$Q_s = (N) \times (50) \times (T_s) \times (\%)$$
$$= (2) \times (50) \times (30) \times (0.03)$$
$$= 90 \text{ gallons of 3\% AFFF concentrate required.}$$

8. *Determine total discharge rate.*

$$D_{total} = D + D_s$$
$$= 786 + 100$$
$$= 886 \text{ gpm}$$

9. *Determine total foam concentrate quantity.*

$$Q_{total} = Q + Q_s$$
$$= 1296 + 90$$
$$= 1386 \text{ gallons of 3\% AFFF concentrate required}$$

SEAL PROTECTION FOR FLOATING ROOF TANKS

floating roof
a tank roof that floats on the surface of a flammable or combustible liquid

A **floating roof** floats on the surface of a flammable or combustible liquid as shown on **Figure 4-11,** rising and falling as the liquid is added to or removed from the tank. The floating roof is similar to a pontoon vessel that allows no space between the bottom of the roof and the surface of the tank for the collection of flammable vapors, preventing ignition of the flammable liquid below the roof.

Table 4-6 *Minimum application rates and discharge times for fixed-roof (cone) tanks containing flammable and combustible liquids requiring alcohol-resistant foams.*

	Minimum Discharge Time (min)	
Application Rate for Specific Product Stored	**Type I Foam Discharge Outlet**	**Type II Foam Discharge Outlet**
Consult manufacturer for listings on specific products	30	55

Note: Most currently manufactured alcohol-resistant foams are suitable for use with Type II fixed foam discharge outlets. However, some older alcohol-resistant foams require gentle surface application by Type I fixed foam discharge outlets. Consult manufacturers for listings on specific products.

Source: NFPA 11 (2005), Table 5.2.5.3.4. Reprinted with permission from NFPA 11, *Low-, Medium-, and High-Expansion Foam,* Copyright © 2005, National Fire Protection Association, Quincy, MA 02169. This reprinted material is not the complete and official position of the National Fire Protection Association on the referenced subject which is represented only by the standard in its entirety.

Table 4-7 *Number of fixed foam discharge outlets for fixed-roof tanks containing hydrocarbons or flammable and combustible liquids requiring alcohol-resistant foams.*

Tank Diameter (or equivalent area)		Minimum Number of Discharge Outlets
m	**ft**	
Up to 24	Up to 80	1
Over 24 to 36	Over 80 to 120	2
Over 36 to 42	Over 120 to 140	3
Over 42 to 48	Over 140 to 160	4
Over 48 to 54	Over 160 to 180	5
Over 54 to 60	Over 180 to 200	6

Source: NFPA 11 (2005), Table 5.2.5.2.1. Reprinted with permission from NFPA 11, *Low-, Medium-, and High-Expansion Foam,* Copyright © 2005, National Fire Protection Association, Quincy, MA 02169. This reprinted material is not the complete and official position of the National Fire Protection Association on the referenced subject which is represented only by the standard in its entirety.

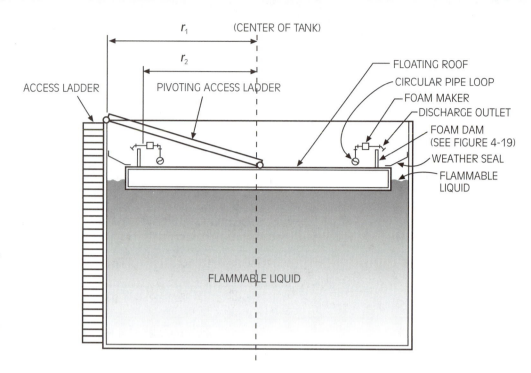

Figure 4-11

Sectional view of a floating roof tank with above-seal low-expansion foam protection using a foam dam.

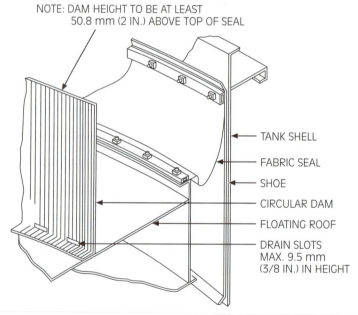

Figure 4-12 *Typical foam dam for floating roof tank protection.*

Source: NFPA 11 (2005), Figure 5.3.5.4.5. Reprinted with permission from NFPA 11, *Low-, Medium-, and High-Expansion Foam,* Copyright © 2005, National Fire Protection Association, Quincy, MA 02169. This reprinted material is not the complete and official position of the National Fire Protection Association on the referenced subject which is represented only by the standard in its entirety.

seal protection
filling or covering the
seal area with low-
expansion foam

The space between the edge of the floating roof and the perimeter of the tank, however, can allow for a possible ignition scenario, and this space is the object of protection for floating roof tanks. The seal that covers this area requires **seal protection**, which sometimes involves building a dam around the perimeter of the floating roof, as shown in **Figure 4-12** and **Figure 4-13**, and filling or covering this ring with low-expansion foam. The dam must be constructed of #10 gage steel, must be 12-inch minimum height, and must be between one and two feet from the edge of the roof.

Top-of-seal protection is shown on **Figure 4-14**. In some cases, it may be necessary for the foam distribution piping to penetrate the seal and protect below the seal, as shown on **Figure 4-15** and **Figure 4-16**. In these cases, a foam dam may not be needed if the seal is capable of holding the foam that has been injected into it. If the seal is combustible, the seal should be protected from above. Floating roof foam makers, as shown on **Figure 4-17** are used for both above seal and below seal applications.

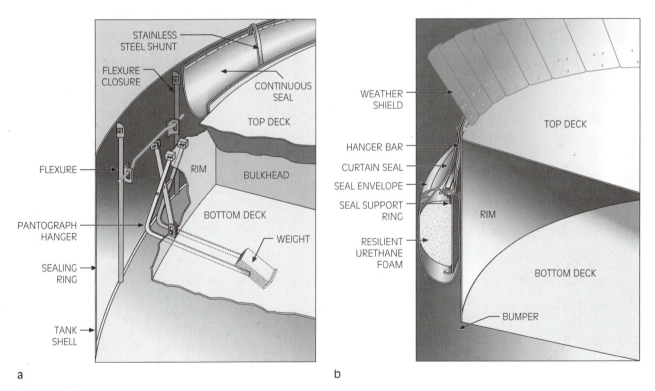

a   b

Figure 4-13   *Floating roof tanks.*

Source: NFPA 11 (2005), Figures 5.3(a), 5.3(b), 5.3(d). Reprinted with permission from NFPA 11, *Low-, Medium-, and High-Expansion Foam*, Copyright © 2005, National Fire Protection Association, Quincy, MA 02169. This reprinted material is not the complete and official position of the National Fire Protection Association on the referenced subject which is represented only by the standard in its entirety.

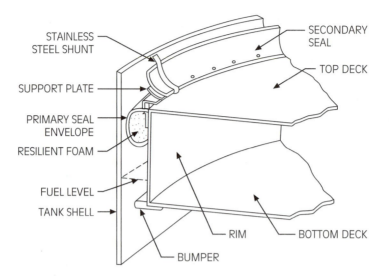

STAINLESS
STEEL SHUNT

SECONDARY
SEAL

TOP DECK

SUPPORT PLATE

PRIMARY SEAL
ENVELOPE

RESILIENT FOAM

FUEL LEVEL

TANK SHELL

RIM

BOTTOM DECK

BUMPER

c

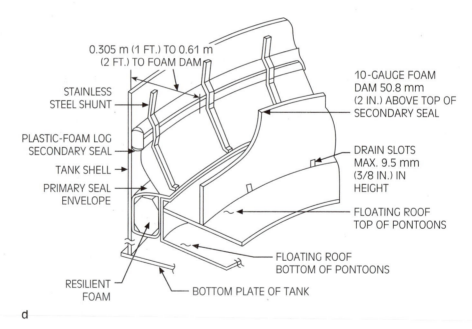

0.305 m (1 FT.) TO 0.61 m
(2 FT.) TO FOAM DAM

STAINLESS
STEEL SHUNT

PLASTIC-FOAM LOG
SECONDARY SEAL

TANK SHELL

PRIMARY SEAL
ENVELOPE

10-GAUGE FOAM
DAM 50.8 mm
(2 IN.) ABOVE TOP OF
SECONDARY SEAL

DRAIN SLOTS
MAX. 9.5 mm
(3/8 IN.) IN
HEIGHT

FLOATING ROOF
TOP OF PONTOONS

FLOATING ROOF
BOTTOM OF PONTOONS

RESILIENT
FOAM

BOTTOM PLATE OF TANK

d

Figure 4-13   *(continued)*

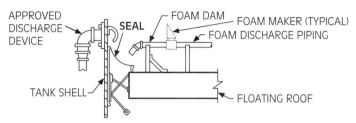

DETAIL A — TOP-OF-SEAL APPLICATION
FOAM DISCHARGE ABOVE MECHANICAL SHOE SEAL

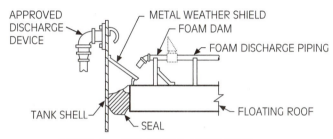

DETAIL B — TOP-OF-SEAL APPLICATION
FOAM DISCHARGE ABOVE METAL WEATHER SEAL

Figure 4-14 *Typical foam system illustrations for top-of-seal fire protection. Both fixed foam (wall-mounted) and roof-mounted discharge outlets are shown for illustrative purposes. Although both methods are shown, only one is needed.*

Source: NFPA 11 (2005), Figure 5.3.5.3.1. Reprinted with permission from NFPA 11, *Low-, Medium-, and High-Expansion Foam,* Copyright © 2005, National Fire Protection Association, Quincy, MA 02169. This reprinted material is not the complete and official position of the National Fire Protection Association on the referenced subject which is represented only by the standard in its entirety.

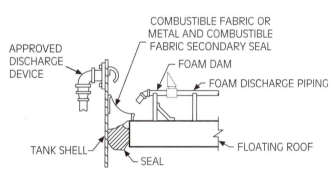

DETAIL C — TOP-OF-SEAL APPLICATION
FOAM DISCHARGE ABOVE SECONDARY COMBUSTIBLE
FABRIC SEAL, OR METAL WITH COMBUSTIBLE FABRIC SECTIONS

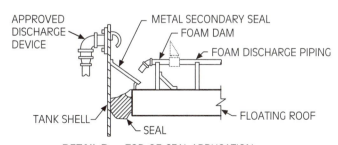

DETAIL D — TOP-OF-SEAL APPLICATION
FOAM DISCHARGE ABOVE METAL SECONDARY SEAL

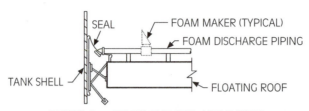

DETAIL A — BELOW–THE–SEAL APPLICATION
FOAM DISCHARGE BELOW MECHANICAL SHOE SEAL— NO FOAM DAM

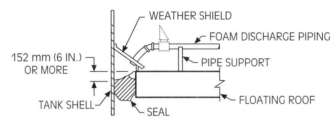

DETAIL B — BELOW-THE-SHIELD APPLICATION
FOAM DISCHARGE BELOW METAL WEATHER SHIELD
TOP OF SEAL 152 mm (6 IN.) OR MORE BELOW TOP OF FLOATING ROOF

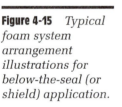

Figure 4-15 *Typical foam system arrangement illustrations for below-the-seal (or shield) application.*

Source: NFPA 11 (2005), Figure 5.3.5.3.5.1. Reprinted with permission from NFPA 11, *Low-, Medium-, and High-Expansion Foam,* Copyright © 2005, National Fire Protection Association, Quincy, MA 02169. This reprinted material is not the complete and official position of the National Fire Protection Association on the referenced subject which is represented only by the standard in its entirety.

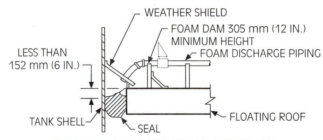

DETAIL C — BELOW-THE-SHIELD APPLICATION
FOAM DISCHARGE BELOW METAL WEATHER SHIELD
TOP OF SEAL LESS THAN 152 mm (6 IN.) BELOW TOP OF FLOATING ROOF

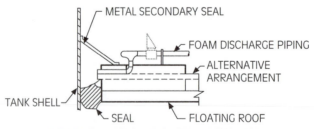

DETAIL D — BELOW-THE- SEAL APPLICATION
FOAM DISCHARGE BELOW METAL SECONDARY SEAL
THIS FOAM APPLICATION METHOD IS NOT ACCEPTABLE IF SECONDARY SEAL
CONSTRUCTED OF ANY COMBUSTIBLE FABRIC SECTIONS.
(REFER TO APPLICATION ABOVE SEAL.)

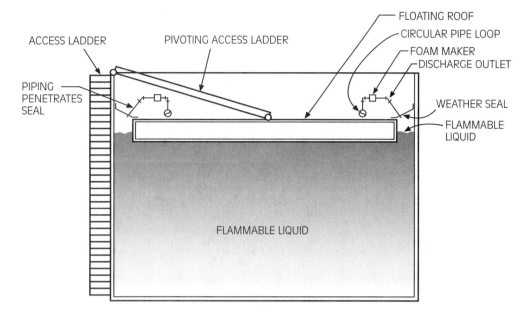

Figure 4-16
Sectional view of a floating roof tank with below-seal low-expansion foam protection.

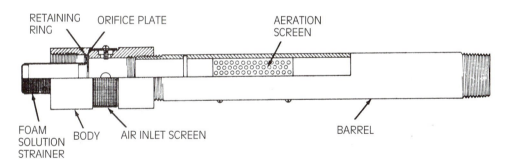

Figure 4-17   *Floating roof foam maker. (Courtesy of Ansul, Inc.)*

DESIGN PROCEDURE FOR FLOATING ROOF TANK SEAL PROTECTION

Low-expansion foam protection of floating roof tank seals is approached using the following methodology.

1. *Calculate foam distribution area.* The foam distribution area is a ring, bounded by the tank wall and the edge of the foam dam for above seal protection, as shown on Figure 4-12, or bounded by the tank wall and the edge of the floating roof for below seal protection, as shown in Figure 4-15.

This ring is calculated by subtracting the circular unprotected roof area from the total tank area:

$$A = \text{(total roof surface area)} - \text{(unprotected roof area)}$$
$$= [(3.1416) \times (r_1{}^2)] - [(3.1416) \times (r_2{}^2)]$$

where r_1 = radius of the outside perimeter of the tank wall
r_2 = radius of the dike wall mounted to the floating roof

2. *Determine application rate and discharge time* in accordance with NFPA 11. Top-of-seal protection is determined by **Table 4-8**, and below-the-seal protection is determined by **Table 4-9**.

3. *Calculate foam discharge rate and concentrate quantity* to protect tank surface area:

Calculate the minimum foam discharge rate

$$D = (A) \times (R)$$

Calculate the foam concentrate quantity

$$Q = (A) \times (R) \times (T) \times (\%)$$

where Q = primary foam concentrate quantity (gallons)
D = foam discharge rate (gpm) (l/min)
A = surface area of protected ring (see step 1) (ft^2) (m^2)
R = application rate (see step 2) (gpm/ft^2) (lpm/m^2)]
T = discharge time (see step 2) (min)
$\%$ = concentrate percentage for foam selected—1%, 3%, or 6% (percentage is represented as a decimal; i.e., 0.01 for 1%, 0.03 for 3%, and 0.06 for 6%).

4. *Determine spacing of discharge outlets.*

Top of seal protection (see Table 4-8)

Below-the-seal protection (see Table 4-9)

5. *Determine number of discharge devices.*

$$N = \frac{C}{S}$$

where N = number of discharge devices
C = circumference of tank = (3.1416) × (tank diameter)
S = maximum spacing between devices (see step 4)

6. *Determine supplementary hose demand and concentrate quantity* per Table 4-3 and Table 4-4.

Table 4-8 *Top-of-seal fixed foam discharge protection for open-top floating roof tanks.*

Seal Type	Applicable Illustration Detail	Minimum Application Rate		Minimum Discharge Time (min)	Maximum Spacing Between Discharge Outlets with			
					305 mm (12 in.) Foam Dam		610 mm (24 in.) Foam Dam	
		L/min · m²	gpm/ft²		m	ft	m	ft
Mechanical shoe seal	A	12.2	0.3	20	12.2	40	24.4	80
Tube seal with metal weather shield	B	12.2	0.3	20	12.2	40	24.4	80
Fully or partly combustible secondary seal	C	12.2	0.3	20	12.2	40	24.4	80
All metal secondary seal	D	12.2	0.3	20	12.2	40	24.4	80

Note: Where the fixed foam discharge outlets are mounted above the top of the tank shell, a foam splashboard is necessary due to the effect of winds.

Source: NFPA 11 (2005), Table 5.3.5.3.1. Reprinted with permission from NFPA 11, *Low-, Medium-, and High-Expansion Foam,* Copyright © 2005, National Fire Protection Association, Quincy, MA 02169. This reprinted material is not the complete and official position of the National Fire Protection Association on the referenced subject which is represented only by the standard in its entirety.

Table 4-9 *Below-the-seal fixed foam discharge protection for open-top floating roof tanks.*

Seal Type	Applicable Illustration Detail	Minimum Application Rate		Minimum Discharge Time (min)	Maximum Spacing Between Discharge (Outlets)
		L/min · m²	gpm/ft²		
Mechanical shoe seal	A	20.4	0.5	10	39 m (130 ft.) — Foam dam not required
Tube seal with more than 152 mm (6 in.) between top of tube and top of pontoon	B	20.4	0.5	10	18 m (60 ft.) — Foam dam not required
Tube seal with less than 152 mm (6 in.) between top of tube and top of pontoon	C	20.4	0.5	10	18 m (60 ft.) — Foam dam required
Tube seal with foam discharge below metal secondary seal*	D	20.4	0.5	10	18 m (60 ft.) — Foam dam not required

*A metal secondary seal is equivalent to a foam dam.

Source: NFPA 11 (2005), Table 5.3.5.3.6.1. Reprinted with permission from NFPA 11, *Low-, Medium-, and High-Expansion Foam,* Copyright © 2005, National Fire Protection Association, Quincy, MA 02169. This reprinted material is not the complete and official position of the National Fire Protection Association on the referenced subject which is represented only by the standard in its entirety.

7. *Calculate supplementary hose demand foam quantity.*

$$D_s = (N) \times (50 \text{ gpm})$$
$$Q_s = (N) \times (50 \text{ gpm}) \times (T_s) \times (\%)$$

where D_s = supplementary discharge rate (gpm) (l/min)
  Q_s = of hose lines (see step 6)
  T_s = hose discharge time (see step 6) (min)
  % = concentration percentage, expressed as a decimal
50 gpm (189.27 l) = minimum NFPA 11 requirement per hose line

8. *Hydraulically calculate the system.*

EXAMPLE 4-3

Floating Roof Tank—Foam Dam and Top-of-Seal Protection

A 100-foot diameter floating roof tank, 60 feet tall, with a 1-foot-wide foam dam that is 12 inches tall, requires a low-expansion foam system. Calculate the minimum concentrate quantity required, using the top of seal, mechanical shoe seal protection method and 3% AFFF foam concentrate.

Solution

1. *Calculate foam distribution area.*

$$A = (\text{total tank area}) - (\text{unprotected roof area})$$
$$= [(3.1416) \times (50)^2] - [(3.1416) \times (49)^2]$$
$$= 7854 - 7543$$
$$= 311 \text{ square feet}$$

2. *Determine application rate and discharge time* in accordance with NFPA 11.

Application density = 0.30 gpm/square foot per Table 4-8
discharge time = 20 minutes per Table 4-8

3. *Calculate foam discharge rate and concentrate quantity* for tank seal protection.
Calculate the minimum foam discharge rate

$$D = (A) \times (R)$$
$$= (311) \times (0.30) = 94 \text{ gpm}$$

Calculate the foam concentrate quantity

$$Q = (A) \times (R) \times (T) \times (\%)$$
$$= (311) \times (0.30) \times (20) \times (0.03)$$
$$= 56 \text{ gallons of 3\% AFFF concentrate required}$$

4. *Determine spacing of discharge outlets.* Outlets are spaced at 40 feet along foam dam circumference for top-of-seal protection with 12-inch dam height, per Table 4-8.

5. *Determine number of discharge devices.*

$$N = \frac{C}{S}$$

$$= \frac{[(3.1416) \times (100')]}{40'}$$

$$= 8 \text{ (rounded up to nearest whole number)}$$

6. *Determine supplementary hose demand and concentrate quantity.* Two hose outlets must flow for 30 minutes for a 100-foot diameter tank, in accordance with Table 4-7.

$$D_s = (N) \times (50 \text{ gpm})$$

$$= (2) \times (50 \text{ gpm}) = 100 \text{ gpm}$$

$$Q_s = (N) \times (50 \text{ gpm}) \times (T_s) \times (\%)$$

$$= (2) \times (50) \times (30) \times (0.03)$$

$$= 90 \text{ gallons of 3\% AFFF concentrate required}$$

7. *Determine total discharge rate.*

$$D \text{ (total)} = D + D_s$$

$$= 94 + 100 = 194 \text{ gpm}$$

8. *Determine total foam concentrate quantity.*

$$Q_{\text{total}} = Q + Q_s$$

$$= 56 + 90 = 146 \text{ gallons}$$

tank farm
an enclosure containing vertical cylindrical tanks, horizontal cylindrical tanks, or spherical tanks, which store flammable or combustible liquids, surrounded by a containment dike

dike protection systems
systems where the dike area is flooded with foam that will float on top of a flammable liquid that spills within the containment dike

DIKE PROTECTION LOW-EXPANSION FOAM SYSTEMS

A **tank farm**, an enclosure containing vertical cylindrical tanks, horizontal cylindrical tanks, or spherical tanks, which store flammable or combustible liquids, surrounded by a containment dike, may be ideal for protection by a low-expansion foam dike protection system. With **dike protection systems**, the dike area is flooded with foam that will float on top of a flammable liquid that spills within the containment dike, as shown in **Figure 4-18**.

A dike protection system also may be recommended as supplemental protection for a fixed cone roof or for a floating roof tank dike area, or as a supplement to a low-expansion foam system designed to apply foam to the exterior surfaces of all cylindrical or spherical flammable or combustible liquid storage tanks contained within the dike.

Portable or mobile foam application may be considered for cases involving 24-hour detection or surveillance and adequate mobile suppression facilities. The discussion here, however, is devoted exclusively to fixed automatic dike protection systems.

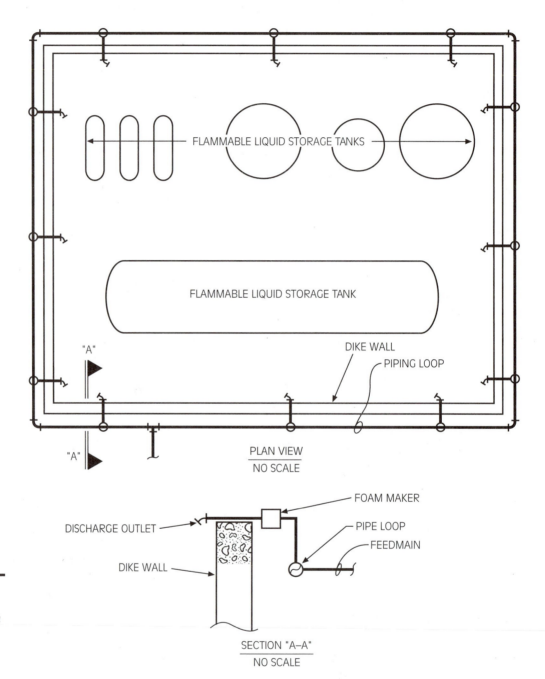

FLAMMABLE LIQUID STORAGE TANKS

FLAMMABLE LIQUID STORAGE TANK

DIKE WALL

PIPING LOOP

"A"

"A"

PLAN VIEW
NO SCALE

FOAM MAKER

DISCHARGE OUTLET

PIPE LOOP

FEEDMAIN

DIKE WALL

SECTION "A–A"
NO SCALE

Figure 4-18 *Plan and sectional view of a dike protection system with low-expansion foam protection.*

Design Procedure for Dike Protection

Low-expansion foam systems for dike protection are designed using the following methodology.

1. *Calculate dike area.*

$$A = \text{(dike length)} \times \text{(dike width)}$$

If a tank is installed with its bottom mounted flush to the floor of the dike, the surface area of the tank may be deducted from the total dike area. If flammable liquid is capable of flowing beneath a vessel, the entire dike surface area is to be calculated.

2. *Determine application rate and discharge times* per NFPA 11. See **Table 4-10**

3. *Calculate foam discharge rate and concentrate quantity.*

    Calculate the minimum foam discharge rate

$$D = (A) \times (R)$$

    Calculate the foam concentrate quantity

$$Q = (A) \times (R) \times (T) \times (\%)$$

where Q = primary foam concentrate rate quantity (gallons) (liters)
  D = foam discharge rate (gpm) (l/min)
  A = dike area (see step 1) (ft^2) (m^2)
  R = application rate (see step 2) (gpm/ft^2) (lpm/m^2)
  T = discharge time (see step 2) (minutes)
  $\%$ = concentrate percentage for foam selected (percentage is expressed as a decimal, i.e., 0.01 for 1%, 0.03 for 3%, and 0.06 for 6%).

Table 4-10 *Minimum application rates and discharge times for fixed foam application on diked areas involving hydrocarbon liquids.*

Type of Foam Discharge Outlets	Minimum Application Rate		Minimum Discharge Time (min)	
	L/min · m^2	gpm/ft^2	Class I Hydrocarbon	Class II Hydrocarbon
Low-level foam discharge outlets	4.1	0.10	30	20
Foam monitors	6.5	0.16	30	20

Source: NFPA 11 (2005), Table 5.7.3.2. Reprinted with permission from NFPA 11, *Low-, Medium-, and High-Expansion Foam,* Copyright © 2005, National Fire Protection Association, Quincy, MA 02169. This reprinted material is not the complete and official position of the National Fire Protection Association on the referenced subject which is represented only by the standard in its entirety.

4. *Determine number of foam discharge devices required.* NFPA 11 specifies that "where fixed discharge outlets installed at a low level are used as the primary protection, they shall be located so that no point in the dike area is more than 9 m (30 ft.) from a discharge outlet where the discharge per outlet is 225 l/min (60 gpm) or less. For outlets having discharge rates higher than 225 l/min (60 gpm), the maximum distance between discharge outlets shall be 18 m (60 ft.)."

$$N = \frac{(2L + 2W)}{S}$$

where N = number of devices. This quotient is rounded up to the nearest whole device

L = length of dike, ft. (m)

W = width of dike, ft. (m)

S = distance between discharge devices, ft. (m)

EXAMPLE 4-4

Dike Protection System

A dike whose width is 60 feet and length is 250 feet, containing several small gasoline Class I hydrocarbon cylindrical storage tanks, requires a low-expansion foam system. Determine the discharge rate, concentrate quantity, and number of nozzles, assuming that low-level foam discharge outlets are used.

Solution

1. *Calculate dike area.*

$$A = (L) \times (W)$$
$$= (250 \text{ ft.}) \times (60 \text{ ft.})$$
$$= 15{,}000 \text{ square feet}$$

Note that because the dike is 60 feet wide, the NFPA 11 provisions of Item #4 ("Determine number of foam discharge devices required") above are met.

2. *Determine application rate and discharge time*

Application rate = 0.10 gpm/square foot, because fixed discharge outlets are used.

Per Table 4-10, Discharge time = 30 minutes, because flash point of the fuel is less than 100°F.

3. *Calculate foam discharge rate and concentrate quantity,* using 3% AFFF foam concentrate.

Calculate the minimum foam discharge rate

$$D = (A) \times (R)$$
$$= (15{,}000) \times (0.10)$$
$$= 1500 \text{ gpm}$$

Calculate the foam concentrate quantity

$$Q = (A) \times (R) \times (T) \times (\%)$$
$$= (15{,}000) \times (0.10) \times (30) \times (0.03)$$
$$= 1350 \text{ gallons 3\% AFFF concentrate}$$

4. *Determine number of foam discharge devices required.*

$$N = \frac{(2L + 2W)}{S}$$
$$= \frac{(250 \text{ ft.} + 250 \text{ ft.} + 60 \text{ ft.} + 60 \text{ ft.})}{60 \text{ ft.}}$$
$$= 11 \text{ devices (rounded upto the nearest whole Device)}$$

Note that the discharge rate per discharge device (1500 gpm ÷ 11 devices) is 136 gpm per device, which justifies the selection of S = 60 ft. for discharge rates exceeding 60 gpm per item 4. Note also that if $S = 30$, the discharge rate is 75 gpm per device, which exceeds 60 gpm per device, once again justifying the selection of $S = 60$ feet.

LOW-EXPANSION FOAM SYSTEMS FOR AIRCRAFT HANGARS

Considerable experience has been gained with respect to the value of low-expansion foam for aircraft fire protection. Aircraft that are being stored or serviced in an aircraft hangar contain a large quantity of flammable fuel, which could leak from the wings and result in a pool fire covering a large floor area. Servicing the aircraft offers numerous opportunities for the spilled fuel to ignite.

NFPA

NFPA 409, Standard on Aircraft Hangars

A low-expansion foam system, designed in accordance with NFPA 409, *Standard on Aircraft Hangars*, smothers flammable liquid pool fires on the floor and also effectively coats the aircraft skin with an effective exposure protection barrier. Hangars are classified by three groupings per NFPA 409, Group I, Group II, and Group III.

Group I Aircraft Hangars

A Group I aircraft hangar has at least one of the following conditions:

- aircraft access door height over 28 feet (8.5 m)
- a single fire area in excess of 40,000 square feet (3716 m^2)
- provisions for housing an aircraft with tail height greater than 28 feet (8.5 cm)
- provisions for housing strategically important military aircraft, as determined by the U.S. Department of Defense

Group II Aircraft Hangars

A Group II aircraft hangar has both of the following conditions:

- an aircraft door height of 28 feet (8.5 m) or less
- a single fire area for specific types of construction in accordance with NFPA 409, as shown on **Table 4-11**.

Group III Aircraft Hangars

A Group III aircraft hangar has both of the following conditions:

- an aircraft access door 28 feet (8.5 m) in height or less
- a single fire area in accordance with NFPA 409, as shown on **Table 4-12**

Aircraft Hangar Foam System Design

An aircraft hanger fire protection design can consist of three discrete, yet co-dependent low-expansion foam systems: ceiling protection, underwing protection, and supplementary hose protection.

1. *Ceiling protection.* **Nonaspirated sprinklers** or **aspirated foam-water nozzles** are installed at the ceiling for primary protection, as shown on **Figure 4-19**. Aspirated foam-water nozzles have an air inlet that allows air to be introduced to the foam solution before it hits the stream deflector, providing a more aerated foam. Nonaspirated sprinklers are foam discharge

nonaspirated sprinklers
foam discharge devices that do not possess an air inlet between the orifice and the deflector

aspirated foam-water nozzles
nozzles possessing an air inlet that allows air to be introduced to the foam solution before it hits the stream deflector

Table 4-11   *Fire areas for Group II aircraft hangers.*

Type of Construction	Single Fire Area (Inclusive)	
	m²	ft²
Type I (443) and (332)	2,787–3,716	30,001–40,000
Type II (222)	1,858–3,716	20,001–40,000
Type II (111), Type III (211), and Type IV (2HH)	1,394–3,716	15,001–40,000
Type II (000)	1,115–3,716	12,001–40,000
Type III (200)	1,115–3,716	12,001–40,000
Type V (111)	743–3,716	8,001–40,000
Type V (000)	465–3,716	5,001–40,000

Source: NFPA 409 (2004), Table 4.1.2. Reprinted with permission from NFPA 409, *Aircraft Hangars,* Copyright © 2004, National Fire Protection Association, Quincy, MA 02169. This reprinted material is not the complete and official position of the National Fire Protection Association on the referenced subject which is represented only by the standard in its entirety.

Table 4-12 *Maximum fire areas for Group III aircraft hangars.*

Type of Construction	Maximum Single Fire Area	
	m²	ft²
Type I (443) and (332)	2,787	30,000
Type II (222)	1,858	20,000
Type II (111), Type III (211), and Type IV (2HH)	1,394	15,000
Type II (000)	1,115	12,000
Type III (200)	1,115	12,000
Type V (111)	743	8,000
Type V (000)	465	5,000

Source: NFPA 409 (2004), Table 4.1.3. Reprinted with permission from NFPA 409, *Aircraft Hangars,* Copyright © 2004, National Fire Protection Association, Quincy, MA 02169. This reprinted material is not the complete and official position of the National Fire Protection Association on the referenced subject which is represented only by the standard in its entirety.

devices that do not possess an air inlet between the orifice and the deflector and are recommended only for film-forming foams, because they require less energy to aerate the solution. Aspirated foam-water nozzles are significantly more expensive than standard automatic foam-water sprinklers.

water oscillating monitors (WOM)

monitors installed at the floor of an aircraft hangar to provide protection below the wing area

2. *Underwing protection.* Automatic or manual foam **water oscillating monitors** (WOM), installed at the floor to provide protection below the wing area, as shown on **Figure 4-20**.

3. *Supplementary hose protection.* Foam hose lines for small fires, mop-up, or supplementary fire-fighting operations.

The three aircraft hangar groupings require differing strategies for foam fire protection. **Figure 4-21** shows a sample layout for protection of a Group I hangar.

Design Method for Aircraft Hangar Protection

Aircraft hangar low-expansion foam system design is accomplished by using the following methodology.

1. *Determine aircraft grouping and appropriate hangar strategy* per NFPA 409, using the strategy that follows:

   Group I—Hangar Strategy Group I hangars, other than those housing unfueled aircraft, shall be in accordance with any one of the following:

   • A foam-water deluge system, plus supplementary underwing protection for single aircraft whose wing area exceeds 3000 ft.² (279 m²)

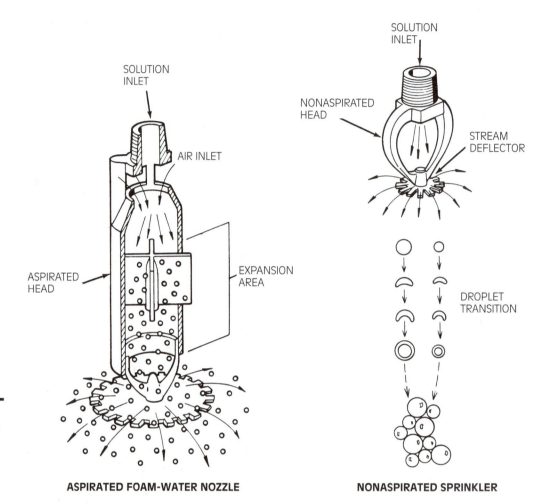

Figure 4-19
Aspirated and non-aspirated foam discharge devices. (Courtesy of Ansul, Inc.)

ASPIRATED FOAM-WATER NOZZLE   **NONASPIRATED SPRINKLER**

- An automatic sprinkler system, plus a low-level foam system
- An automatic sprinkler system, plus a low-level high expansion foam system

Group II—Hangar Strategy Group II hangars, other than those housing unfueled aircraft, shall be in accordance with any one of the following:

- NFPA 409 requirements for Group I hangars.
- An automatic sprinkler system, plus a low-level, low-expansion foam system.
- An automatic sprinkler system, plus an automatic high expansion foam system.
- A closed-head foam-water system.

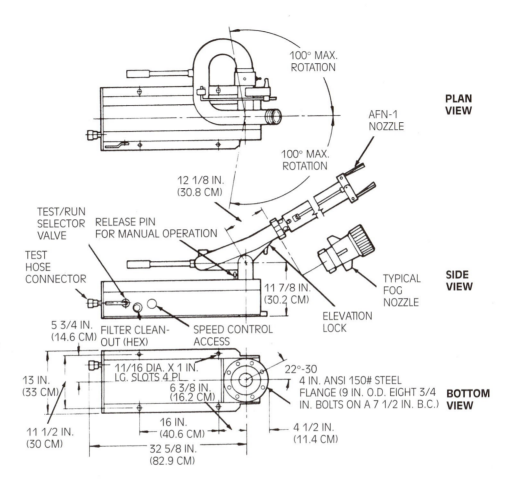

100° MAX.
ROTATION

**PLAN
VIEW**

AFN-1
NOZZLE

100° MAX.
ROTATION

12 1/8 IN.
(30.8 CM)

TEST/RUN
SELECTOR
VALVE

RELEASE PIN
FOR MANUAL OPERATION

TEST
HOSE
CONNECTOR

11 7/8 IN.
(30.2 CM)

TYPICAL
FOG
NOZZLE

**SIDE
VIEW**

ELEVATION
LOCK

5 3/4 IN.
(14.6 CM)

FILTER CLEAN-
OUT (HEX)

SPEED CONTROL
ACCESS

13 IN.
(33 CM)

11/16 DIA. X 1 IN.
LG. SLOTS 4 PL.

6 3/8 IN.
(16.2 CM)

22°-30

4 IN. ANSI 150# STEEL
FLANGE (9 IN. O.D. EIGHT 3/4
IN. BOLTS ON A 7 1/2 IN. B.C.)

**BOTTOM
VIEW**

16 IN.
(40.6 CM)

4 1/2 IN.
(11.4 CM)

11 1/2 IN.
(30 CM)

32 5/8 IN.
(82.9 CM)

Figure 4-20 *Water
oscillating monitor
(WOM) for
underwing
protection in
aircraft hangars.
(Courtesy of Ansul,
Inc.)*

Group III—Hangar Strategy Group III hangars, other than those housing
unfueled aircraft, shall be in accordance with any one of the following:

- Foam systems are not required.
- The AHJ may opt to protect Group III hangars with the Group II strategy
  when aircraft operations are perceived as hazardous.

2. *Determine application time per NFPA 409.*

- Foam concentrate capable of supplying 10 minutes of foam-water application is required, with an equal amount as a reserve supply required,
  based upon all systems in the vicinity of an anticipated fire actuating.
- 45 minutes of water application is required, based upon all systems in
  the vicinity of an anticipated fire actuating and including an additional
  500 gpm for water flow from fire hydrants if hydrants are installed in
  the vicinity of the aircraft hangar.

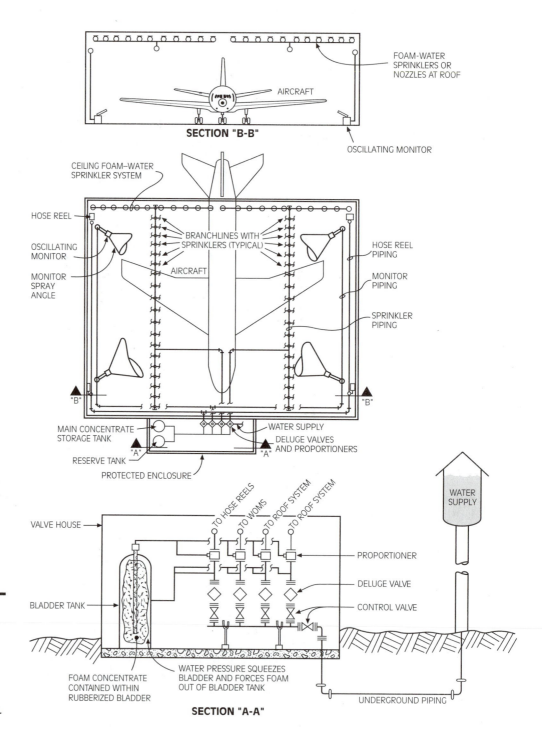

SECTION "B-B"

FOAM-WATER SPRINKLERS OR NOZZLES AT ROOF

AIRCRAFT

OSCILLATING MONITOR

CEILING FOAM–WATER SPRINKLER SYSTEM

HOSE REEL

OSCILLATING MONITOR

MONITOR SPRAY ANGLE

BRANCHLINES WITH SPRINKLERS (TYPICAL)

AIRCRAFT

HOSE REEL PIPING

MONITOR PIPING

SPRINKLER PIPING

"B"

"B"

MAIN CONCENTRATE STORAGE TANK

RESERVE TANK

"A"

"A"

WATER SUPPLY

DELUGE VALVES AND PROPORTIONERS

PROTECTED ENCLOSURE

VALVE HOUSE

TO HOSE REELS

TO WOMS

TO ROOF SYSTEM

TO ROOF SYSTEM

WATER SUPPLY

PROPORTIONER

DELUGE VALVE

CONTROL VALVE

BLADDER TANK

FOAM CONCENTRATE CONTAINED WITHIN RUBBERIZED BLADDER

WATER PRESSURE SQUEEZES BLADDER AND FORCES FOAM OUT OF BLADDER TANK

UNDERGROUND PIPING

SECTION "A-A"

Figure 4-21 *Plan view of an aircraft hangar with low-expansion foam roof protection, underwing protection, and hosereel protection.*

3. *Determine roof system design density.*

Density Selection for Group I Hangars

- Foam-water deluge systems
  - Area shall not exceed 15,000 ft.^{2} (1394 m^2) per system.
  - Sprinkler spacing shall not exceed 130 ft.^{2} (12 m^2) per sprinkler.
  - Maximum distance between sprinklers or branch lines shall not exceed 12 ft. (3.7 m).
  - Design density shall be hydraulically calculated to ensure a minimum of 0.20 gpm/ft.^{2} (8.1 L/min/m^2) for air-aspirating discharge devices.
  - Design density shall be hydraulically calculated to ensure a minimum of 0.16 gpm/ft.^{2} (6.5 L/min/m^2) for non-air-aspirating discharge devices.

- Supplementary underwing protection systems for single aircraft whose wing area exceeds 3000 ft.^{2} (279 m^2):
  - Systems shall be designed in accordance with NFPA 11.
  - One method of protecting such areas is by use of water oscillating monitors (WOMs).

- Closed-head automatic sprinkler protection system:
  - Systems shall be either wet pipe or preaction.
  - Systems shall not exceed 52,000 ft.^{2} (4831 m^2) in total area.
  - Sprinkler spacing shall not exceed 130 ft.^{2} (12 m^2) per sprinkler.
  - Maximum distance between sprinklers or branch lines shall not exceed 12 ft. (3.7m).
  - Design density shall be hydraulically calculated to ensure a minimum of 0.17 gpm/ft.^{2} (6.8 L/min/m^2) over the hydraulically most remote 15,000 ft.^{2} (1394 m^2).
  - Quick response sprinklers with a temperature rating of 175°F (79.4°C) shall be used.

- Low-level low-expansion foam systems:
  - Systems shall be designed to cover the entire floor area.
  - Where AFFF is used, the design density is 0.10 gpm/ft.^{2} (4.1 L/min/m^2).
  - Where protein-based or fluoroprotein-based foam is used, design density is 0.16 gpm/ft.^{2} (6.5 L/min/m^2).

- Low-level high-expansion foam systems:
  - Systems shall be designed to cover the entire floor area.
  - Application rate shall be 3 ft.^{3}/min/ft.^{2} (0.9 m^3/min/m^2).

4. *Estimate roof protection discharge rate.*

$$D = (A) \times (R)$$

where D = foam solution discharge rate, gpm (l/min)
   A = hangar floor area, square feet (m^2)
   R = application rate per step 3, gpm per square foot
      (l/min · m^2)

5. *Estimate concentrate quantity for roof protection.*

$$Q = (A) \times (R) \times (T) \times (\%)$$

where Q = foam concentrate quantity, gallons (liters)
   T = foam discharge time = 10 minutes
   % = concentrate percentage, expressed as a decimal (i.e., 0.01
      for 1% concentrates, 0.03 for 3% concentrates, 0.06 for 6%
      concentrates). NFPA 409 requires a reserve supply of
      concentrate, equal to the primary supply:

Total concentrate supply = (2) × (Q)

6. *Determine aircraft wing area.* The aircraft wing area may prevent the effective application of low-expansion foam from the roof system to the area below the wings. This is a serious problem for aircraft with large wing areas because the wings hold large quantities of fuel and an ignition of spilled fuel below a large wing can place the fuel stored within the wing in extreme jeopardy. For this reason, water oscillating monitors (WOMs) are designed to apply a foam-water blanket beneath the wings. **Table 4-13** lists several commonly encountered aircraft with their associated wing areas. Total wing area for the hangar is the sum of all wing areas expected to be housed in the hangar. If the aircraft is not listed on Table 4-13, the wing area must be obtained from the owner or client.

7. *Determine underwing water oscillating monitor location.* Monitors should be located so full coverage beneath the wings can be obtained. Ideally, for an aircraft hangar housing one aircraft whose exact position is known, at least two monitors would be located so they can be perpendicular to the fuselage of the aircraft and provide unobstructed protection beneath the wings. Ascertaining the precise location for an aircraft proposed for storage in an aircraft hangar may not be feasible, especially because the usage of the hangar may change with time. For such cases, WOM coverage may have to be provided over the entire floor area. WOM coverage is illustrated in **Figure 4-22**.

8. *Determine water oscillating monitor coverage area.* A water oscillating monitor will project foam in a fraction of a circle, of known radius as shown on **Figure 4-22**. A photo of a WOM is shown on **Figure 4-23**. The radius is a function of the ability of the monitor to distribute foam and

Table 4-13 *Gross wing area and overall height for selected aircraft.*

Aircraft	Gross Wing Area		Overall Height	
	m²	ft²	m	ft–in.
Tupolev TU-154	201.5[†]	2169	11.4[†]	37–4
Boeing 757	185.2[†]	1994	13.5[†]	44–6
Tupolev TU-204	182.4[†]	1963	13.9[†]	45–7
Boeing 727-200	157.9[†]	1700	10.4[†]	34–0
Lockheed L-100J Hercules	162.1[†]	1745	11.6[†]	38–3
Yakovlev Yak-42	150.0[†]	1614	9.3[†]	32–3
Boeing 737-600, -700, -800, -900	125.0[†]	1345	12.5[†]	43–3
Airbus A-318, A-319, A-320, A-321	122.6[†]	1319	11.8[†]	38–8
Boeing MD 80	112.3[†]	1209	9.0[†]	29–7
MD 90			9.3[†]	30–7
Gulfstream V	105.6[†]	1137	7.9[†]	25–10
Boeing 737-300, -400, -500	105.4[†]	1135	11.1[†]	36–6
Tupolev TU-334, TU-354	100.0[†]	1076	9.4[†]	30–9
BAC 1-11-500	95.8[†]	1031	7.5[†]	24–6
NAMC YS-11	94.8[†]	1020	8.9[†]	29–5
Fokker 100, 70	93.5[†]	1006	8.5[†]	27–10
BAC 1-11-300, -400	93.2	1003	7.5[†]	24–6
Boeing 717	93.0[†]	1001	8.8[†]	29–1
DC-9-30	93.0[†]	1001	8.4[†]	27–6
Boeing 737-200	91.0[†]	980	11.3[†]	37–0
Gulfstream IV	88.3[†]	950	7.4[†]	24–5
DC 9-10	86.8[†]	934	8.4[†]	27–6
BAe 146, RJX-70, -85, -100	77.3[†]	832	8.6[†]	28–3
Fokker 50, 60	70.0[†]	753	2.7[†]	27–3
Canadair RJ-700	68.6[†]	738	7.6[†]	24–10
Dash 8 Q400	63.0[†]	679	7.5[†]	24–7
ATR 72	61.0[†]	656	7.6[†]	25–1
Airtech CN-235	59.1[†]	636	8.2[†]	26–10
Saab 2000	55.7[†]	600	7.7[†]	25–4
Canadair RJ-100, -200	54.5[†]	587	6.2[†]	20–5
ATR 42	42.5[†]	586	7.6[†]	24–10
Dash 8 Q100, Q200	54.3[†]	585	7.5[†]	24–7
Embraer ERJ-135, -145	51.1[†]	550	6.9[†]	22–1
Cessna 750	48.9[†]	527	5.8[†]	18–11
Cessna 680	47.9[†]	516	5.5[†]	19–2
Saab 340	41.8[†]	450	6.9[†]	22–1
Embraer EMB-120	39.4[†]	424	6.3[†]	20–10
Bell Boeing V-22	39.5[†]	382	6.6[†]	21–9
Britten-Norman BN2	30.2[†]	325	4.2[†]	13–8
Cessna 650	28.9[†]	312	5.1[†]	16–9
Beech 1900	28.8[†]	310	4.7[†]	15–6
Beech King Air C90	27.3[†]	294	4.3[†]	14–3

*Aircraft with wing areas in excess of 279 m² (3000 ft².).

[†]Data from *Jane's All the World's Aircraft.*

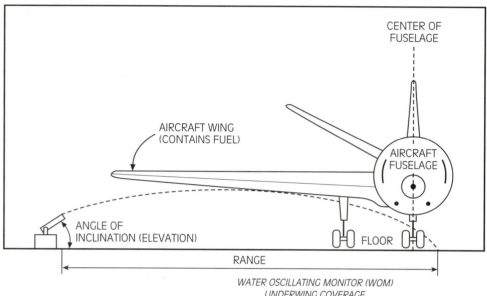

WATER OSCILLATING MONITOR (WOM)
UNDERWING COVERAGE
(SIDE VIEW)

THE WOM MUST BE AIMED TO COVER THE UNDERWING AREA WITHOUT INTERFERENCE FROM THE WING. THE ANGLE OF INCLINATION MUST BE GREAT ENOUGH TO PROVIDE COVERAGE TO THE CENTER OF THE FUSILAGE, WITHOUT INTERFERENCE BY THE WING. THE DESIGNER MUST REFER TO MANUFACTURER'S DATA FOR DETERMINATION OF THE RANGE OF THE WOM AS A FUNCTION OF THE ANGLE OF INCLINATION (ELEVATION). A CRITICAL FEATURE IS THE LIMITATION TO THE ANGLE OF INCLINATION CAUSED BY THE AIRCRAFT WING, WHICH MAY REDUCE THE RANGE.

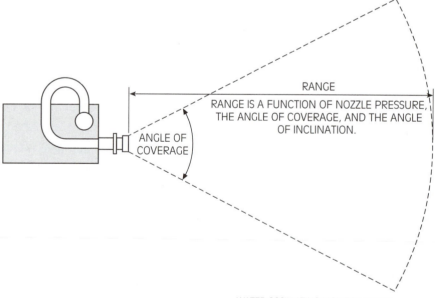

WATER OSCILLATING MONITOR (WOM)
PLAN VIEW (LOOKING DOWN FROM ABOVE)

WOM COVERAGE AREA LOOKS LIKE A PIECE OF PIE, WHEN VIEWED FROM THE CEILING AND TRAUSA PARABOLA WHEN VIEWED FROM THE SIDE. AS YOU HAVE LEARNED WHEN WATERING YOUR LAWN, THE SMALLER THE ANGLE OF COVERAGE, THE GREATER THE RANGE.

Figure 4-22　*Water oscillating monitor (WOM).*

Figure 4-23 *WOM protection of the underwing area of an aircraft. (Courtesy of National Foam, Inc.)*

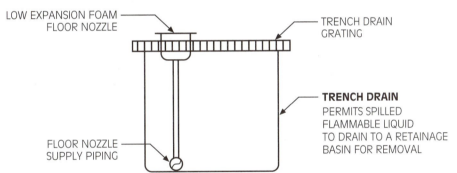

LOW EXPANSION FOAM
FLOOR NOZZLE

TRENCH DRAIN
GRATING

TRENCH DRAIN
PERMITS SPILLED
FLAMMABLE LIQUID
TO DRAIN TO A RETAINAGE
BASIN FOR REMOVAL

FLOOR NOZZLE
SUPPLY PIPING

Figure 4-24 *Floor-mounted underwing nozzles; upon activation, floor nozzles raise to distribute foam below aircraft wings.*

varies with monitor vendor and monitor inlet pressures. The monitor manufacturer must be consulted to verify that the monitor selected will provide throw sufficient to protect the entire underwing area. Once the throw or radius is obtained, the arc angle must be determined. This can be done by drawing the arc on the hangar plan and measuring the angle required to cover the wing area. The area of monitor coverage is determined by calculating what fraction of a full circle is being covered by the monitor:

$$\text{Monitor area} = [(3.1416) \times (r^2)] \times \frac{(\text{angle of coverage})}{360}$$

When monitors cover the entire floor area, the monitor area is determined by dividing the floor area by the number of monitors.

9. *Water oscillating monitor system discharge time and application rate.*

   • Discharge time is 10 minutes foam per NFPA 409.

   • Application rate is 0.10 gpm per square foot (4.1 l/min · m^2) per NFPA 409 when AFFF is used.

10. *Determine water oscillating monitor discharge rate and concentrate quantity.*

$$D = (A) \times (R) \times (N)$$
$$Q = (A) \times (R) \times (N) \times (T) \times (\%)$$

where D = foam discharge rate, gpm (l/min)
A = area of monitor coverage per step 8, in square feet (m^2)
R = application rate, 0.10 gpm per square foot (4.1 l/min · m^2)
N = number of monitors installed
T = discharge time, 10 minutes
$\%$ = concentrate percentage expressed as a decimal (e.g., 0.01 for 1% concentrate, 0.03 for 3% concentrate, 0.06 for 6% concentrate). NFPA 409 requires a reserve supply of foam equal to the primary supply.

Total concentrate supply = (2) $\times$ (Q)

Note that an option to WOMs are nozzles installed at the floor of the aircraft hangar, usually within trench drains, as shown on **Figure 4-24**. NFPA 409 requires trench drains on aircraft hanger floors, that direct spilled fuel to retainage basins, reducing fire severity within the hanger. Aircraft hangar floors are required to pitch a minimum of 0.5% to the trench drains.

11. *Determine the supplementary hose discharge requirements.* The designer must determine the location of racked or reeled hose stations and determine hose lengths, such that hose protection is available for all portions of the hangar, including the aircraft interior. NFPA 409 requires, as a minimum:

    a. two hand hose lines
    b. 60 gpm (227 l/min) per hose line
    c. 20-minute foam operation

12. *Determine hose discharge rate and concentrate requirement.*

$$D = (N) \times (R) \qquad Q = (N) \times (R) \times (T) \times (\%)$$

where D = hose foam discharge rate, gpm (liters/min)
N = number of hose lines flowing
R = discharge rate per hose, 60 gpm (227 l/min)
T = discharge time, 20 minutes
$\%$ = concentrate percentage, expressed as a decimal (e.g., 0.01 for 1% concentrate, 0.03 for 3% concentrate, 0.06 for 6% concentrate).

EXAMPLE 4-5

Aircraft Hangar Low-Expansion Foam System Application

A Group I aircraft hangar, housing one Lockheed L-500 Galaxy aircraft, requires a foam system. Hangar dimensions are 300 feet long × 250 feet wide, with a 29-foot aircraft access door. Nonaspirated sprinkler heads and 3% AFFF foam will be used. Determine the minimum low-expansion foam system demand and minimum foam concentrate requirement.

Solution

1. *Foam discharge rate from the roof system:*

$$D = (A) \times (R)$$
$$= (300' \times 250') \times (0.16 \text{ gpm/sq.ft.})$$
$$= 12,000 \text{ gpm foam solution rate, assuming sprinkler distribution of foam.}$$

Note that because maximum system size is 15,000 square feet per deluge system, a minimum of five deluge systems are required. Also note that 12,000 gpm is a minimum estimate and does not include gallonage for frictional losses and balancing of overdischarge.

2. *Minimum concentrate quantity for the roof system using 3% AFFF.*

$$Q = (A) \times (R) \times (T) \times (\%)$$
$$= (12,000 \text{ gpm}) \times (10 \text{ min.}) \times (0.03)$$
$$= 3600 \text{ gallons 3% AFFF foam concentrate}$$

Total minimum main and reserve concentrate supply = $(2) \times (Q)$ = 7200 gallons

3. *Aircraft wing area of A Lockheed L-500 galaxy is 6200 square feet per Figure 4-25.*

4. *Two water oscillating monitors* are centrally located, perpendicular to the fuselage, each covering an angle of 90°. The minimum monitor radius is one-half of the hangar width, or 125 feet.

5. *Water oscillating monitor coverage area.*

$$A = [(3.1416) \times (r^2)] \times (90/360)$$
$$= [(3.1416) \times (125^2)] \times 0.25$$
$$= 12,272 \text{ square feet per monitor}$$

6. *Water oscillating monitor discharge rate and concentrate quantity.*

$$D = (A) \times (R) \times (N)$$
$$= (12,272) \times (0.10) \times (2)$$
$$= 2454 \text{ gpm for two monitors}$$

$$Q = (A) \times (R) \times (N) \times (T) \times (\%)$$
$$= (12,272) \times (0.10) \times (2) \times (10 \text{ min}) \times (0.03)$$
$$= 736 \text{ gallons of 3% AFFF concentrate}$$

7. *Supplementary hose requirements,* assuming the use of a 100-foot hose with a 30-foot hose spray, with four hose stations in the hangar;

$$D = (N) \times (R)$$
$$= (4) \times (60 \text{ gpm})$$
$$= 240 \text{ gpm}$$

$$Q = (N) \times (R) \times (T) \times (\%)$$
$$= (4) \times (60 \text{ gpm}) \times (20 \text{ min}) \times (0.03)$$
$$= 144 \text{ gallons 3\% AFFF concentrate required}$$

8. *Estimate total minimum system demand and minimum concentrate requirement.*
Total system demand

$$\text{Total demand} = (\text{roof demand}) + (\text{monitor demand}) + (\text{hose demand})$$
$$= (12,000 \text{ gpm}) + (2454 \text{ gpm}) + (240 \text{ gpm})$$
$$= 14,694 \text{ gpm minimum system demand}$$

Total concentrate quantity

$$\text{Total concentrate} = (\text{roof}) + (\text{monitor}) + (\text{hose})$$
$$= (3600 \text{ gals}) + (736 \text{ gals}) + (144 \text{ gals})$$
$$= 4480 \text{ gallons 3\% AFFF concentrate}$$

$$\text{Total concentrate (main + reserve)} = 8960 \text{ gallons}$$

LOADING RACK PROTECTION

NFPA

NFPA 16, Standard for the Installation of Foam-Water Sprinkler and Foam-Water Spray Systems

A loading rack is the critical point where flammable or combustible liquids are pumped from or to storage tanks (discussed earlier in this chapter), to or from a truck or rail car. The entire manufacturing process of flammable and combustible liquids should be protected by fixed fire protection systems, as shown on **Figure 4-25**.

Low-expansion foam protection of the truck loading rack, shown on **Figure 4-26**, is designed in accordance with NFPA 11 and NFPA 16, *Standard for the Installation of Foam-Water Sprinkler and Foam-Water Spray Systems.*

Hazards Associated with Loading Racks

Considering storage tank protection and dike protection, loading rack protection may be the most sensitive operation involved in the manufacture, storage, and transfer of a flammable or combustible liquid. Pumping the liquid to the truck or rail car requires pressurization of the flammable or combustible liquid, creating the possibility of a pressure spray fire if ignited.

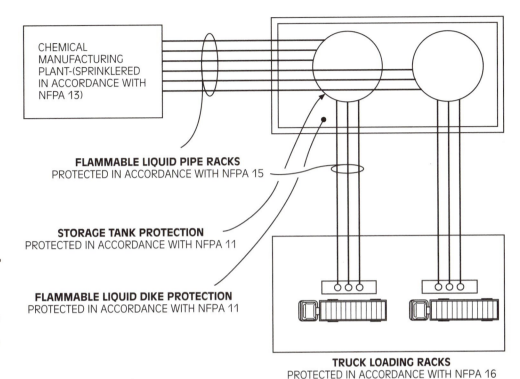

CHEMICAL
MANUFACTURING
PLANT-(SPRINKLERED
IN ACCORDANCE WITH
NFPA 13)

FLAMMABLE LIQUID PIPE RACKS
PROTECTED IN ACCORDANCE WITH NFPA 15

STORAGE TANK PROTECTION
PROTECTED IN ACCORDANCE WITH NFPA 11

FLAMMABLE LIQUID DIKE PROTECTION
PROTECTED IN ACCORDANCE WITH NFPA 11

TRUCK LOADING RACKS
PROTECTED IN ACCORDANCE WITH NFPA 16

Figure 4-25
Flammable liquid manufacturing process; each phase of the process should be protected by a fixed fire protection system.

Figure 4-26 *Truck loading rack. (Courtesy of National Foam, Inc.)*

The weak links in the process are the pump, which could overheat and fail, and the hose, which could burst or become disengaged from its connector to the truck or rail car. The number of ignition sources available during this process may be cause for concern, with the possibility of smoking, overheated pumps, electrostatic charges, or ignition sources on the truck or rail car, such as the battery.

Fire Protection Strategy for Loading Racks

Foam-water sprinklers or nozzles are installed at the roof of the loading rack. If a loading rack does not have a roof, a structure will be needed to support low-expansion foam system piping. Foam-water sprinklers usually are spaced no more than 10 feet apart, for a maximum coverage area of 100 square feet per sprinkler. **Figure 4-27** illustrates roof foam-water coverage.

The object of the roof protection is to provide complete protection of the drainage area. The drainage area is a curbed area designed to contain spilled flammable liquid as it flows toward floor drains. The hazardous liquid is directed from the floor drains to a retainage basin for recovery and disposal. The drainage area may not directly coincide with the roof area, so care must be taken to ensure coverage of the drainage area from the roof sprinklers, as shown on **Figure 4-28**.

Figure 4-27 *Truck loading rack with low-expansion foam protection. (Courtesy of National Foam, Inc.)*

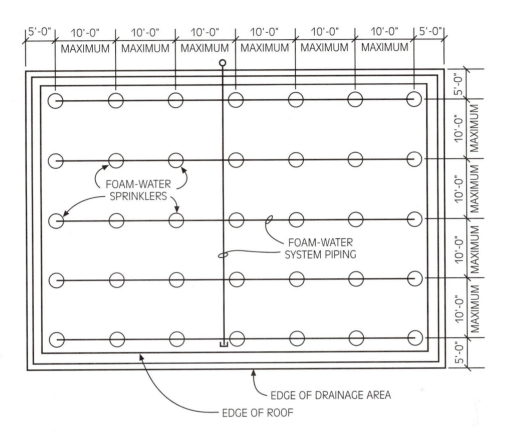

Figure 4-28 *Truck loading rack spacing of roof sprinklers.*

Additional nozzles are aimed directly at the point of connection of the hose to the truck or rail car, and beneath the truck or rail car to enable the sweeping of liquid away from the truck or rail car, as shown on **Figure 4-29**.

Loading Rack System Design Procedure

A methodical approach to the design of low-expansion foam protection of loading racks, as shown on Figure 4-29, is presented as follows.

1. *Define the hazard area.* The hazard area is always the drainage area, not the roof area.
2. *Determine the type of foam to be used* (3% AFFF, 6% AFFF, etc.), and determine the primary discharge rate associated with the fuel protected, per **Table 4-14**.
3. *Determine the discharge time.* The discharge time for low-expansion foam application with truck loading racks is 10 minutes. Where loading racks are protected by monitor nozzles, the minimum discharge time is 15 minutes, per **Table 4-15**.

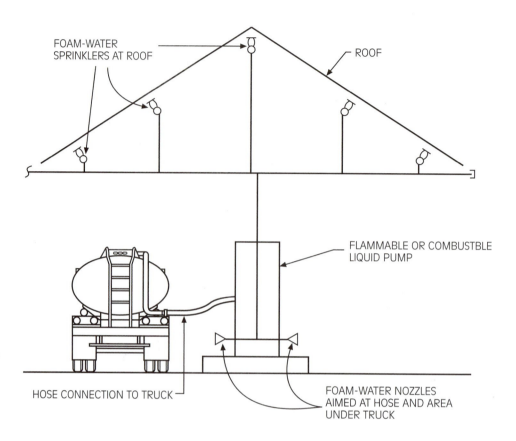

Figure 4-29 *Truck loading rack-ground sweep nozzles.*

4. *Calculate the primary foam concentrate quantity.*

$$Q = A \times D \times T \times \%$$

where Q = primary foam concentrate quantity from roof system, in gallons
   A = drainage area, length times width
   D = application density rate, from Table 4-14. Where loading racks are protected by monitor nozzles, the minimum application rate is determined by Table 4-15.
   T = discharge time, 10 minutes
   $\%$ = low-expansion foam percentage, from Table 4-14

5. *Determine the number of ground sweep nozzles.* The number of nozzles is determined by field survey, and is usually two nozzles between each pump.

6. *Select nozzle and design attributes.* The designer must reference a book of manufacturer's data to select a specific nozzle. An example of such a nozzle is a wide-angle nozzle, spraying a 120° pattern, discharging 29 gpm at 30 psi.

Table 4-14   *Truck loading rack application rates for various fuels and low-expansion foams. (Courtesy of National Foam, Inc.)*

Fuel Protected	Foam Concentrate	Application Rate	
		gpm/ft^2	(Lpm/m^2)
Hydrocarbons	Protein	0.16	(6.5)
(Water Insoluble)	Fluoroprotein	0.16	(6.5)
	AFFF	0.16	(6.5)
Alcohols			
Methanol	ARC	0.17	(6.9)
	3×3	0.16	(6.5)
	3×6	0.16	(6.5)
Ethanol	ARC	0.17	(6.9)
	3×3	0.16	(6.5)
	3×6	0.16	(6.5)
Isopropanol	ARC	0.17	(6.9)
	3×3	0.16	(6.5)
	3×6	0.16	(6.5)
Ketones	ARC	0.17	(6.9)
	3×3	0.16	(6.5)
	3×6	0.16	(6.5)
Carboxylic Acids	ARC	0.17	(6.9)
	3×3	0.16	(6.5)
	3×6	0.16	(6.5)
Aldehydes	ARC	0.16	(6.5)
	3×3	0.16	(6.5)
	3×6	0.16	(6.5)
Esters	ARC	0.17	(6.9)
	3×3	0.16	(6.5)
	3×6	0.16	(6.5)
Ethers	ARC	0.16	(6.5)
	3×3	0.16	(6.5)
	3×6	0.16	(6.5)

Note: 3 × 6 concentrate is 6% concentrate, and 3 × 3 concentrate is 3% concentrate; ARC is alcohol resistant concentrate.

7. *Determine the ground sweep foam quantity.*

$$Q_g = N \times D \times T \times \%$$

where Q_g = ground sweep foam quantity, in gallons
N = number of ground sweep nozzles on the system
D = gpm discharging from each nozzle (in the example nozzle given in step 6, 29 gpm is the discharge rate)
T = discharge time, 10 minutes
$\%$ = low-expansion foam percentage, from Table 4-14

Table 4-15 *Minimum application rates and discharge times for loading racks protected by foam monitor nozzle systems.*

Foam Type	Minimum Application Rate		Minimum Discharge Time (min)	Product Being Loaded
	L/min · m²	gpm/ft²		
Protein and fluoroprotein	6.5	0.16	15	Hydrocarbons
AFFF, FFFP, and alcohol-resistant AFFF or FFFP	4.1	0.10	15	Hydrocarbons
Alcohol-resistant foams	Consult manufacturer for listings on specific products		15	Flammable and combustible liquids requiring alcohol-resistant foam

Source: NFPA 11 (2005), Table 5.6.5.3.1. Reprinted with permission from NFPA 11, *Low-, Medium-, and High-Expansion Foam*, Copyright © 2005, National Fire Protection Association, Quincy, MA 02169. This reprinted material is not the complete and official position of the National Fire Protection Association on the referenced subject which is represented only by the standard in its entirety.

8. *Determine total foam concentrate quantity.*

$$Q_t = Q + Q_g$$

where Q_t = total foam concentrate quantity
Q = primary foam concentrate quantity, from roof system
Q_g = foam concentrate quantity from ground sweep nozzles

9. *Determine the spacing of the roof sprinklers.* Divide the length of the drainage area by 10 feet and round up to the nearest whole number, then do the same for the width of the drainage area. As an example, a drainage area of 128 feet long by 102 feet wide would have 13 sprinklers spaced equidistantly along its length, and 11 sprinklers spaced equidistantly along its width, for a total of 143 sprinklers.

HYDRAULIC CALCULATION OF FOAM SYSTEMS

After the foam quantity estimates have been made in accordance with this chapter, the designer draws a detailed CAD layout of the system and performs a comprehensive computerized hydraulic calculation of the low-expansion foam system.

A number of nuances are unique to the hydraulic calculation of foam systems, as opposed to the calculation of automatic sprinkler systems. For example, a calculation result for an automatic sprinkler system that shows system demand significantly in excess of available water supply is considered an advantageous attribute, whereas in foam systems, such a result could mean that the desired foam percentage is not being obtained, or that the desired foam discharge duration is not being met. Foam systems require calculations that are carefully balanced at each junction node, and balanced with respect to the water supply, to ensure that the required percentage and duration of foam is being met.

When hydraulically calculating a foam system, perform the calculation both to the supply and to the demand. A supply calculation flows the water supply to the flowing nodes and tells you what density is flowing from each flowing node, and the flowing densities are compared to the NFPA minimum density requirements. A demand calculation begins at the hyrdaulically most remote flowing node, using the minimum density prescribed by the appropriate NFPA standard, and tells you whether this minimum density, given the pipe sizes shown, will produce a demand above or below the water supply curve. If the demand point is above the water supply curve, pipe sizes must be increased, and the demand must be recalculated until it falls below the water supply curve. In each case, perform calculations to ensure that the proper amount of foam concentrate is provided. In most cases, the demand point will need to be very close below the water supply curve to ensure the correct discharge time and foam concentrate percentage. For a detailed methodology for hydraulically calculating an automatic sprinkler system or low-expansion foam system, please refer to *Design of Water-Based Fire Protection Systems*, also published by Delmar Learning.

Foam systems that use proportioners must have foam demands and water supplies carefully balanced at the inlet and discharge ports of the proportioner. This will ensure proper operation of the proportioner and proper injection of the correct percentage of concentrate.

SUMMARY

Low-expansion foam systems are designed for a wide variety of applications involving flammable liquids, with flash points below 100°F, and combustible liquids, with flash points at or above 100°F. Low-expansion foam has an expansion ratio of up to 20:1 and consists of a mixture of foam concentrate, water, and air. A foam system may use protein, fluoroprotein, AFFF, or alcohol-resistant foam.

Fixed low-expansion foam systems are commonly used for subsurface injection or surface application of flammable or combustible storage tanks. Low-expansion foam systems also are frequently used to provide dike protection, seal protection for floating roof tanks, protection for aircraft hangars, and for other hazardous areas where flammable or combustible liquids are present.

REVIEW QUESTIONS

1. Classify the following fuels as flammable or combustible and identify class:

    a. acetone

    b. butyl alcohol

    c. creosote oil

    d. mineral oil

    e. methyl alcohol

    f. fish oil

    g. gasoline

2. A fixed cone roof flammable liquids storage tank has a diameter of 185 feet, is 100 feet tall, and stores 100 octane gasoline. Draw a 3% AFFF low-expansion subsurface injection foam system for this tank, and calculate the minimum amount of low-expansion foam concentrate required for the protection of this tank. Also determine the minimum amount of foam solution required and the minimum amount of water that must be available for this design.

3. Draw a 3% AFFF low-expansion foam surface application system for the tank described in problem 2, and calculate the minimum amount of low-expansion foam concentrate that would be required for the protection of this tank. Also determine the minimum amount of foam solution required and the minimum amount of water that must be available for this design.

4. Draw a 6% AFFF low-expansion top-of-seal protection foam system for a floating roof tank whose diameter is 200 feet, 80 feet tall, with a 2-foot wide foam dam that is 12 inches high. Calculate the minimum amount of low-expansion foam concentrate that is required for the protection of this tank. Also determine the minimum amount of foam solution required and the minimum amount of water that must be available for this design.

5. A dike measuring 300 feet by 300 feet surrounds a tank farm. Assume a fuel with a flash point less than 100°F, and assume the use of fixed discharge devices. Draw a 6% AFFF low-expansion foam system, and calculate the minimum amount of low-expansion foam concentrate that is required for the protection of the dike area. Also determine the minimum amount of foam solution required and the minimum amount of water that must be available for this design.

6. An aircraft hangar has a door height of 32 feet and measures 200 feet × 200 feet. Draw a 3% AFFF low-expansion foam system for this hangar, showing all required systems, using aspirated foam-water sprinklers. A Boeing 747 is stored in the hangar. Assume two WOMs with a 90° discharge angle and a 100' throw are used. Assume four hand hose lines operate. Calculate the minimum amount of low-expansion foam concentrate that is required for the protection of the aircraft hangar. Also determine the minimum amount of foam solution required and the minimum amount of water that must be available for this design.

DISCUSSION QUESTIONS

1. For Group I aircraft hangars, why are the required design densities for air-aspirated nozzles and non-air-aspirating nozzles different?

2. What advantages would floor drain-mounted foam discharge devices have as opposed to WOMs in an aircraft hangar?

ACTIVITIES

1. Arrange to visit a power generating facility, and take note of the methods used for the storage of flammable and combustible liquids. Determine whether any cone roof storage tanks are protected by low-expansion foam. Do any cone roof storage tanks exist with no low-expansion foam protection? Consult the plant or corporate fire protection engineer, and determine the criteria used to decide whether a tank receives foam protection. Do any dikes have low-expansion foam systems installed?

2. Visit an aircraft hangar at a commercial airport or on a military facility. Classify the hangar by taking measurements of the wing areas of aircraft housed by the hangar. If an existing foam system is installed, draw the system, and consult the base fire protection engineer to determine whether the criteria used for the design of the system match NFPA criteria.

Chapter

5

MEDIUM- AND HIGH-EXPANSION FOAM SYSTEM DESIGN

Learning Objectives

Upon completion of this chapter, you should be able to:

- Determine appropriate situations for the use of low-expansion, medium-expansion, and high-expansion foam.

- Discuss the differences in the application and the methods of extinguishment for low-expansion, medium-expansion, and high-expansion foam.

- Perform a calculation of a high-expansion foam system to determine the rate of discharge and number of high-expansion foam generators required.

- Lay out a high-expansion foam system, showing foam fences, generator locations, and piping locations.

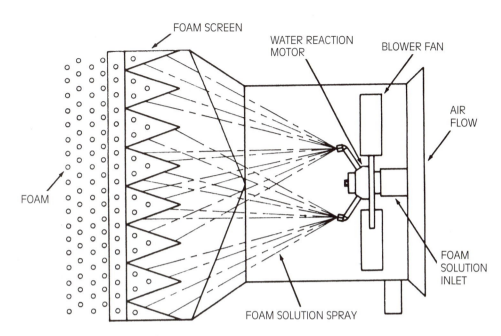

FOAM SCREEN

WATER REACTION MOTOR

BLOWER FAN

AIR FLOW

FOAM

FOAM SOLUTION INLET

FOAM SOLUTION SPRAY

Figure 5-1 *High-expansion foam generator. (Courtesy of Ansul, Inc.)*

high-expansion foam system

a system of air-filled bubbles created by the mechanical expansion of a foam solution by air and water, with foam-to-solution ratio of between 200 to 1 and 1000 to 1

NFPA

NFPA 11, Standard for Low-, Medium-, and High-Expansion Foam

high-expansion foam generators

devices that add air to the foam solution spray, creating foam

Medium-expansion foam has an expansion ratio of 20:1 to 200:1. A **high-expansion foam system** is defined as a system that delivers a blanket of air-filled bubbles created by the mechanical expansion of a foam solution by air and water, with a foam-to-solution ratio of between 200 to 1 and 1000 to 1. High-expansion foam systems are designed in accordance with NFPA 11, *Standard for Low-, Medium-, and High-Expansion Foam.* Some devices are capable of higher expansion ratios.

To achieve expansion ratios within the prescribed range, one or more **high-expansion foam generators**, resembling large fans, are used as the discharge devices. A foam–water solution, consisting of between 2.5% and 3% foam concentrate, is sprayed into the air stream of a fan and projected through a foam screen. The fan adds air to the foam solution spray, creating foam, which flows through a screen and distributes a thick blanket of high-expansion foam.

High-expansion foam generators operate electrically or are powered by water pressure. Electric generators require a reliable power source with emergency backup power, whereas water-powered generators require no external sources of power, providing a level of reliability often preferred by specifying engineers. **Figure 5-1** shows a water-powered high-expansion foam generator.

LOCAL APPLICATION MEDIUM- AND HIGH-EXPANSION FOAM SYSTEMS

The majority of medium- and high-expansion foam systems are total flooding systems, in which foam fills a volume to a specified height. Local application systems discharge foam directly onto a fire or spill hazard. NFPA permits the use of local foam application for extinguishing flammable and combustible liquid fires, liquefied natural gas (LNG) fires, and fires involving ordinary combustibles. Such applications assume that walls or other volumetric containment features do not exist, but these features may be added to enhance the effectiveness of the protection for three-dimensional hazards.

For local application systems, NFPA 11 specifies minimum design criteria, which include the following:

- Sufficient foam must be discharged to cover the hazard to a depth of 2 feet (0.6 m).
- The required depth must be attained within 2 minutes.
- The system shall operate for a duration of at least 12 minutes.

ELECTRICAL CLEARANCES FOR MEDIUM- AND HIGH-EXPANSION FOAM SYSTEMS

Because foam contains water, the application of foam could transmit electricity if the designer does not take the care to prevent such an occurrence, keeping in mind that it may be possible to automatically disconnect electrical power in advance of foam discharge. Further, the proximity of foam components to live electrical components must be coordinated. **Table 5-1** shows electrical clearances for medium and high-expansion foam equipment to live uninsulated electrical components.

Table 5-1 applies at altitudes of 3300 feet (1000 m) or less. For altitudes greater than 3300 feet (1000 m), the clearance is required to be increased at the rate of 1% for each 330 feet (100 m) of altitude above 3300 feet (1000 m). Further, at voltages greater than 161 kV, a relationship between the design basic insulation level (BIL) kV and the nominal voltages is not uniform among electrical components, and close coordination is needed to determine accurate clearances.

MEDIUM-EXPANSION FOAM

NFPA 11 does not provide specific design requirements for medium-expansion foam systems. Instead, it provides some guidelines for use by designers:

- The required depth of medium-expansion foam over a protected hazard shall vary as a function of expansion ratio. Coordination with the foam manufacturer is therefore required.
- The depth of medium-expansion foam shall be determined by tests. NFPA 11 refers to Annex H, which details performance tests for Class A, Class B,

Table 5-1 *Clearance from medium- and high-expansion foam equipment to live uninsulated electrical components.*

Nominal Line Voltage (kV)	Nominal Voltage to Ground (kV)	Design BIL[1] (kV)	Minimum Clearance[2]	
			mm	in.
To 15	To 9	110	178	7
23	13	150	254	10
34.5	20	200	330	13
46	27	250	432	17
69	40	350	635	25
115	66	550	940	37
138	80	650	1118	44
161	93	750	1321	52
196–230	114–132	900	1600	63
		1050	1930	76
		1175	2210	87
		1300	2489	98
287–380	166–220	1425	2769	109
		1550	3048	120
500	290	1675	3327	131
		1800	3607	142
		1925	3886	153
500–700	290–400	2100	4267	168
		2300	4674	184

[1]Basic insulation level (BIL) values are expressed as kilovolts (kV), the number being the crest value of the full wave impulse test that the electrical equipment is designed to withstand.
[2]For voltages up to 69 kV, the clearances are taken from NFPA 70.

and LNG fires. Annex H is not a part of the NFPA requirements but may provide some guidance for designers. Most foam manufacturers have a database of foam test results for specific commodities, which may provide the basis for design criteria for designers.

• The rate of discharge of medium-expansion foam shall be determined by tests. Annex H and manufacturers' data may provide some guidance.

• The quantity of medium-expansion foam shall be determined by tests conducted by an independent testing laboratory. These tests may be arranged for a specific proposed use, or may be made available by the manufacturer or testing laboratory for hazards similar to the ones proposed.

APPLICATIONS FOR HIGH-EXPANSION FOAM SYSTEMS

For hazards in which smothering of a three-dimensional fire or oxygen deprivation of a three-dimensional fire is the primary objective, a high-expansion foam system is advantageous. A three-dimensional object requiring high-expansion foam is one that requires foam to be totally flooded and completely covered to an elevation above the highest level of the object requiring protection. Examples of a three-dimensional object that may require high-expansion foam include a liquified natural gas (LNG) pump, and a rack or shelf containing highly flammable or combustible contents.

Low-expansion foam and foam–water systems, discussed in Chapter 4, are used primarily for two-dimensional pool fires where a thin blanket of foam is sufficient to float on top of a flammable or combustible liquid. Low-expansion foam systems demand sizable amounts of water and create a considerable amount of foam–water solution that requires disposal. This problem becomes complicated when flammable liquids or materials having a detrimental environmental impact are involved. These materials are diluted by the foam–water solution, and will run off to the nearest drain, requiring a drainage and retainage system that can handle and store flaming materials that can contain environmentally damaging material.

A way to minimize this adverse environmental impact is to specify a high-expansion foam system, shown on **Figure 5-2**. A high-expansion foam system

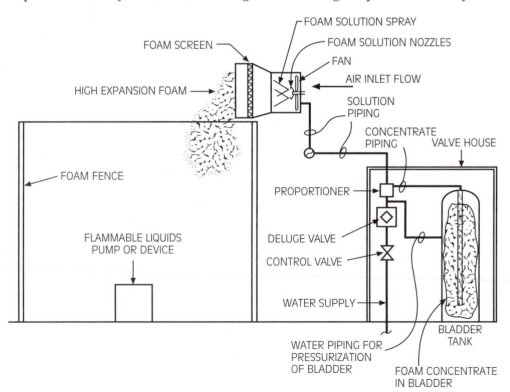

Figure 5-2 *Sectional view of a flammable liquid pump protected by a high-expansion foam system.*

requires much less water and is significantly more efficient when a containment barrier is provided. Containment of high-expansion foam ordinarily is accomplished by building a foam fence around the hazard, which prevents runoff. A foam fence can be as simple as a standard chain-link fence with strips of plastic woven between the links. The foam-holding integrity requirements of the foam fence are a function of the environmental danger involved in the runoff. Concerns relative to water damage are decreased because the amount of water required to make high-expansion foam is relatively small compared to low-expansion foam systems.

High-expansion foam systems are permitted to protect liquified natural gas (LNG) applications, and application rates are determined by testing in accordance with NFPA 11, Annex G.

OTHER APPLICATIONS FOR HIGH-EXPANSION FOAM SUPPRESSION SYSTEMS

High-expansion foam systems are used primarily for the extinguishment of flammable liquid fires, but should be specified with at least some degree of caution with respect to personnel safety. Two examples of high-expansion foam suppression are presented to demonstrate the logic that is considered in specifying high-expansion foam for the protection of flammable liquid hazards.

HIGH-EXPANSION FOAM SYSTEMS FOR ROBOTIC FLAMMABLE LIQUID RACK STORAGE

A robotic flammable liquids storage scenario in Singapore, presents an opportunity to demonstrate the process used by a fire protection engineer in the specification of a high-expansion foam system for an unusual application.[1] An 80-foot-(24.4. m) tall flammable liquids storage facility featured a robotically operated rack storage system in which an automatic robot scanned bar codes on 55-gallon (208.2 l) barrels of flammable liquid and used that information to direct the barrel to a specific rack at a prespecified level. Although this process seems fool-proof, accidents do happen, and the possibility exists for a barrel to drop and initiate a pool fire.

The first consideration involves personnel. Although the drawings and specifications present the building as a robotic facility, service and maintenance personnel may be asked to enter the facility periodically for repairs and troubleshooting. The prospect for a person to be caught in the middle of a flammable liquids pool fire or to be buried in 80 feet of high-expansion foam is not to be taken lightly. The following are actions that a fire protection engineer may consider in such situations:

- Establishment of a procedure for lockout of the flammable liquid storage area from individuals who enter the facility accidentally. The lockout system involves accounting for all personnel at all times, keeping all doors of the robotic facility locked in the closed position at all times, and engaging heightened fire watches and rescue operations when a person enters the facility, including video monitoring and electronic communication systems.

- Establishment of a protocol for entering the facility by authorized personnel, such as an individualized code or a special key.

- Establishment of a procedure for keeping track of personnel who enter the facility, such as coded entry, sign-in sheets, video surveillance, and special notification procedures.

- Establishment of a protocol for periodically training personnel who may be required to enter the facility, or personnel in charge of tracking those who enter.

- Establishment of a procedure for protecting individuals while in the facility, such as portable breathing assistance, a leader line that a person can follow to return to safety, the initiation of audible and visual alarms and perhaps voice warnings, portable communication equipment, and possibly conversion of the suppression system from automatic to manual operation.

- Establishment of a procedure for rescuing trapped personnel, with periodic testing and training.

Only after personnel considerations have been solved can a high-expansion foam system be specified and designed. Once personnel considerations are accommodated successfully, the fire protection engineer must determine and solve the performance objectives of the system. Some of the performance objectives to contemplate include the following:

- With the possibility of a cataclysmic explosion ensuing from a pool fire and its effect on neighboring barrels exposed to elevated temperatures, the time interval for actuating the system must be decided. Flame detectors, discussed later in this book, would enhance rapid initiation of the system. Field or laboratory testing, or computer models may be needed to determine the amount of time that is appropriate to initiate suppression before a pool fire endangers adjacent barrels.

- The time required for extinguishment of the pool fire must be ascertained.

- The rising high-expansion foam provides exposure protection for the stacked barrels and provides cooling for barrels that may have been heated by the pool fire. A rate for the rise of high-expansion foam has to be decided, and the time for the high-expansion foam to reach the top barrel has to be determined. Testing may be needed to ascertain these values.

- The high-expansion foam suppression system has to be field tested to ensure that the performance objectives have been met.

- A methodology for collecting and disposing of the high-expansion foam has to be established. [In the example being discussed, 80 feet (24.4 m) of high-expansion foam may take several days to return to solution form.] If sprinklers are present in the facility, they may be used to knock down the foam by temporarily replacing closed sprinklers with sprinklers with open orifices, then replacing the sprinklers after successful testing and disposal. Arrangements have to be made to recover the foam solution and pump it safely into a tanker truck or other container.

HIGH-EXPANSION FOAM SYSTEMS FOR AIRCRAFT HANGARS

NFPA 409 permits high-expansion foam systems to be specified in lieu of the low-expansion foam systems discussed in the previous chapter. **Figures 5-3**, **5-4**, **5-5**, and **5-6** show high-expansion foam systems operating in aircraft hangars.

Figure 5-3 *High-expansion foam generator discharging in an aircraft hangar (wall-mounted). (Courtesy of Ansul, Inc.)*

Figure 5-4 *High-expansion foam generator discharging in an aircraft hangar (ceiling discharge). (Courtesy of Ansul, Inc.)*

Figure 5-5 *Ceiling discharge of high-expansion foam generators in an aircraft hangar. (Courtesy of Ansul, Inc.)*

Figure 5-6 *Aircraft protected by high-expansion foam. (Courtesy of Ansul, Inc.)*

Historically, overhead AFFF low-expansion foam systems have been used to protect flammable liquid spill fire hazards in aircraft hangars, supplemented by water oscillating monitors (WOMS), which can become susceptible to false trips under certain circumstances. The Environmental Protection Agency (EPA) has increased regulatory requirements related to the proper collection and disposal of AFFF,[2] which brings an added layer of challenge for the designer and has opened up avenues for alternative protection schemes.

High-expansion foam may be a possible answer to EPA concerns relative to low-expansion foam because the expansion ratio of high-expansion foam means that there will be considerably less concentrate per unit volume than low-expansion foam. High-expansion foam is primarily a volumetric filling agent, but it is not necessary to completely fill a volume such as an aircraft hangar. As noted previously, the foam height has to be only 10% higher than the hazard height. If the primary hazard is the flammable liquid, theoretically one could consider the hazard height as negligible and employ the 2-ft. minimum high-expansion foam height above the combustible. If the hazard includes aircraft that may be trapped within a spill fire, the hazard height would have to be high enough to provide exposure protection for aircraft without foam intrusion into aircraft or without creating a potentially dangerous condition for personnel.

Some designers use a proposed high-expansion foam depth of 1 meter (3.28 ft.) because it addresses the concerns mentioned previously, and because it replicates the expected average depth produced by a low-expansion foam system. To accomplish this performance objective, measures will have to be taken to ensure that aircraft hangar doors are maintained in a closed position, or to ensure that electronic interfaces are present to automatically close aircraft doors before system actuation.

One concern associated with high-expansion foam is the rate at which the foam travels from its contact point at the floor, to the furthest extent of required protection, which can be calculated based on an approximate velocity of 2 ft./sec on a dry floor, and 3 ft./sec over a flammable liquid surface.[3] This is a concern particularly when high-expansion foam generators are located along the outside wall, as shown in Figure 5-3, and foam is required to travel to the building centerline to achieve complete protection.

To compensate for this dynamic, a designer may elect to adjust the distance between high-expansion foam generators based upon concerns relative to travel time, and may mandate that the generators be located at the roof of the hangar rather than positioned along the perimeter of the exterior walls. The downside of the roof-mounted arrangement is the possibility for foam to enter an aircraft cockpit via open canopies.

Consider the case of an aircraft trapped within a spill fire in the center of a large aircraft hangar, with high-expansion foam generators mounted along the perimeter of the facility. If aircraft exposure protection is an important performance objective for a high-expansion foam system, compare the scenario described previously to the exposure protection distributed along the surface area of the craft when foam–water low-expansion foam sprinklers are used, and use this comparison to make judgments relative to the spacing of high-expansion foam generators at the roof, as shown on Figures 5-4 and 5-5. It also may be possible to position either roof-mounted high-expansion foam generators to avoid discharge onto aircraft with open canopies, or to pre-arrange the positioning of aircraft where canopy location is coordinated to be noncolinear with high-expansion foam generator discharge.

Another crucial design consideration that must be specified is the depth of high-expansion foam required to protect the hazard. Because we are dealing with a two-dimensional pool fire, our initial concern is to coat the surface of the flammable liquid with high-expansion foam. But we also may be responsible for protecting aircraft from flame exposure, so we should consider exposure protection of the aircraft with high-expansion foam. Further, we must take personnel into careful consideration, so filling the facility to the roof may be unwise, hazardous, and unnecessary. It may be sufficient to meet the performance objectives of property protection and life safety protection by filling the facility with high-expansion foam to the aircraft wing or aircraft fuselage, as shown on Figure 5-6.

Figure 5-7 *High-expansion foam generators filling a dike for a flammable liquids storage tank. (Courtesy of Ansul, Inc.)*

HIGH-EXPANSION FOAM SYSTEMS AS DIKE PROTECTION

Chapter 4 introduced low-expansion foam as a suppression methodology for flammable liquid dikes, but successful use of high-expansion foam has been employed as well. **Figures 5-7** and **5-8** show a successful test of a high-expansion foam system protecting a flammable liquids dike.

HIGH-EXPANSION SYSTEM EXTINGUISHMENT MECHANISMS

High-expansion foam systems are suitable for the protection of Class A ordinary combustibles and are ideal for the protection of Class B combustible liquids. Extinguishment is accomplished by smothering, cooling, insulating, and penetrating.

- *Smothering* is accomplished by building a barrier or foam fence that allows the foam blanket to collect and rise above the combustible substance. The fresh air flow required to replace the combustion gases lost in the vertical plume is inhibited, and the deepening blanket eventually cuts off the oxygen supply and halts the combustion process.

- *Cooling* occurs when the water in the high-expansion foam turns to steam, coats the surface, and absorbs heat from the combustible. The conversion of water to steam reduces the available oxygen at the flame surface. This

Figure 5-8 *High-expansion foam filling a flammable liquids dike. (Courtesy of Ansul, Inc.)*

phenomenon, in conjunction with blanket smothering, brings the quantity of available oxygen below the level required for sustained combustion.

- *Insulation* of the hazard occurs when the foam blanket covers combustibles to the fire, providing exposure protection and preventing the spread of the fire.

- *Penetration* of the combustible is facilitated when the water in the foam soaks into Class A combustibles, providing cooling of burning materials and insulation of unburned materials. High-expansion foam, therefore, can be used successfully for Class A deep-seated fires.

The combination of these high-expansion foam extinguishment mechanisms provides an effective fire-protection strategy for the extinguishment of Class A and Class B combustibles. A flammable or combustible liquids containment, retainage, and removal strategy can be the solution to the vexing environmental problems that occur when environmentally hazardous flammable materials mix with foam–water solution.

DESIGN OF TOTAL FLOODING HIGH-EXPANSION FOAM SYSTEMS

total flooding method
a method of fire suppression that involves completely filling a room or enclosure volume with a fire protection agent

The **total flooding method** of fire suppression involves completely filling a room or enclosure volume with a fire protection agent to achieve suppression. Total flooding is the method of choice for most high-expansion foam systems because

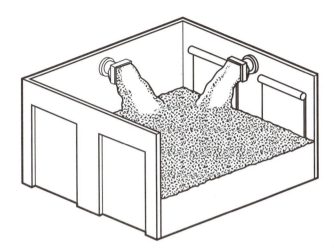

Figure 5-9 *Total flooding method. (Courtesy of Ansul, Inc.)*

the presence of containment barriers surrounding a flammable or combustible object enhances the ability of the high-expansion foam to achieve the elevation required to blanket and smother combustion at the highest elevation of the object. It may be difficult or inefficient in some cases to achieve complete suppression using the local application methodology. Installing barriers to retain high-expansion foam may result in the use of less high-expansion foam to achieve suppression, which makes cleanup and disposal somewhat easier.

The ideal situation to achieve total flooding of a combustible would be if the combustible were contained within a small interior room, as shown on **Figure 5-9**. Containment in such cases becomes solely a function of the existing room volume, assuming that the room is amenable or adaptable to holding foam, with normally-closed or electronically-closable openings and few sources of leakage.

Larger interior areas require a foam barrier to be installed around the hazard. The cost of a foam fence is minimal compared to the cost of permanent walls or the cost of filling a very large facility. Access doors or other openings must be designed to close automatically upon actuation of the system. The foam barrier containment system should be sized as small as possible to isolate the hazard area from other areas, and to enhance the velocity at which the foam blanket rises.

Personnel Considerations for High-Expansion Foam

All high-expansion foam systems must be designed with full consideration of personnel who may become trapped within the enclosure. To illustrate, personnel considerations would certainly be less of a concern on an aircraft carrier than in a home for the elderly. Evaluation of personnel, personnel training, and personnel fitness is key to evaluation of personnel considerations for suppression system specification in a facility.

Self-contained breathing apparatus should be made available within an enclosed space, but canister-type gas masks cannot be considered because high-expansion foam and the chemicals contained within the canister are not compatible. A method for temporarily shutting down the system by a trapped occupant may have to be considered in extreme cases, but this arrangement is not recommended. Interlocks should be installed on the system, capable of shutting down electrical equipment contained within the enclosure and thereby preventing the transmission of electric current through the foam solution.

Although a trapped occupant might be able to clear an area around the nose and mouth sufficient to accomplish egress in a high-expansion foam discharge, being buried in foam may cause disorientation or irrational behavior, and opening doors during foam discharge may allow high-expansion foam to escape, possibly compromising the system's effectiveness. For this reason, careful consideration of the health, fitness, and training of personnel that may be potentially exposed to high-expansion foam must become a fundamental part of the design of any high-expansion foam system. For example, more exits than NFPA 101 (the life safety code) requires may be specified, and an electronic methodology to prevent re-entry is advisable.

High-Expansion Foam Components

High-expansion foam concentrate ordinarily is stored in a bladder tank in a manner similar to low-expansion foam. When pressurized by water, the bladder squeezes the foam concentrate from the bladder tank and directs it to a proportioner that mixes the solution in the proper ratio, usually producing a 1%, 3%, or 6% solution. A deluge valve, supervised by a detection system and a solenoid valve that actuates opening of the normally-closed deluge valve, controls the flow of water and initiates simultaneous compression of the bladder and proportioning of foam solution.

The foam solution is distributed to high-expansion foam generators by way of a piping system whose materials are selected to be compatible with the foam solution and the environment. The manufacturer should be consulted with respect to compatibility and reactivity concerns between piping and concentrate. Some foams may react with standard black or galvanized pipe and may require the installation of stainless steel pipe and fittings, especially in pipes carrying 100% foam concentrate. Piping installed outside or in an area subject to ambient corrosion must have materials selected accordingly. A strainer for the water piping system is required.

The deluge valve, control panel, and bladder tank must be physically separated from the hazard, not subject to foam flooding, and protected from environmental and mechanical injury. The separation distance between the hazard and the valve house is unspecified but must be sufficient to permit operation of control valves in the valve house without hazard to personnel. The deluge valve room should have a door that provides direct access from outside a building.

Determination of High-Expansion Foam Quantity

NFPA 11 requires that high-expansion foam depth be 10% higher than the highest hazard, or 2 feet above the hazard, whichever is greater. Flammable liquids ordinarily are protected by a containment area that is 2 feet higher than the hazard, but testing and the recommendation of the manufacturer may require a higher containment area.

Where the hazard height is 20 feet or greater, the minimum required high-expansion foam volume is determined by:

$$V = [(room\ length) \times (room\ width) \times (hazard\ height)] \times (1.10)$$

The quantity 1.10 represents a height 10% above the height of the combustible and must be at least 2 feet (0.6 m), as described above.

$$V = high\text{-}expansion\ foam\ volume$$

In cases where the hazard height is less than 20 feet (6.1 m):

$$V = [(room\ length) \times (room\ width) \times (hazard\ height)] + (2\ ft.)$$

A conservative method for determining high-expansion foam volume would result if no deductions were made for machinery or other solid objects located within the containment area. Additions to the minimum required volume should be considered if unclosable openings exist, or if a delay is anticipated for the closing of automatically closable openings.

NFPA 11 requires continuous operation of the high-expansion foam system for 25 minutes, or to generate four times the submergence volume, whichever is less, but in no case less than enough for 15 minutes of operation.

Duration of High-Expansion Foam Application

Duration of application varies with the classification of the combustible, construction of the enclosure, and presence or absence of automatic fire sprinklers at the roof. Water spray from automatic sprinklers impinges on foam bubbles and tends to dilute or knock down the foam blanket, and their presence requires extended discharge times. The density of the combustibles, along with the possibility for a deep-seated fire, creates a situation in which extended discharge times are required.

Table 5-2 can be used to select a submergence time in minutes, assuming a 30-second delay from detection to commencement of discharge. Using Table 5-2, for example, a combustible liquid with a flash point of 250°F in a sprinklered building of light or unprotected steel construction requires a submergence time of 4 minutes.

Discharge Rate

With low-expansion foam, we used a foam discharge rate in gpm/ft^2 (lpm/m^2). This is appropriate since our objective was to cover a two-dimensional area with

Table 5-2 *Maximum submergence time for high-expansion foam measured from start of foam discharge (in minutes).*

Hazard	Light or Unprotected Steel Construction		Heavy or Protected or Fire-Resistive Construction	
	Sprinklered	Not Sprinklered	Sprinklered	Not Sprinklered
Flammable liquids [flash points below 38°C (100°F)] having a vapor pressure not exceeding 276 kPa (40 psia)	3	2	5	3
Combustible liquids [flash points of 38°C (100°F) and above][a]	4	3	5	3
Low-density combustibles (i.e., foam rubber, foam plastics, rolled tissue, or crepe paper)	4	3[b]	6	4[b]
High-density combustibles (i.e., rolled paper kraft or coated banded)	7	5[b]	8	6[b]
High-density combustibles (i.e., rolled paper kraft or coated unbanded)	5	4[b]	6	5[b]
Rubber tires	7	5[b]	8	6[b]
Combustibles in cartons, bags, or fiber drums	7	5[b]	8	6[b]

[a]Polar solvents are not included in this table. Flammable liquids having boiling points less than 38°C (100°F) might require higher application rates. See NFPA 30.
[b]These submergence times might not be directly applicable to storage piled above 4.6 m (15 ft.) or where fire spread through combustible contents is very rapid.

Source: NFPA 11 (2005), Table 6.12.7.1. Reprinted with permission from NFPA 11, *Low-, Medium-, and High-Expansion Foam*, Copyright © 2005, National Fire Protection Association, Quincy, MA 02169. This reprinted material is not the complete and official position of the National Fire Protection Association on the referenced subject which is represented only by the standard in its entirety.

low-expansion foam. With high-expansion foam, our objective is volumetric filling, so we therefore deal with a new term, CFM or cubic feet per minute (ft³/min or m³/min). NFPA 11 provides a simple formula for the determination of a discharge rate required for the high-expansion foam protection of combustible:

$$R = [(V/T) + R_s] \times (C_n) \times (C_l)$$

where R = total rate of high-expansion foam discharge in cubic feet per minute (CFM) [ft³/min or m³/min]

V = submergence volume, in cubic feet (m³); the computed height of 10% above the highest combustible, or 2 ft. (0.6 m) higher than the highest combustible, whichever is greater

T = submergence time, as determined by Table 5-2, is selected given the type of combustible, the construction of the foam barrier, and the presence of sprinklers

R_s = rate of foam breakdown by the sprinkler system, which can be estimated by the formula:

$$R_S = (S) \times (Q)$$

where S = foam breakdown in cubic feet per minute per gallons per minute of sprinkler discharge; (m³/min) × (L/min) NFPA 11A recommends 10 cfm per gpm (0.0748 m³/min) × (L/min) of sprinkler discharge

Q = calculated discharge of the sprinkler system using the number of sprinklers calculated in the area occupied by the high-expansion foam system as determined by the sprinkler system hydraulic calculations, in gpm (L/min)

C_n = the compensation for normal foam shrinkage, as recommended by the foam manufacturer; a C_n value of 1.15 is acceptable for most foams

C_l = a compensation factor for foam leakage, as determined by evaluation of the foam containment enclosure; a factor of 1.0 assumes that a tight containment area is present, whereas a factor of 1.2 assumes some normal leakage; the designer may consider an adjustment of this factor for wind or other containment concerns

Number of Generators Required

Once the rate of discharge (R) is computed, the number of high-expansion foam generators (N) is determined by dividing the high–expansion foam system discharge rate (R) by the rating of the generator (R_G) in cubic feet per minute (m³/min), rounded up to the nearest whole integer:

$$N = \frac{R}{R_G}$$

See Table 5-3 for the generator rating (R_G), the minimum solution flow to the generator required to produce the foam rating indicated, and the minimum pressure required to operate the generator.

Hydraulic calculations of the high-expansion foam system to determine pipe diameters can be performed using the minimum solution flow demand obtained from the previous calculation, balancing each generator with other generators at hydraulic confluence nodes, creating a total solution flow for all generators designed for the system.

EXAMPLE 5-1

High-Expansion Foam System

A room contains a combustible liquid with a flash point of 150°F. An automatic sprinkler system is installed at the ceiling of the room, producing a calculated automatic sprinkler system demand of 500 gpm. The room is of light or unprotected steel construction, with dimensions 100 ft. long, 100 ft. wide, and 30 ft. high, with a storage height of 20 ft. Determine the rate of high-expansion foam discharge, and perform a layout of the system.

Solution

$$R = \text{rate of discharge in cubic feet per minute.}$$

$$R = [(V/T) + R_s] \times (C_n) \times (C_l)$$

$$V = \text{submergence volume, in cubic feet.}$$

$$V = (100') \times (100') \times (20') \times (1.10)$$

$$= 220,000 \text{ ft.}^3$$

Note that the storage height, 20 feet, was increased by 10%, or a factor of 1.10, in accordance with NFPA 11. Note also that 10% of 20 ft. is exactly 2 ft., as required by NFPA 11.

$$T = \text{submergence time, minutes; per Table 5-2, } T = 4 \text{ min.}$$

$$R_s = (S) \times (Q)$$

$R_s = $ rate of foam breakdown by sprinklers, cubic feet per minute (cfm).

$S = $ foam breakdown in cfm per gpm of sprinkler discharge; as recommended by NFPA 11, $S = 10$ cfm/gpm

$Q = $ total discharge from sprinklers, in gallons per minute (gpm) = 500 gpm.

$R_s = (10 \text{ cfm/gpm}) \times (500 \text{ gpm}) = 5000$ cfm

$C_n = $ compensation for normal foam shrinkage; as recommended by NFPA 11A, $C_n = 1.15$

$C_l = $ compensation for leakage; assuming that normal leakage will occur, $C_l = 1.20$.

$$R = (220,000/4 + 5000) \times (1.15) \times (1.20)$$

$$= 82,800 \text{ cfm}$$

From **Table 5-3**, the designer has decided to select a Jet-X-5A generator capable of producing 5,700 cfm at 50 psi, with a minimum solution flow of 61 gpm and an expansion ratio of 700 to 1. The generator has a 1½" national pipe thread (n.p.t.) piping connection.

$$N = \frac{R}{R_G}$$

where $N = $ minimum number of generators required (rounded up to the nearest whole number)

$R = $ calculated total rate of discharge = 82,800 cfm

$R_g = $ discharge rating of the generator selected = 5700 cfm

$N = (82,800/5700) = 15$ generators, rounded up to the nearest whole generator

d. Lay out the high-expansion foam system, using a scale of $1/8'' = 1' - 0''$, showing the hazard area, generator locations, the piping system, and the deluge valve riser and tank location.

DISCUSSION QUESTIONS

1. How does high-expansion foam differ from low-expansion foam, and which hazards or occupancies are amenable for each type of system?

2. Describe the impact and influence of the EPA on foam system design.

ACTIVITIES

1. Visit a facility where a high-expansion foam system is installed, and meet with the fire protection engineer or responsible fire official.

a. Determine the criteria used for deciding whether high-expansion foam is used on a given hazard.

b. Determine the safeguards designed into the system that are intended to protect personnel.

c. Evaluate the design criteria relative to NFPA 11 criteria. If you find any difference(s), determine the reason for the difference.

d. Perform a layout of the system, and calculate the system in accordance with NFPA 11, as covered in this chapter.

e. Witness a test of the existing system, and compare results to your calculations.

NOTES

1. "High-Expansion Foam—An Unusual Application," Robert M. Gagnon, PE, *FPE Magazine*, Issue 1, Society of Fire Protection Engineers, Bethesda, Maryland.

2. "Status Report on Environmental Concerns Related to Aqueous Film-Forming Foam (AFFF)," Joseph L. Scheffey, PE, and Christopher P. Hanauska, PE, 2002 Federal Aviation Administration Technology Transfer Conference Proceedings, May 2002.

3. "Fighting Fires in Oil Storage Tanks Using Base Injection of Foam, Part 1," P. Nash and John Whittle, *Fire Technology*, Volume 14, Number 1, February 1978, Pages 15–27.

Table 5-3 *Jet-X high-expansion foam generators—performance characteristics. (Courtesy of Ansul, Inc.)*

Model No.	Generator Inlet Pressure		Foam Outlet		Solution Flow		Expansion
	(psi)	(kPa)	(cfm)	(cmm)	(gpm)	(Lpm)	
Jet-X-2	50	345	1,140	32	30	114	275:1
	75	517	1,770	50	38	144	340:1
	100	690	2,170	61	44	166	360:1
Jet-X-2A	50	345	2,240	63	35	132	465:1
	75	517	3,200	91	42	159	555:1
	100	690	3,735	106	50	189	545:1
Jet-X-5	50	345	5,350	151	36	136	1105:1
	75	517	6,720	190	45	170	1115:1
	100	690	7,225	204	50	189	1080:1
Jet-X-5A	50	345	5,700	161	61	230	700:1
	75	517	7,500	212	75	283	750:1
	100	690	8,000	226	87	329	685:1
Jet-X-15	50	345	13,535	383	157	594	645:1
	75	517	18,960	537	184	696	770:1
	100	690	16,875	478	210	794	600:1
Jet-X-15A (UL)	50	345	13,880	393	119	450	870:1
	75	517	17,410	493	145	548	900:1
	100	690	19,545	553	169	639	865:1
Jet-X-15A (FM)	50	345	12,985	368	105	397	925:1
	75	517	17,985	509	128	484	1050:1
	100	690	17,100	484	150	567	855:1
Jet-X-15A (LNG)	50	345	12,625	357	180	681	525:1
	75	517	14,495	410	220	832	495:1
	100	690	18,240	516	260	984	525:1

Model	A		B		C		D		E
	in.	(cm)	in.	(cm)	in.	(cm)	in.	(cm)	NPT
Jet-X-2	22.25	57	22.25	57	25	63	3.0	8	1 in.
Jet-X-2A	25.00	63	25.00	63	25	63	3.0	8	1 in.
Jet-X-5	42.00	107	42.00	107	40	101	7.0	18	1.5 in.
Jet-X-5A	42.00	107	42.00	107	40	101	7.0	18	1.5 in.
Jet-X-15	64.25	163	64.25	163	46	117	8.5	22	2 in.
Jet-X-15A (UL)	64.00	163	64.00	163	46	117	8.5	22	2 in.
Jet-X-15A (FM)	64.00	163	64.00	163	46	117	8.5	22	2 in.
Jet-X-15A (LNG)	64.00	163	64.00	163	46	117	8.5	22	2 in.

A layout of the high-expansion foam system can be performed. For symmetry, 16 generators will be used, four along each wall at the 30-ft. elevation. The generators can be connected to a piping loop that runs along the perimeter of the outside of the building. Because it is known that the generator selected requires 50 psi at 61-gpm solution flow for each generator, a hydraulic calculation can be performed to determine the pipe sizes of the piping system supplying the generator arrangement.

For the purpose of hydraulic calculation, pipes are differentiated from one another by a C factor, which describes the relative roughness of the internal pipe wall. A C factor appropriate to the type of piping selected for this project must be used in the calculations. For instance, plastic pipe is smoother on the inside than steel pipe and would have a different C factor than steel pipe. Because the connection on the generator is 1½ in., the pipe branching off to each generator must be a minimum of 1½ in. Because the solution flow for each generator is small, 61 gpm in this case, especially when compared to the gpm demanded by an open-nozzle aircraft hangar, which could exceed 10,000 gpm, high-expansion foam system piping is relatively small.

The piping loop feeding the generators must be sized to support the demand of the generators served by the loop. In this case, 16 generators each demand a minimum of 61 gpm in solution flow. The deluge valve, proportioner, and bladder tank must be placed in a heated enclosure, protected from mechanical injury, accessible for maintenance or emergency operation, not subject to damage from a fire in the hazard area, and protected from flooding by the high-expansion foam system.

SUMMARY

High-expansion foam systems are used primarily for the volumetric total flooding of three-dimensional objects, using a foam with an expansion ratio of between 200 to 1 and 1000 to 1. The foam is expanded by a high-expansion foam generator that resembles a large fan. High-expansion foam extinguishes fire by smothering, cooling, insulating, and penetrating.

NFPA 11 requires that high-expansion foam be flooded to an elevation exceeding 10% above the highest combustible, or 2 ft. above the hazard, whichever is higher. The rate of high-expansion foam discharge must consider foam breakdown that might occur if a sprinkler system is activated above the area protected by high-expansion foam.

REVIEW QUESTIONS

1. A room is 150 ft. long, 150 ft. wide, and 30 ft. high, and contains rubber tires stored to a height of 22 ft. The room is protected by an automatic sprinkler system, and the calculated sprinkler discharge over the hazard area is 375 gpm. The room is tightly sealed, is constructed of fire-resistive construction, and no leakage from the room is expected.

   a. Calculate the rate of high-expansion foam discharge, in cubic feet per minute.

   b. Using Jet-X-5A Generators at 50 psi, determine the number of generators required to deliver the rate of discharge calculated.

   c. Using the number of Jet-X-5A Generators obtained in (b) and using the cumulative discharge of all Jet-X-5A generators at their listed flow, calculate the time required for the high-expansion foam to hit the ceiling of the room.

   d. Lay out the high-expansion foam system, using a scale of 1/8″ = 1′ – 0″, showing the hazard area, generator locations, the piping system, and the deluge valve riser and tank location.

2. A room is 75 ft. long, 225 ft. wide, and 15 ft. high, and contains a flammable liquid stored in drums to a height of 12 ft. The room is protected by an automatic sprinkler system, and the calculated sprinkler discharge over the hazard area is 615 gpm. The room is not tightly sealed, is constructed of unprotected steel construction, and some leakage from the room is expected.

   a. Calculate the rate of high-expansion foam discharge, in cubic feet per minute.

   b. Using Jet-X-5A Generators at 50 psi, determine the number of generators required to deliver the rate of discharge calculated.

   c. Using the number of Jet-X-5A Generators obtained in (b) and using the cumulative discharge of all Jet-X-5A generators at their listed flow, calculate the time required for the high-expansion foam to hit the ceiling of the room.

   d. Lay out the high-expansion foam system, using a scale of 1/8″ = 1′ – 0″, showing the hazard area, generator locations, the piping system, and the deluge valve riser and tank location.

3. A room is 225 ft. long, 200 ft. wide, and 20 ft. high, and contains banded rolled paper stored to a height of 12 ft. The room is no protected by an automatic sprinkler system The room is tightly sealed, is constructed o fire-resistant construction, and no leakag from the room is expected.

   a. Calculate the rate of high-expansion foar discharge, in cubic feet per minute.

   b. Using Jet-X-2 Generators at 50 psi, dete mine the number of generators required t deliver the rate of discharge calculate

   c. Using the number of Jet-X-2 Generato obtained in (b) and using the cumulati discharge of all Jet-X-2 generators at the listed flow, calculate the time it takes f the high-expansion foam to hit the ce ing of the room.

   d. Lay out the high-expansion foam syste using a scale of 1/8″ = 1′ – 0″, showi the hazard area, generator locations, t piping system, and the deluge valve ris and tank location.

4. A room is 100 ft. long, 100 ft. wide, and 15 high, and contains combustibles in fib drums a to a height of 12 ft. The room is p tected by an automatic sprinkler system, a the calculated sprinkler discharge over t hazard area is 1015 gpm. The room is n tightly sealed, is constructed of unprotect steel construction, and some leakage fro the room is expected.

   a. Calculate the rate of high-expansion foa discharge, in cubic feet per minute.

   b. Using Jet-X-2 Generators at 50 psi, dete mine the number of generators requir to deliver the rate of discharge calculate

   c. Using the number of Jet-X-2 Generato obtained in (b) and using the cumulati discharge of all Jet-X-2 generators at th listed flow, calculate the time requir for the high-expansion foam to hit t ceiling of the room.

WATER MIST SYSTEMS

Learning Objectives

Upon completion of this chapter, you should be able to:

- Evaluate water mist as a potential halon replacement.
- Discuss the applications for water mist systems.
- Compare the performance objectives of a water mist spray to the spray of a large-drop sprinkler.
- List water mist system types and configurations.
- Discuss the reasons why zoning of water mist systems can increase system effectiveness for systems with limited water supplies.

NFPA

NFPA 750, Standard for the Installation of Water Mist Fire Protection Systems

water mist system
a distribution system connected solely to a water supply or alternatively to a water supply and an atomizing media (air or nitrogen), that is equipped with one or more nozzles capable of delivering water mist intended to control, suppress, or extinguish fires, as defined by NFPA 750

fire control
a reduction in thermal exposure, threat to occupants, or fire-related characteristics

fire suppression
a sharp reduction in heat release rate and prevention of regrowth

fire extinguishment
complete suppression of the fire

temperature control
performance objective aimed at reducing room temperatures during combustion to allow safe egress and reduced damage

When the first NFPA standard on water mist systems was introduced in 1996, water mist technology reached a new and critical stage. NFPA 750, *Standard for the Installation of Water Mist Fire Protection Systems,* created an atmosphere for expanding the use of water mist and broadening of the number of applications for this unique system. A **water mist system** is a distribution system connected solely to a water supply or alternatively to a water supply and an atomizing media (air or nitrogen), that is equipped with one or more nozzles capable of delivering water mist intended to control, suppress, or extinguish fires, as defined by NFPA 750. Water mist systems have the potential to serve as replacement systems for occupancies formerly projected by halon systems. Water mist systems are similar in cost to clean-agent systems.

WATER MIST PERFORMANCE OBJECTIVES

In selecting a water mist system, it is essential to specify and design the system to meet one or more performance objectives specific to the hazard being protected. The NFPA has reorganized its standards to emphasize performance objectives, design objectives, and fire test protocols as focal points for system selection and design. Water mist performance objectives recognized by NFPA 750 are:

- **fire control**—a reduction in thermal exposure, threat to occupants, or fire-related characteristics
- **fire suppression**—a sharp reduction in heat release rate and prevention of regrowth
- **fire extinguishment**—complete suppression

Water mist design application parameters include:

- compartment geometry
- compartment ventilation
- fire hazard classification—fuel type, Class A, Class B, Class C, or combination fires
- fire location—combustible elevation; location against walls or corners of space
- obstructions and shielding—floor-mounted obstructions (tables); ceiling-mounted obstructions (signs, beams)
- reliability—based on operating experience (a usable table is in Annex D of NFPA 750, included as Table 6-2), or predictive techniques (simplified equations, fault tree analysis, and Markov analysis).

In addition to fire control, fire suppression and fire extinguishment, water mist has been found to be effective for:

- **temperature control**, in which the room temperatures are reduced during combustion to allow safe egress and reduced damage; and
- **exposure protection**, in which combustibles adjacent to the fire are wetted to delay their ignition.

exposure protection
the wetting of combustibles adjacent to the fire to delay their ignition

water mist
a water spray with water droplets of less than 1000 microns at the minimum operation pressure of the discharge nozzle

WATER MIST DROPLETS

NFPA 750 defines **water mist** as a water spray with water droplets of "less than 1000 microns at the minimum operation pressure of the discharge nozzle." Size of water droplets is a function of the discharge pressure through an orifice of fixed diameter, and is a key contributing factor in the ability of a water spray to evaporate and cool a flame for certain configurations of combustibles. The total absorption of heat per unit time increases greatly as drop size decreases, given a fixed water volume, because the available surface area for heat transfer increases, as illustrated in **Figure 6-1**.

As detailed in the Annex of NFPA 750, research has shown that water mist protecting Class A fires requires larger water mist droplets, close to 1000 microns in diameter, whereas protection of Class B fires requires smaller droplets. Although the NFPA 750 definition includes the total range of droplets, the droplet size should match the commodity.

Small droplets absorb heat quickly and evaporate readily, creating steam, whereas larger droplets are less prone to do so. Large droplets are much more likely to provide coating of exposed combustibles, which is an important performance objective of Class A fires. Large drops are more prone to agitate the surface of a flammable liquid fire, which explains why smaller droplets are more conducive to extinguishment of Class B fires.

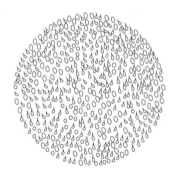

ONE GALLON OF WATER DIVIDED INTO VERY SMALL DROPLETS

- MORE DROPLETS
- LARGER WATER SURFACE AREA EXPOSED TO HEAT
- MORE DROPS WILL EVAPORATE AND TURN TO STEAM
- STEAM ABSORBS MORE HEAT PER UNIT TIME FROM THE FLAME, REDUCING THE FLAME TEMPERATURE

ONE GALLON OF WATER DIVIDED INTO VERY LARGE DROPLETS

- FEWER DROPLETS
- SMALLER WATER SURFACE AREA EXPOSED TO HEAT
- MORE DROPS WILL REMAIN UNEVAPORATED AND COAT THE SURFACE OF THE COMBUSTILBES
- A COMBUSTIBLE COATED WITH WATER IS COOLED BY THE WATER

Figure 6-1 *Droplet size.*

Droplet Size Performance Objectives

It makes sense that if droplet sizes of differing discharge devices can be identified, a fire protection system designer armed with this information could make important decisions relative to the performance objective of the system. Water mist may be selected if it is decided that creating heat-absorbing vapor is the primary performance objective, or where water supplies are limited, or where water damage may be a concern. Combustibles that may be protected successfully using this performance objective include ordinary combustibles, electronics and telecommunications equipment, and flammable liquids.

Extending our previous discussion of water mist droplets specified for Class A and Class B fires, an NFPA 13 standard sprinkler system with its larger droplet sizes may be appropriate and cost-effective with the vast majority of Class A fires. An exception could be the protection of historic or irreplaceable Class A storage. Although extinguishment of flammable liquid pool fires is addressed in NFPA 15, it may be advisable to use water mist on small flammable liquid containers such as diptanks, and NFPA 15 water spray systems on larger flammable liquid risks such as a chemical manufacturing and processing plant.

Water Mist Pressure and Droplet Size

low-pressure water mist systems
systems in which the pressures encountered by the system piping are 175 psi or less

intermediate-pressure water mist systems
systems in which the pressures encountered by the system piping are between 175 psi and 500 psi

high-pressure water mist systems
systems in which the pressures encountered by the system piping are 500 psi or greater

Pressure, either by fire pumps or by pressurized air or gas, is a strong contributor to the size of water mist droplets, given a nozzle of fixed diameter and orifice characteristics. In addition, the pressure at the nozzle is a significant contributor to the ability of a nozzle to project water mist droplets at extended distances from the nozzle. Generally, the lower the pressure at the nozzle, the lower is the fluid velocity and resultant projected range, and the closer the nozzle is required to be to the axis of the plume centerline. NFPA 750 has established three classifications for system pressurization:

1. **Low-pressure water mist systems**, in which the pressures encountered by the system piping are 175 psi (12.1 bar) or less, the same pressure range used for most standard sprinkler systems

2. **Intermediate-pressure water mist systems**, in which the pressures encountered by the system piping are between 175 psi (12.1 bar) and 500 psi (34.5 bar)

3. **High-pressure water mist systems**, in which the pressures encountered by the system piping are 500 psi (34.5 bar) or greater

System pressure becomes extremely important in specifying material characteristics for water mist systems, such as piping, fittings, nozzles, and valves. System components are required to be capable of meeting or exceeding the highest pressures expected to be encountered on a water mist system.

WATER MIST SYSTEM CONFIGURATIONS

Commonly encountered water mist systems include the following:

- high-pressure gas-driven water mist system with stored water, as shown on **Figure 6-2**, which features pressure-rated water cylinders, compressed gas cylinders, open nozzles, and a detection system

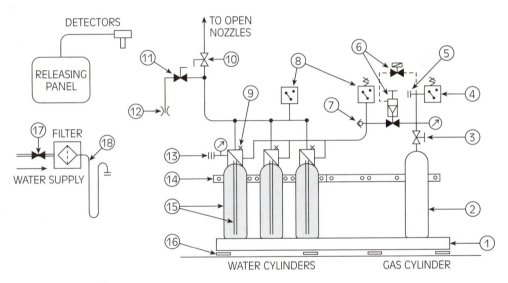

Legend

1. Steel base and frame
2. Compressed gas cylinder (driving medium)
3. Cylinder control valve
4. Pressure switch, supervise cylinder pressure
5. Burst disc
6. Solenoid operated master release valve
7. Microleakage valve
8. Pressure switches, alarm if system trips
9. Vent port, for filling water cylinders (fill until water discharges from open port)
10. Primary system or sectional control valve
11. Test connection and drain
12. Test orifice (alternative to full discharge)
13. Cylinder discharge header with filling port
14. Cylinder rack with restraints
15. Pressure-rated water cylinders with dip tube
16. Optional load cells
17. Water supply valve, normally closed
18. Filter and hose with adaptor fitting for filling cylinders

Figure 6-2   *Schematic representation of a high-pressure gas-driven system with stored water (pre-engineered system).*

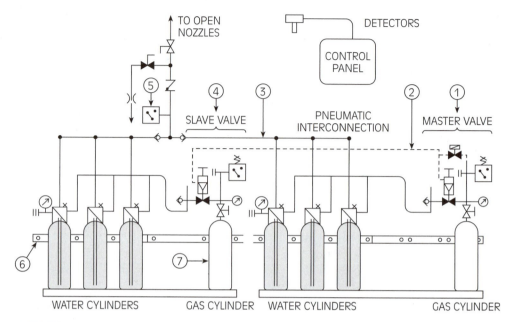

Legend

1. Solenoid operated master release valve (with local manual release)
2. Pneumatic tubing interconnecting "master" valve to "slave" valve
3. Discharge header
4. Pneumatically activated slave valve (no local manual release)
5. Pressure switch, alarm when system trips
6. Cylinder rack with restraints
7. Compressed gas cylinders (driving medium)

Figure 6-3 *Schematic representation of a high-pressure gas-driven system with multiple accumulator units for extended duration.*

Source: NFPA 750 (2006), Figure A.11.1.6(b). Reprinted with permission from NFPA 750, *Water-Mist Fire Protection Systems,* Copyright © 2006, National Fire Protection Association, Quincy, MA 02169. This reprinted material is not the complete and official position of the National Fire Protection Association on the referenced subject which is represented only by the standard in its entirety.

- high-pressure gas-driven water mist system with multiple accumulator units for extended duration, as shown on **Figure 6-3**, which is similar to the high-pressure gas-driven water mist system with stored water, but with an additional bank of water and gas cylinders
- low-pressure twin-fluid water mist system, as shown on **Figure 6-4**, which features a low-pressure rated water tank, a bank of compressed gas cylinders, water and air pipes to each low-pressure twin-fluid nozzle, and a detection system
- single-fluid water mist system, as shown on **Figure 6-5**, which features a pressure-rated water storage tank, a compressed gas cylinder, and a water line to the nozzles

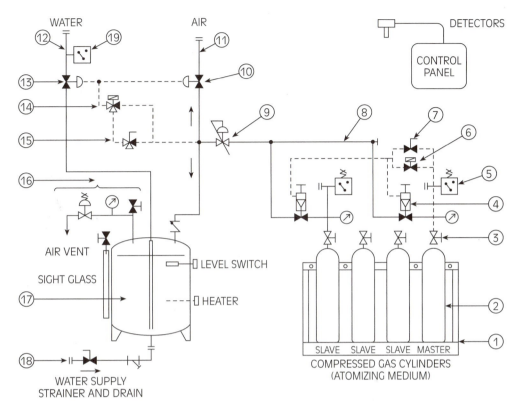

WATER

AIR

DETECTORS

CONTROL PANEL

AIR VENT

SIGHT GLASS

LEVEL SWITCH

HEATER

WATER SUPPLY
STRAINER AND DRAIN

SLAVE   SLAVE   SLAVE   MASTER

COMPRESSED GAS CYLINDERS
(ATOMIZING MEDIUM)

Legend

1. Steel base and frame
2. Compressed gas cylinders (atomizing medium)
3. Cylinder control valve
4. Pneumatic cylinder release valve
5. Pressure supervisory switch with burst disc
6. Solenoid operated master release valve
7. Manually operated master relief valve
8. 1/2 in. high pressure tubing manifold
9. Air pressure control valve (high to low pressure)
10. Air-actuated globe valve (cycle air line)

11. Air-line to twin-fluid nozzles (low pressure)
12. Water line to twin-fluid nozzles (low pressure)
13. Air-actuated globe valve (cycle water line)
14. Low pressure solenoid valves (for operating air-actuated globe valves)
15. Manual release valve (opens globe valves)
16. Pressure gauge, pressure relief valve, and vent valve
17. Low pressure rated water tank
18. Drain and refill connection and strainer
19. Pressure switch, alarm on discharge

Figure 6-4   *Schematic representation of a low-pressure twin-fluid water mist system.*

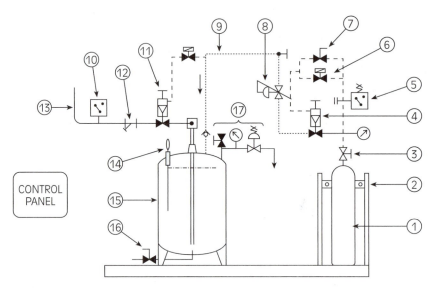

Legend

1. Compressed gas cylinder
2. Steel frame and cylinder restraints
3. Cylinder control valve
4. Pneumatic cylinder release valve
5. Pressure switch disc with burst disc
6. Solenoid operated master release valve
7. Manually operated master release valve
8. Pressure regulating valve, high to low
9. Air-line tubing

10. Pressure switch, alarm on discharge
11. Primary system control valve
12. Strainer on discharge valve
13. Water line to nozzles
14. Water level indicator (dipstick)
15. Pressure rated water storage tank (30 bar)
16. Drain and refill connection with strainer
17. Pressure gauge, pressure relief valve, and vent valve

Figure 6-5   *Schematic representation of a single-fluid water mist system.*

- pump-driven water mist system, as shown on **Figure 6-6**, which features a connection to a water supply, a connection to a regulated air supply, an electric or diesel fire pump, solenoid-released water mist selector valves, and a detection system

- positive-displacement pump assembly with unloader valves on each pump and a pressure relief valve on the discharge manifold, as shown on **Figure 6-7**, which features a connection to a freshwater or seawater supply, a connection to a regulated air supply, positive displacement pumps with unloader valves, solenoid-actuated water mist sectional control valves, and a detection system

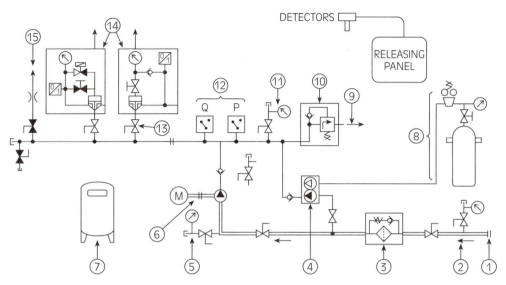

Legend

1. Connection to water supply
2. Low pressure gauge
3. Filters or screens with bypass
4. Standby pressure maintenance pump (pneumatic or electric)
5. NPSH gauge (+/−)
6. Pump and driver (electric or diesel)
7. Pump controller
8. Air supply and regulator for pneumatic pressure maintenance pump (4), includes plant air and compressors

9. Unloader valve discharge line to drain or break tank
10. Unloader valve rated for 100% of pump flow capacity
11. Pressure gauge
12. Pressure (P) and flow (Q) switches/transmitters connected to pump assembly
13. Manual isolation valve
14. Sectional control valves to mist systems (solenoid release)
15. Test connection with flow meter

Figure 6-6   *Schematic diagram of a pump-driven water mist system.*

Source: NFPA 750 (2006), Figure A.11.1.6(e). Reprinted with permission from NFPA 750, *Water-Mist Fire Protection Systems,* Copyright © 2006, National Fire Protection Association, Quincy, MA 02169. This reprinted material is not the complete and official position of the National Fire Protection Association on the referenced subject which is represented only by the standard in its entirety.

- break tank connection in supply to positive displacement pumps, for marine systems, as shown on **Figure 6-8**, which features a freshwater or seawater supply, a break tank, and piping to a positive-displacement pump assembly with unloader valves on each pump and a pressure relief valve on the discharge manifold, as shown on Figure 6-7

- gas pump unit for machinery spaces and gas turbine enclosures, as shown on **Figure 6-9**, which features a water tank or a connection to a water supply, a bank of gas or nitrogen cylinders, a gas pump unit, and a bank of water mist selector valves

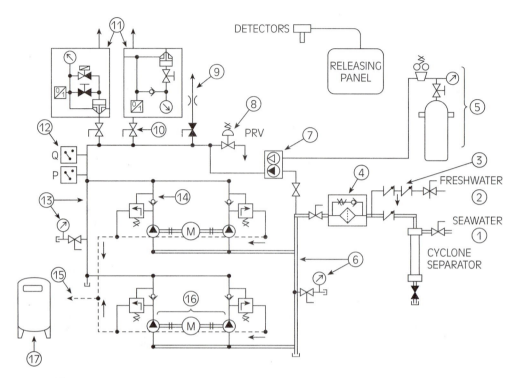

Legend

1. Seawater supply with cyclone separator
2. Freshwater supply
3. Backflow device (optional)
4. Filters or screens with bypass
5. Air supply and regulator for standby pressure maintenance pump, plant air or compressor
6. Suction manifold with NPSH gauge (+/–)
7. Pneumatic standby pressure pump
8. Pressure relief valve (optional)
9. Test connection with flow meter
10. Isolation valve for sectional valves

11. Solenoid actuated sectional control valves
12. Pressure (P) and flow (Q) switches/ transmitters connected to controller
13. Discharge manifold with pressure gauge
14. Unloader valve (one per pump)
15. Unloader valve discharge bypass line (to drain or break tank)
16. Positive displacement pumps (two per motor)
17. Pump controller

Figure 6-7 *Schematic representation of a positive displacement pump assembly with unloader valves on each pump and a pressure relief valve on discharge manifold.*

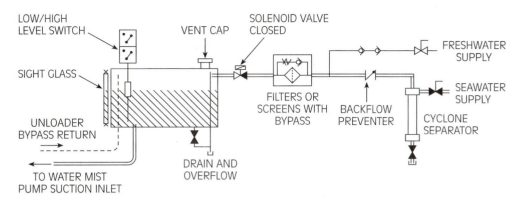

Figure 6-8 *Schematic diagram of a break tank connection in supply to positive displacement pumps (marine systems).*

Source: NFPA 750 (2006), Figure A.11.1.6(g). Reprinted with permission from NFPA 750, *Water-Mist Fire Protection Systems,* Copyright © 2006, National Fire Protection Association, Quincy, MA 02169. This reprinted material is not the complete and official position of the National Fire Protection Association on the referenced subject which is represented only by the standard in its entirety.

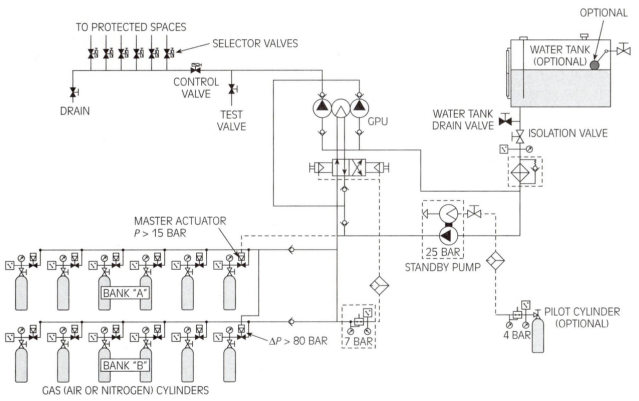

Figure 6-9 *Gas pump unit for machinery spaces and gas turbine enclosures.*

Source: NFPA 750 (2006), Figure A.11.1.6(h). Reprinted with permission from NFPA 750, *Water-Mist Fire Protection Systems,* Copyright © 2006, National Fire Protection Association, Quincy, MA 02169. This reprinted material is not the complete and official position of the National Fire Protection Association on the referenced subject which is represented only by the standard in its entirety.

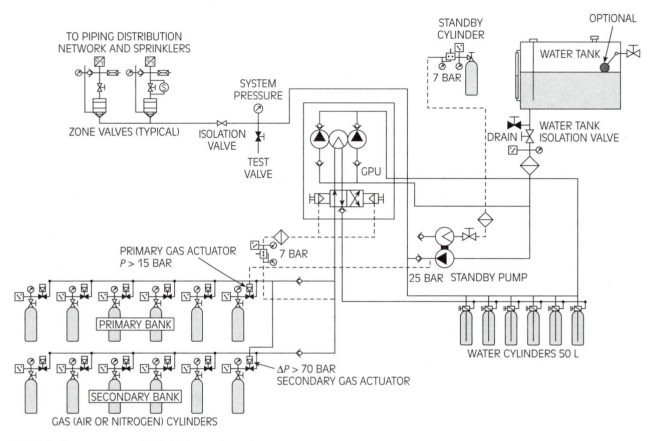

Figure 6-10 *Gas pump for light hazard applications.*

Source: NFPA 750 (2006), Figure A.11.1.6(i). Reprinted with permission from NFPA 750, *Water-Mist Fire Protection Systems,* Copyright © 2006, National Fire Protection Association, Quincy, MA 02169. This reprinted material is not the complete and official position of the National Fire Protection Association on the referenced subject which is represented only by the standard in its entirety.

local application method
a method of fire suppression in which the agent is applied directly onto the hazard

- gas pump for light hazard applications, as shown on **Figure 6-10,** which features an arrangement similar to that shown on Figure 6-9, but with water cylinders and water mist zone valves

WATER MIST SYSTEM DESIGN

total flooding method
a method of fire suppression that involves completely filling a room or enclosure volume with a fire protection agent

In a manner similar to gaseous agent suppression systems, water mist systems can be designed to employ the **local application method,** in which the agent is applied directly onto the hazard, or designed for the **total flooding method,** which involves completely filling a room or enclosed volume with a fire protection agent, as shown in **Figure 6-11.** Local application systems may be considered in very large compartments where total flooding is not an achievable option.

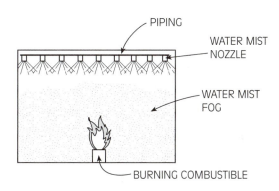

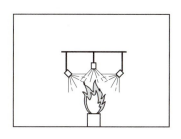

Figure 6-11 *Water mist system configuration.*

TOTAL COMPARTMENT APPLICATION

THE ENTIRE COMPARTMENT IS FLOODED WITH WATER MIST DROPLETS.

LOCAL APPLICATION SYSTEM

WATER MIST IS PROJECTED DIRECTLY ONTO AN EXPECTED FIRE.

zoned application systems

systems in which a volume is protected by several distinct suppression zones, each with its own detection system

Systems in room volumes too large to be designed practically for a total compartment flooding application may be considered as a candidate for a **zoned application system**, as shown on **Figure 6-12**. In this system, a volume is protected by several distinct suppression zones, each with its own detection system. Although the zoned scenario shown for the aircraft is ideal, some systems with larger volumes, especially those with excessive ceiling heights, may not be amenable to the zoned concept.

A designer can program a fire alarm control unit to predetermine the number of zones that should actuate upon receipt of a detection signal from any zone, based on the amount of water available and the expected duration of discharge. For example, a detector located in one detection zone could actuate the suppression zone in question, plus the two adjacent suppression zones, for a total of three or more suppression zones in actuation. The zoned arrangement saves water and provides opportunities for the possible protection of aircraft in the future.

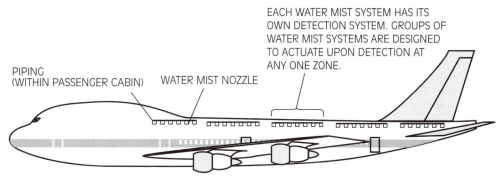

Figure 6-12 *Zoned water mist system.*

AIRCRAFT ARE BEING TESTED FOR A ZONED APPLICATION SCENARIO.

Water mist systems may be considered as an alternate to a gaseous suppression system, and may be an acceptable replacement for a halon system, because there is no risk of ozone depletion or global warming potential with a water mist system.

Types of Water Mist System

Designers of water mist systems may consider any of the following types of systems designed in accordance with NFPA 750.

deluge water mist systems

systems with open nozzles that discharge water mist simultaneously from all nozzles on the system

- **deluge water mist systems,** with open nozzles that discharge water mist simultaneously from all nozzles on the system
- **wet pipe water mist systems,** having nozzles with an individual actuating device, in which each nozzle actuates individually, and with piping filled with water

wet pipe water mist systems

systems having nozzles with an individual actuating device, where each nozzle actuates individually, and with piping filled with water

Table 6-1 *Darcy-weisbach and associated equations for pressure loss in intermediate- and high-pressure systems.*

	SI Units	**U.S. Customary Units**
Darcy–Weisbach equation	$\Delta p_m = 2.252 \dfrac{fL\rho Q^2}{d^5}$	$\Delta p = 0.000216 \dfrac{fL\rho Q^2}{d^5}$
Reynolds number	$\mathrm{Re} = 21.22 \dfrac{Q\rho}{d\mu}$	$\mathrm{Re} = 50.6 \dfrac{Q\rho}{d\mu}$
Relative roughness	Relative roughness $= \dfrac{\varepsilon}{d}$	Relative roughness $= \dfrac{\varepsilon}{D}$

where:
Δp_m = friction loss (bar gauge)
L = length of pipe (m)
f = friction factor (bar/m)
Q = flow (L/min)
d = internal pipe diameter (mm)
ε = pipe wall roughness (mm)
ρ = weight density of fluid (kg/m^3)
μ = absolute (dynamic) viscosity [centipoise (cP)]

where:
Δp = friction loss (psi gauge)
L = length of pipe (ft)
f = friction factor (psi/ft)
Q = flow (gpm)
d = internal pipe diameter (in.)
D = internal pipe diameter (ft)
ε = pipe wall roughness (ft)
ρ = weight density of fluid (lb/ft^3)
μ = absolute (dynamic) viscosity [centipoise (cP)]

Source: NFPA 750 (2006), Table 9.2.1. Reprinted with permission from NFPA 750, *Water-Mist Fire Protection Systems,* Copyright © 2006, National Fire Protection Association, Quincy, MA 02169. This reprinted material is not the complete and official position of the National Fire Protection Association on the referenced subject which is represented only by the standard in its entirety.

preaction water mist systems

systems having nozzles with an individual actuating device, where each nozzle actuates individually, with piping filled with air

dry pipe water mist systems

systems with air-filled piping, having nozzles with an individual actuating device, where each nozzle actuates individually

- **preaction water mist systems**, having nozzles with an individual actuating device, in which each nozzle actuates individually, with piping filled with air; designed to prevent water flow until a reliable detection signal is received
- **dry pipe water mist systems**, systems with air-filled piping, with nozzles having an individual actuating device, in which each nozzle actuates individually; water flows into the piping when a nozzle actuates

Water mist systems also can be either pre-engineered systems or engineered systems. Pre-engineered water mist systems are modular packages in which nozzles, piping, fittings, media storage, system layout and flow rates, and the fire alarm releasing unit are selected from a chart that relates hazard type and size to system volume. Engineered water mist systems require calculations to determine media storage volume, system layout, system flow rates, nozzle selection, and pipe sizes.

Water Mist System Calculation Criteria

NFPA 750 recognizes that at the present time no uniform design method or criteria exists for the design and calculation of water mist systems. Each manufacturer has a proprietary design method associated with the system, and the listing of the components is obtained for the system in accordance with the performance objectives for that specific system. When a water mist system is designed or specified, the water mist criteria for that system must be met in accordance with its listing.

NFPA 750 requires that water mist systems classified as intermediate- or high-pressure, single-fluid, single-liquid phase, are to be calculated using the Darcy-Weisbach equation shown on **Table 6-1**. For low pressure water-filled pipe water mist systems, the Hazen-Williams formula shall be used, as given in **Figure 6-13**.

Advice and reference material for the performance of flow calculations are included in NFPA 750. These may be used once the system is selected and the performance criteria of the manufacturer are determined. In many cases the manufacturer either has a pre-engineered system for a given hazard, or will offer a contract package that includes engineered system design.

WATER MIST SYSTEM EFFECTIVENESS AND OPERATING EXPERIENCE

Although water mist systems are still in their infancy, the NFPA 750 Annex provides some information relative to operating experience. **Table 6-2** lists systems for a variety of hazards, with some interesting testing and operating data.

(1) For SI units:

$$P_m = 6.05 \frac{Q_m^{1.85}}{C^{1.85} d_m^{4.87}} \times 10^5$$

where:
P_m = frictional resistance (bar/m of pipe)
Q_m = flow (L/min)
d_m = actual internal diameter of pipe (mm)
C = friction loss coefficient

(2) For U.S. customary units:

$$P_f = \frac{4.52 Q^{1.85}}{C^{1.85} d^{4.87}}$$

where:
P_f = frictional resistance (psi/ft of pipe)
Q = flow (gpm)
d = actual internal diameter of pipe (in.)
C = friction loss coefficient

Figure 6-13 *Friction losses for water-filled pipe determined on the basis of the Hazen-Williams formula.*

Source: NFPA 750 (2006), Table 9.3.2. Reprinted with permission from NFPA 750, *Water-Mist Fire Protection Systems,* Copyright © 2006, National Fire Protection Association, Quincy, MA 02169. This reprinted material is not the complete and official position of the National Fire Protection Association on the referenced subject which is represented only by the standard in its entirety.

WATER MIST APPLICATIONS

In addition to its strong potential as a halon replacement, potential uses for water mist technology are diverse, and this diversity creates a solid basis for the specification and design of water mist systems for numerous unique problems, such as protection of ships, planes, the space shuttle, or a circus that travels by rail and, to remain insured, requires an automatic fixed fire protection system.

Some applications, most notably shipboard applications including living spaces, engine rooms, and mechanical spaces, are well past the experimental stage and are accruing an impressive record of reliable performance. Current applications for water mist also include turbine enclosures and museum protection. Other applications, such as aircraft applications, remain experimental but provide an exciting potential for an increased level of safety.

Table 6-2   *Water mist operating experience.*

Protected Hazard	No. of Systems	Time in Operation	Fires Success	Fail	Accidental Operation	Application Flood	Local	Detection/ Actuation	Listed Yes	No	Acceptance Test (Y/N)	Notes
600 HP engine	1	Mid-1996	N/A	N/A	No		X	IR flame		X	Y—discharge	1
Compressor lube oil system	1	Feb. 1996	N/A	N/A	No		X	Smoke		X	Y—discharge	2
Engine test cells			1	No	1	X?		Heat detection				3
7 combustion turbines (on oil platforms)	7	Spring 1993	Several	No	Unknown	X?						4
5 diesel fire pumps (on oil platforms)	5	Spring 1993	N/A	N/A	Unknown							
5 diesel generators (on oil platforms)	5	Spring 1993	N/A	N/A	Unknown							
2 deep fat fryers (on oil platforms)	2	Spring 1993	N/A	N/A	Unknown							
6 cruise ships	?	Unknown	None	None	1							5
Lube oil systems for 6 combustion turbine driven compressors (natural gas pipeline)	6	1996 (2) 1997(2) 1998(2)	None	None	None	Combination						
Thermal oil system on fiberboard press	1		2	None	None		X?		X			6

Notes:

1. At some time after acceptance test, white residue was noted in pipe. Laboratory tests determined it to be zinc oxide. Piping was flushed but residue remained.

2. During acceptance test, system failed to operate due to low pressure. Investigation determined that a ¼ in. copper tube had separated from a brass fitting. On second discharge test, 3 of 25 nozzles were plugged, possibly due to use of pipe joint compound. System remained out of service 22 months after installation due to concern for accidental operation during cutting and welding operations. Chalk-like material had formed on interior surface of pipe; believed to be oxide.

3. Fire activation: A hydraulic line break occurred on an engine under test. System operated properly and extinguished the fire. Accidental activation: System operated 50 hours into engine test, near end of test cycle. One heat detector may have been located too close to exhaust stack.

4. Systems are tested periodically with limited discharge of a few seconds to verify the system is operational.

5. Six vessels have water mist protection for engine rooms, dining areas, ballrooms, and escape routes. Engine room systems are manually controlled. One accidental operation, in dining room.

6. Oil spray fire burned 2–3 hours, fought unsuccessfully by plant personnel using hose streams. Water mist system was manually activated and extinguished fire within 20 seconds. When press was placed back in service, a second fire developed at an undetected crack in thermal oil system piping. This fire was also extinguished by the water mist system.

Electronic/Telecommunications Applications

In a large telecommunications switchgear facility, significant floor area is occupied by telecommunication enclosures that are combustible but are largely unaffected by automatic sprinkler protection at roof level. Telecommunications switchgear equipment, consisting primarily of vertically mounted circuit boards, traditionally was internally protected by halon systems or carbon dioxide systems. Carbon dioxide, applied in concentrations of 34% or greater, is dangerous in areas occupied by personnel, whereas halon systems are no longer an option, as discussed previously.

A total flooding water mist system is able to negotiate some obstructions within the switchgear, depending upon placement of the nozzle, which is installed within the equipment, with standard automatic sprinkler protection at the ceiling. A water mist system also could be considered for other types of electronic equipment.

Gas Turbine Applications

Another application that has been affected directly by the phaseout of halon is the fire protection of gas turbines. Total flooding water mist systems are cycled to initiate suppression two or more times to enhance extinguishment, and tests have shown that conditions are survivable for personnel who might become trapped within the turbine enclosure. Water mist does not cause damage to the turbine bearings.

Gas-Well Blowout Applications

Water mist nozzles spraying vertically and parallel to a gas flame have been demonstrated to be effective in extinguishing the flame using very small amounts of water, which lessens or eliminates danger to personnel involved in fire fighting efforts. Such tactics may be considered for process industries using flammable solvents. An issue of concern has been the possibility of freezing and the measures that would have to be taken to address the potential for freezing.

Exposure Protection Applications

Vessels containing flammable liquids traditionally use water spray protection in accordance with NFPA 15. A water spray that completely envelops the vessel and eliminates dry spots may be considered in some cases where water supply is insufficient. Intermittent water mist application with very low flow rates and strategically placed nozzles will provide adequate protection of hot surfaces with a small expenditure of water spray. In cases where exposure protection area is large and water supply limitations are not a factor, NFPA 15 water spray systems most likely would be specified.

Life Safety Applications

Awareness of the droplet size of sprinkler water in the early 1950s prompted development of the standard spray sprinkler as a replacement for the "old style" sprinkler. During development of the standard automatic sprinkler, it was noted that the smaller droplets provided more available surface area for heat absorption, vaporization, and plume cooling. It was decided to space the "teeth" on the sprinkler deflector more closely than the old style sprinkler, thereby breaking the water spray into finer drops. Automatic sprinkler protection has been based on this principle since the early 1950s.

Some cases benefit from smaller droplets than are emitted by automatic-sprinklers. Further, water supply is severely limited in some circumstances, such as on a maritime vessel. Water mist has been used successfully on living quarters on maritime vessels, and this usage has been expanded to land-based living quarters with water supplies that are insufficient to support standard automatic sprinkler systems.

AUTOMATIC SPRINKLER SYSTEMS AS A HALON REPLACEMENT

Some system designers are considering a standard automatic sprinkler system as a replacement for a halon system. Sprinkler systems present no personnel safety concerns, nor do they have any effect on the ozone layer. The only concern in specifying a well-designed sprinkler system is relative to property damage in cases where electronic devices or irreplaceable objects are protected.

NFPA

NFPA 13, Standard for the Installation of Sprinkler Systems

These concerns may be addressed by installing a double-interlocked preaction system in accordance with NFPA 13, *Standard for the Installation of Sprinkler Systems*. A double-interlocked preaction system is filled with air instead of water, which requires a detector as well as a sprinkler to actuate before water flows to the actuated sprinkler.

SUMMARY

The Montreal Protocol prohibits the manufacture of halogenated agents in countries participating in the agreement. Water mist systems may be considered as halon system replacements when they are designed in accordance with NFPA 750. Water mist systems are designed to address preconceived performance objectives, including fire control, fire suppression, fire extinguishment, compartment geometry, compartment ventilation, fire hazard classification, fire location, obstructions and shielding, reliability, temperature control, and exposure protection.

A designer selects among low-pressure water mist systems, intermediate-pressure water mist systems, and high-pressure water mist systems, and will design a high-pressure gas-driven water mist system with stored water, a high-pressure gas-driven water mist system with multiple accumulator units for extended duration, a low-pressure twin-fluid water mist system, a single-fluid water mist system, a pump-driven water mist system, a positive-displacement pump assembly with unloader valves on each pump and a pressure relief valve on the discharge manifold, a break tank connection in supply to positive displacement pumps for marine systems, a gas pump unit for machinery spaces and gas turbine enclosures, or a gas pump for light hazard applications.

REVIEW QUESTIONS

1. Give an example of a situation in which a local application water mist system is preferred and one in which a total flooding water mist system is preferred.

2. What is the difference between wet pipe, dry pipe, deluge, and preaction water mist systems?

3. Will water mist droplets be larger or smaller if the pressure is increased for water flowing through a fixed orifice? Why?

4. For a large passenger aircraft or a large room on a passenger ship, why would it be necessary to install a zoned application water mist system? For a system with a fixed amount of stored water, why would a zoned application water mist system be more effective than a total flooding water mist system?

5. List the performance objectives of water mist system design, and describe how water mist systems meet each objective.

DISCUSSION QUESTIONS

1. Assuming that you are the chief engineer at an electric power plant, do some Internet research and provide some examples of applications in which water mist could be considered.

2. Provide examples of water mist applications for historical preservation of priceless and irreplaceable artifacts.

ACTIVITIES

1. Visit a manufacturer of water mist systems. Determine the methodology that determines vessel pressure requirements and pre-engineered capacities.

2. Visit a testing laboratory and witness a test of a water mist system. Determine the protocol for testing a variety of water mist systems.

Chapter 7

ULTRA HIGH-SPEED EXPLOSION SUPPRESSION SYSTEMS AND ULTRA HIGH-SPEED WATER SPRAY SYSTEMS

Learning Objectives

Upon completion of this chapter, you should be able to:

- List and explain the difference in performance objective between an explosion suppression system and an ultra high-speed water spray system, and identify the NFPA standard applicable to each performance objective.

- Discuss the difference between a deflagration and a detonation.

- List and discuss methods other than an explosion suppression system that may reduce the likelihood of an explosion.

- List applications for explosion suppression systems.

- Identify the extinguishment methodologies for explosion suppression systems.

- Evaluate an explosion pressure profile for a commodity, comparing a suppressed profile to an unsuppressed profile. Determine whether an explosion suppression system would be valuable for a vessel of a given yield strength.

- Describe the sequence of a suppressed explosion.

NFPA

NFPA 69, Standard on Explosion Prevention Systems

explosion suppression systems
systems used for the protection of vessels or other enclosures where overpressurization is the primary concern

NFPA

NFPA 15, Standard for Water Spray Fixed Systems for Fire Protection

ultra high-speed water spray systems
systems used for the protection of explosive hazards, where water is the media used, and where overpressurization is not the primary concern

explosion
a rapid release of combustion energy that increases pressure in a vessel, container, or building and results in its eventual rupture

NFPA 69, *Standard on Explosion Prevention Systems*, is used to design **explosion suppression systems** for vessels and other enclosures where overpressurization is the primary concern, such as within vessels. In the 1996 edition of NFPA 15, *Standard for Water Spray Fixed Systems for Fire Protection*, a chapter was added for **ultra high-speed water spray systems** protecting explosive hazards, where water is the media used, and where overpressurization is not the primary concern, such as within rooms occupied by personnel.

DETONATIONS AND DEFLAGRATIONS

To understand the components of the explosion suppression system, readers must understand the types of reactions that result in explosions. An **explosion** is a rapid release of combustion energy that increases the pressure within a container, resulting in the eventual rupture of a vessel, container, or building. This release of energy is not instantaneous. An explosion actually is a rapidly spreading fire that creates a corresponding increase in pressure when a process or hazardous substance is in a tightly enclosed area.

The two types of reactions that create this energy release are detonations and deflagrations. A **detonation** is a reaction involving a fuel that contains its own oxidizer, such as a high-explosive. This reaction produces a flame front that burns faster than the speed of sound. Explosion suppression systems may not be effective in halting or suppressing a flame front traveling at that rate. Detonations normally are associated with munitions that are completely assembled, such as within an artillery shell, although they may occur during the munitions manufacturing process. The probability of a detonation in the munitions manufacturing process can be greatly reduced by limiting the extent of munitions handled in a process, by separating the munitions being handled from the main munitions supply, and by providing the most rapid suppression system possible.

A **deflagration** results from ignition of a flammable fuel–oxidant mixture in an enclosed area, forming a flame front that moves into the unburned material more slowly than the speed of sound, with a concurrent rapid increase of pressure within the container. This creates a scenario conducive for effective extinguishment by an explosion suppression system. In a deflagration the speed of the flame front accelerates with the passage of time. In an enclosed container a deflagration could result in a rapid increase in pressure that can exceed the designed pressure rating of a vessel, resulting in rupture.

For explosion suppression systems, therefore, an explosion can be summarized as being a combustion process in a poorly vented or protected container resulting in a flame front that expands at a velocity that exceeds the speed of sound and creates a pressure that exceeds the rated pressure of a container, resulting in rupture. Many industrial-process containers are rated for a normal operating pressure of less than 5 psig, with bursting failure occurring in the 80 to 120 psig range.

detonation
a reaction in which the flame front expands at a rate greater than the speed of sound

deflagration
a reaction in which the flame front moves into the unburned material at less than the speed of sound

Deflagrations can involve a manufactured airborne fuel such as ignitable dust or gas, or a non-airborne commodity such as a pile of powdered munition, or a liquid, such as nitroglycerine, that, when ignited, creates a flame front that moves more slowly than the speed of sound. The essential elements of a deflagration are a sufficient concentration of fuel, air, or other oxidant, and a source of ignition.

With a high-speed detection system and control unit, an explosion suppression system can be effective in halting, controlling, and extinguishing the flame front in a deflagration. By operating in the earliest stages of a deflagration, an explosion suppression system has ample time to distribute the agent, halt the deflagration, and protect the vessel.

DESIGN APPROACHES

point protection
application of suppressant directly onto an expected point of hazard

area protection
application of suppressant over the entire floor surface area of a room or enclosure

Explosion suppression systems and ultra high-speed water spray systems protect a hazard either by **point protection** or **area protection**. Point protection, shown on **Figure 7-1**, is the application of suppressant directly onto an expected point of hazard, such as a saw that cuts solid munitions. Area protection, shown on **Figure 7-2**, is the application of suppressant over the entire floor surface area of a room or enclosure.

Figure 7-1 *Point protection (pilot-operated system shown). (Courtesy of Superior Automatic Sprinkler Corp.)*

Figure 7-2 *Area protection (pilot-operated system shown). (Courtesy of Superior Automatic Sprinkler Corp.)*

EXPLOSION SUPPRESSION SYSTEMS

The first explosion suppression systems were introduced by the British in the late 1940s to protect aircraft engines. With aircraft fuel stored primarily in the wings, loss of an engine imperiled the mission. Assuming that an aircraft was designed to continue the mission with the loss of one engine, suppression of an engine fire could save the crew and the craft. This technology was extended to industrial use in the early 1950s.

Explosion suppression systems consist of a cylinder filled with a suppressing agent, a detection system, and a system of control circuitry. Ordinarily they can be differentiated visually from ultra high-speed deluge fire suppression systems by their lack of system piping, by their uniquely shaped agent containers, and by the wide variety of agents that can be used with these systems. Explosion suppression systems are designed to protect enclosed vessels or containers in which overpressurization is the primary concern.

To ensure that the components are being used properly, some manufacturers provide a pre-engineered system design for a client's specific needs and do not sell individual explosion suppression system components without an engineering analysis of the client's hazard. After a laboratory test or site inspection, if necessary, the manufacturer's representative recommends the agent choice, detector type, agent capacity, and suppression agent container location.

Industrial explosions can affect much more than machines. People can fall victim to industrial processes, as witnessed by many tragic incidents. A grain-processing complex in Louisiana exploded in 1977, killing 36. In Galveston in that same year, a grain terminal explosion killed 18. In 1978, a flour mill explosion in Bremen, Germany, killed 12. In 1982, a cereal silo explosion in Metz, France, killed 12. Between 1980 and 1986, there were 154 reported agricultural dust explosions. Between 1977 and 1985, there were 26 reported coal dust explosions.

Alternatives or Enhancements to Explosion Suppression Systems

As an enhancement to the effectiveness of an explosion suppression system, improvements in flammable-commodity process design can be considered. NFPA 69 includes explosion suppression systems as one method of explosion protection. Other means of guarding against explosions include revisions to the hazardous commodity manufacturing process. Process design improvements that can be considered in an explosion prevention risk analysis fall into three categories:

1. *Control of oxidant concentration.* Purge gas can be introduced into a volume containing hazardous material or dust to reduce the level of available oxidant in the volume to below that required for combustion. Air consists of 21% oxygen and 79% nitrogen. Fire protection engineers use the flammability diagrams shown on **Figure 7-3** to regulate flammability potential.

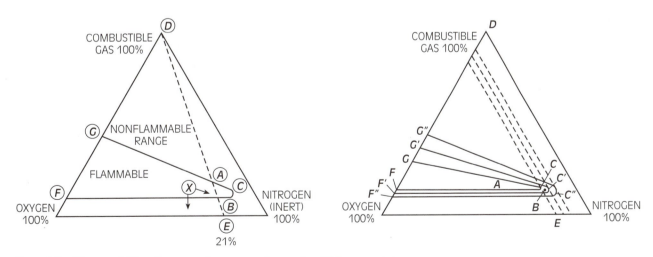

Figure 7-3 *Flammability diagrams for a sample combustible gas mixture.*

Note that regulation of the oxygen concentration from its current level of 21% (line DE) affects the flammability potential of a combustible gas. The flammable range is described by triangle FCG; point A is the upper flammable limit; and point B is the lower flammable limit of the gas mixture.

2. *Control of combustible concentration.* Ventilation or air dilution is done to maintain the concentration of airborne combustible at a level of 25% below its **lower flammable limit (LFL)**, the minimum concentration of airborne combustibles required for ignition. Airborne dusts or gases can be extremely dangerous when an ignition source is introduced to the combustible at a concentration above its lower flammable limit. By limiting the amount of hazardous material permitted in a volume, the risk of ignition decreases.

3. *Deflagration pressure containment.* Some vessels can be designed with stronger metals and tighter seals to withstand the anticipated pressure from an internal deflagration. The NFPA 69 formula below can be used to calculate an explosion protection design pressure for a vessel:

$$P_r = \frac{(1.5) \times [R(P_i + 14.7) - 14.7]}{F_u}$$

$$P_d = \frac{(1.5) \times [R(P_i + 14.7) - 14.7]}{F_y}$$

where P_r = design pressure to prevent rupture of the container resulting from internal deflagration, in psig

P_d = design pressure to prevent deformation of the container resulting from internal deflagration, in psig

P_i = maximum initial pressure at which the combustible atmosphere exists, in psig

R = ratio of maximum deflagration pressure to maximum initial pressure, in absolute pressure units. R is 9 for most gas/air mixtures, 10 for most organic dust/air mixtures, and 13 for most St-3 dust/air mixtures.

F_u = ratio of **ultimate stress** of the vessel to allowable stress of the vessel. Ultimate stress is the stress at which the vessel is likely to rupture. F_u is 4.0 for vessels of low carbon steel and low alloy stainless steel.

F_y = ratio of the **yield stress** of the vessel to allowable stress of the vessel. The yield stress is the stress at which the vessel is likely to deform. F_y is 2.0 for vessels of low carbon steel and low alloy stainless steel.

Passive safety devices such as adjustable pressure release door safety latches can provide pressure relief to protect a vessel from rupture. It may not be economically feasible to provide the strength of enclosure required to suppress a deflagration in a vessel.

lower flammable limit (LFL)
the minimum concentration of airborne combustibles required for ignition

ultimate stress
the stress at which a vessel is likely to rupture

yield stress
the stress at which a vessel is likely to deform

NFPA

NFPA 68, Guide for Venting of Deflagrations

explosion isolation
the automatic closing of a valve to limit the pressure rise from spreading to a predetermined area

Another enhancement to an explosion suppression system is explosion venting. These passive vents are available at a relatively low cost compared to the cost of strengthing the enclosure, and they sometimes are used in lieu of an explosion suppression system. Venting, however, relieves pressure without any active effort to control the continued burning of the fuel in the container. Care must be taken to avoid placing explosion venting louvers where an operator or passerby could be injured by an explosion relief blast. NFPA 68, *Guide for Venting of Deflagrations*, is used when designing an explosion venting system.

Explosion isolation, either physical or with suppressant, is recommended for ducts or areas adjacent to the vessel protected. Physical isolation of an explosion involves the automatic closing of a valve to limit the pressure rise from spreading to a predetermined area. Electrically initiated and pneumatically operated explosion isolation knife-gate valves physically isolate and contain explosions within predetermined boundaries of a hazardous process. This isolation process is enhanced by the addition of a suppressing agent into each area that has been physically isolated. The system consists of explosion detection, a control unit, and explosion isolation valves.

Typical applications and design of explosion isolation of an explosion suppression system include

- fast-acting automatic-closing valve assemblies, shown on **Figure 7-4**, which isolate a vessel of interest from an explosion source

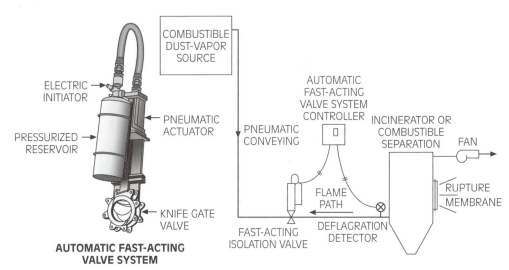

Figure 7-4   *Typical application and design of fast-acting automatic-closing valve assembly.*

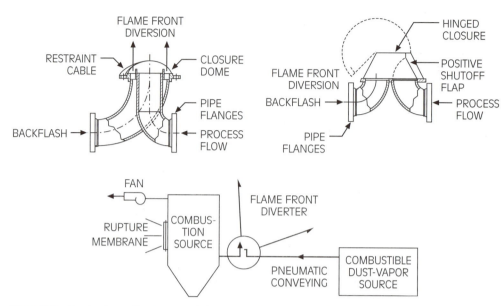

Figure 7-5 *Typical application and design of flame front diverters.*

Source: NFPA 69 (2002), Figure A.9.5. Reprinted with permission from NFPA 69, *Explosion Prevention Systems*, Copyright © 2002, National Fire Protection Association, Quincy, MA 02169. This reprinted material is not the complete and official position of the National Fire Protection Association on the referenced subject which is represented only by the standard in its entirety.

- flame front diverters, shown on **Figure 7-5**, which prevent flames from entering a critical vessel

How Explosion Suppression System Agents Work

Suppressing agents are introduced into an enclosed chamber at the incipient stages of a deflagration to stop the continuation of the combustion reaction. Clean agents, ammonium phosphate, sodium bicarbonate, and sodium chloride interact with the combustion reaction and halt it in three basic ways:

1. The suppressant absorbs energy produced by the combustion reaction.
2. Critical chain reactions are inhibited, halting propagation of the combustion reaction.
3. The concentration of air and dust is affected by the addition of suppressant, creating a concentration outside of the combustible range.

Analysis of these three methods can determine the effectiveness of a suppressant, although some suppressants may be more effective than others with respect to one or more of the three methods. The advantages of the many

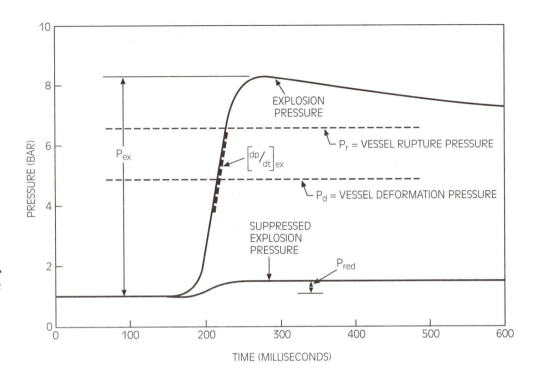

Figure 7-6 *Explosion pressure profile for a combustible undergoing deflagration in an enclosed vessel.*

explosion pressure profile

a graph depicting the effect of suppression upon a specific combustible

suppression agents discussed in this book can be used to provide the best extinguishing mechanism for the combustible involved.

Figure 7-6 shows an **explosion pressure profile**, a graph depicting the effect of suppression on a specific combustible. Note how the introduction of suppressant lowers the pressure to a small fraction of its unsuppressed level. Note also that the unsuppressed explosion pressure of the combustible exceeds the vessel rupture pressure and the suppressed explosion pressure is less than the vessel deformation pressure. The case for installing explosion suppression is demonstrated clearly on this sample explosion pressure profile.

Explosion pressure profiles are created by testing the deflagration of a combustible commodity in a testing vessel capable of withstanding very high pressures. The maximum explosion pressure can be determined and compared to the suppressed pressure in the presence of an explosion suppression system. The profile then can be compared to the yield stress and the ultimate stress of the vessel that would contain the explosive commodity.

Manufacturers of explosion suppression systems regularly perform experimental commodity testing and compile and make available to potential users a computerized database of all experiments and full-scale tests. Other information that can be compiled includes venting, inerting, agent analysis, and detection evaluation. Arrangements can be made with manufacturers or at most testing laboratories for testing materials or processes that are not contained within the database.

An explosion chamber is used for small-scale dust and gas testing, conducted in accordance with ASTM E1226, the standard testing method for pressure and rate of pressure rise for combustible dusts, and the International Organization for Standardization Code ISO 6184, for determining explosion indexes of combustible dusts in air. The types of tests that can be conducted in this chamber include the dust explosibility test, the minimum explosive concentration (MEC) test, the maximum allowable oxygen concentration (MOC) test, and the minimum ignition energy (MIE) test.

The dust explosibility test provides, for a wide range of concentrations, maximum pressure, maximum rate of pressure rise, and dust deflagration index for a specific hazard. The results are used to determine the severity of the explosion and to establish explosion venting requirements. The explosion rate of pressure rise, multiplied by the cube root of the volume of the container, has been found to be a constant.

The MEC test determines the minimum dust concentration that could result in an explosion; the MOC test determines the oxygen concentration necessary to foster an explosion; and the MIE test determines the sensitivity of a dust to electrostatic ignition. These tests are used to determine an explosion prevention strategy for a specific process.

Once the explosion parameters and severity have been established, larger scale testing can be arranged. Proposed explosion suppression systems can be tested with respect to the specific material and process requirements. These enlarged volumes can enable the study of actual explosions, investigation of explosion protection schemes, determination of models for larger-scale hazards, and determination of the correct explosion suppression system for the hazard. The worst case—the one set of conditions that could produce the greatest rate of pressure increase—is chosen for protection, providing a margin of safety should conditions of lesser severity occur.

Not all explosions can be suppressed. Each process must be examined carefully and, if necessary, tested to determine an appropriate protection and suppression scheme. The type of process, methods, construction, routing, vessel geometry, duct design, ignition sources, product hazard intensity, product toxicity, and additional protection mechanisms all can determine how the process is to be protected and suppressed.

A suppressant, when injected into an explosive environment, extinguishes the flame front by chemical action and cooling, and prevents burning of any remaining explosible mixture by removing heat, which causes the temperature of the mixture to fall below the autoignition temperature. The suppressant creates a barrier that prevents heat transfer within the mixture. Monoammonium phosphate powder, sodium bicarbonate powder, and other agents are available for use with suppressant containers.

Isolation is available by way of explosion isolation valves, but a combination of a chemical suppression system and mechanical isolation system is recommended. The isolation valve, when supplemented by a suppression system, may allow some explosive mixture to pass, but the chemical barrier provided by the suppression system should extinguish what may have escaped.

Explosion Suppression System Components

An explosion suppression system consists of an agent storage container, a detection system, and a control unit, as shown in **Figure 7-7**. Note that the suppression system also features explosion isolation valves, shown on Figure 7-4.

Explosion Suppression System Agent Storage Container

The storage container for an explosion suppression system is refillable, enclosing a fixed amount of suppressing agent, and is pressurized with a gas, such as nitrogen. The container can be mounted directly to the vessel it protects, on a wall within a room, or above a ceiling, with piping protruding into a room. The container has a control valve, initiated by the firing of a low-power explosive device that becomes fully open one millisecond from initiation, releasing the agent. High-speed photography has determined that the agent exits the container at a velocity of 150 ft. per sec (45.72 m/s), with differing agents traveling at roughly the same speed.

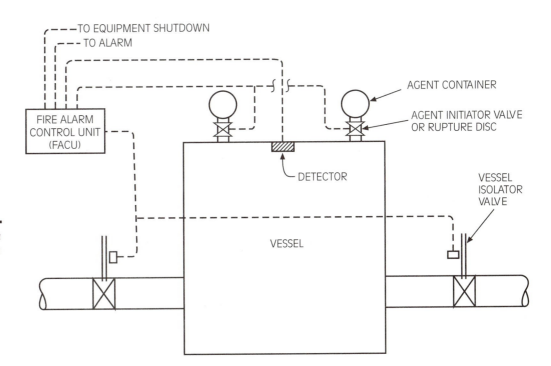

Figure 7-7 *Explosion system schematic; a detection signal is sent to the control unit, and the agent initiator valves and isolator valves are actuated simultaneously.*

Multiple containers can be located within a room or vessel to ensure dispersal of gas to a specific concentration. An agent dispersal nozzle is attached directly to the valve with a nipple or short piece of pipe, or, if needed for local application, a piping dispersal system. Explosion system response time for the suppression system is affected negatively by lengthy runs of pipe from the storage container to the hazard. It is recommended that containers be positioned as close to the hazard as is physically possible. An advantage of this type of container is that it can be mounted well above the floor, allowing for more plant operations space at the floor.

Some manufacturers can supply explosion suppression containers that use stored water as the suppression agent. Such a system may be used either for explosion suppression in accordance with NFPA 69, in which overpressurization is the primary concern, or as an ultra high-speed water spray system in accordance with NFPA 15, in which overpressurization is not the primary concern.

Because of the dispersal distances involved, several smaller containers are recommended, rather than one large container. The best response times are achieved when containers and detectors are attached directly to a vessel with little or no piping.

Detection Systems for Explosion Suppression Systems

Detectors used on explosion suppression systems can be selected from the following:

- ultraviolet flame detectors
- pressure-sensing detectors
- products of combustion detectors
- infrared flame detectors

Thermal detection is not recommended for explosive hazards because of its significantly slower response time. Ultraviolet flame detectors are recommended for vapor explosion suppression, with a response time in milliseconds. Pressure-sensing detectors are well suited for dust hazards in enclosed spaces. These detectors are not susceptible to failure caused by dust loading on the surface of the component, unlike many flame detectors that use a "cone of vision" to detect fires in their incipient phase. If venting is present, pressure-sensing detectors may not be appropriate.

Chapters 11 through 15 thoroughly discuss detector selection and fire alarm and releasing systems.

Fire Alarm Control Units (FACU) for Explosion Suppression Systems

Fire alarm control units (FACU) that can accommodate several explosion suppression zones are available. The unit supervises the detection circuits, has terminals for an audible alarm system, contains built-in standby batteries that operate the system in the event of main power failure, has terminals for equipment shut-down, provides for manual activation, and can initiate automatic actuation of the containers after receiving a detection signal. A power unit provides the firing current necessary to actuate the initiator on the suppressant container.

Applications for Explosion Suppression Systems

Explosion suppression systems can be used in a variety of applications, including the following.

Aerosol Fill Rooms A common application for total flooding explosion suppression systems is the aerosol fill room used in industrial occupancies to fill spray cans with paint or some other potentially flammable mixture. The filling gas used to propel the flammable commodity is butane. The combination of the flammable or combustible liquid and the filling gas creates a severe hazard. The objective of a system protecting this commodity is to limit pressure in the enclosure.

Diptanks Another common hazard protected by explosion suppression systems is a flammable liquid diptank, which serves as a dipping vessel to coat or strip an object, and can be protected with a local-application explosion suppression distribution system.

Dust Collectors A history of dust explosions in grain facilities goes back to the 16th century, and many disastrous incidents in 1977 initiated concentrated study of the problem. Fires in fabric filter dust collectors, with their high velocity air stream, are amenable to protection by an explosion suppression system.

Grain Elevators Extensive testing for the National Grain and Feed Association conclusively proved the value of explosion suppression systems in protecting against grain dust explosions. An explosion suppression system, installed at the head and boot of a bucket grain elevator, successfully inerts a dusty atmosphere and prevents flame propagation to the gallery, silos, and associated equipment. Systems also can be installed in the legs to prevent explosions originating at points external to the bucket elevator.

Other Dust Hazards Additional applications for explosion suppression systems are varied. In addition to grain hazards, dust hazards applications include plastics, coal, rubber, wood, paper, stearates, resins, pharmaceuticals, cellulose, fertilizer, sulfur, acrylics, charcoal, epoxies, aluminum, food additives, chocolate, corn, herbicides, dyes, milk powders, ink toners, talc, insecticides, fungicides, and many other airborne solids.

Vaporous Hazards Applications for vaporous hazards include isobutane, acetone, ethylene, gasoline, kerosene, propane, shellac, toluene, xylene, heptane, hexane, methane, bentone, benzene, butane, ethane, cyclohexane, ethyl alcohol, glycol, hydraulic fluid, aviation fuel, lubricants, naphtha, transformer oil, pentane, rocket fuels, and many other vapors.

Industrial Processes Processes that can be protected successfully are pulverizing (hammermills, shredders, cage mills, separators, grinders, ball mills, flakers, granulators), conveying (cyclones, screw conveyors, fans, bucket elevators, pneumatic ducts), processing (flavoring cylinders, powder paint booths, aerosol fill rooms, blenders, mixers, pipe coating, transformer cooling, ink toning, sanders, cookers, formers), and storing (bag collectors, bins, hoppers, flammable liquid storage areas, fluid bed dryers, tanks).

Other Hazards Other hazards protected are shredders, with dry chemical as the agent; dust collectors, with pressure detectors as the initiating devices; ignitible metal hazards, with sodium chloride as the agent; wood products and dust hazards, with water spray as the agent; and food processing equipment, pharmaceuticals, and agricultural products, with an ultraviolet detection system.

Designers of special hazards systems must be familiar with explosion suppression systems. They also have to be aware of applications where this proven technology would be effective.

Selection of an Explosion Suppression System Agent

A number of factors must be considered when selecting and specifying an explosion suppression system. They include the following:

Personnel Protection The position of personnel in proximity to a vessel that could undergo rupture could be the determining factor in selecting an explosion suppression system. Possible exposure of the agent to personnel determines the type of agent chosen for the hazard.

Effectiveness-to-Weight Ratio Another consideration in selecting an agent is the agent's effectiveness-to-weight ratio. Carbon dioxide ordinarily is not found in explosion suppression systems because of the large concentrations involved, normally in excess of 34% of the container volume. By contrast, some clean agents ordinarily require only 1 pound per 10 cubic feet of volume for effective extinguishment.

Agent Compatibility An additional consideration is the compatibility of the agent with the commodity protected. The agent must be physically compatible with the commodity, and additional consideration may be given to clean-up concerns or the possibility of reuse of the commodity in case of agent discharge, and getting the operation up and running after agent discharge.

Sequence of a Suppressed Explosion by an Explosion Suppression System

An explosion can be studied in several steps, as shown in **Figure 7-8**:

1. A combustible mixture ignites.
2. Pressure increases in the vessel as the flame front expands.
3. An explosion suppression detector senses the energy created by the expanding flame front and notifies the control unit.
4. The control unit initiates agent discharge and isolation valves are closed.
5. The suppression system halts the advancing flame front.

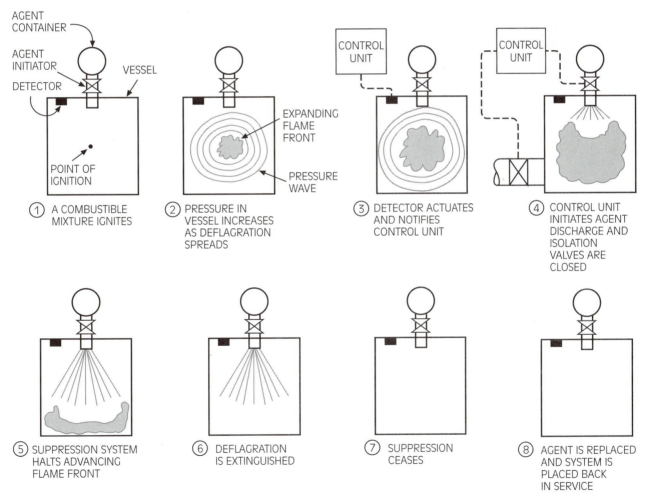

Figure 7-8   *Sequence of a suppressed explosion.*

6. The deflagration is extinguished.

7. Suppression ceases.

8. The agent is replaced and the system is placed back in service.

The time interval between each step in the suppression process is a function of the velocity of the expanding flame front, which varies with differing commodities. Detection time must be selected to ensure that the agent is discharged in sufficient time to prevent pressures in the vessel from exceeding the yield pressure of the vessel.

ULTRA HIGH-SPEED WATER SPRAY SYSTEMS

In addition to explosion suppression systems, the other suppression methodology for suppression deflagrations is an ultra high-speed water spray system, designed in accordance with NFPA 15. Ultra high-speed water spray systems are used to protect hazards where overpressurization is not the primary concern.

Applications for Ultra High-Speed Water Spray Systems

Whereas explosion suppression systems would ordinarily be expected to protect enclosed volumes, ultra high-speed water spray systems would be the most likely protection method for a room or a point of protection, with the munitions manufacturing process being the predominant application. The following applications are representative of the uses for ultra high-speed water spray systems:

- rocket fuel manufacturing or processing rooms
- solid propellant manufacturing or handling rooms
- paint spray can filling rooms
- ammunition manufacturing rooms
- pyrotechnics manufactuiring rooms
- maintenance, renovation, or demilitarization of ordnance items
- the manufacture of other volatile solids, chemicals, dusts, or gases

Munitions Manufacturing and Ultra High-Speed Water Spray Systems

Ultra high-speed water spray systems protect many hazardous applications, but munitions manufacuring applications predominate. The following operations represent the types of munitions operations ordinarily encountered:

- *weighing:* an operation to prevent ignition in the presence of static charges produced by the dust from powdered munitions
- *pressing:* placing solid munitions under pressure for formation into a solid of given geometry

- *pelletizing:* pressing solid munition into pellet-sized shapes
- *propellant loading:* placing solid munitions into a container, such as a shell
- *melting:* placing solid munitions in a kettle and bringing to a critical temperature by steam or hot water
- *extrusion:* pressing a solid propellant through an orifice to create a desired shape
- *mixing:* stirring components of solid or liquid munitions to compose a mixture
- *blending:* bringing together two munitions components in a container
- *screening:* sifting solid munitions (which creates a dust hazard)
- *sawing:* cutting solid munitions to a predetermined size, possibly creating sparks
- *granulating:* breaking solid munitions into small pieces
- *drying:* placing solid munitions in ovens or heating cabinets
- *pouring:* depositing solid or liquid munition into a container
- *machining:* forming or grinding solid munition into a specific shape

Agent Selection for Ultra High-Speed Water Spray Systems

Although explosion suppression systems employ a wide variety of agents, water is almost universally used as the suppressing agent in ultra high-speed water spray systems because it is readily available and provides the cooling, insulation, and dilution needed to extinguish deflagrations. Water knocks down suspended dust and cools the expanding flame front. Many munitions contain their own oxygen supply to promote combustion in tightly sealed enclosures, such as an artillery shell. The presence of oxidizers eliminates smothering agents such as carbon dioxide and clean agents. Cooling is the predominant mode of extinguishment, and water is therefore the most effective agent to accomplish this objective.

Squib-Actuated Ultra High-Speed Water Spray Systems

squib-actuated ultra high-speed water spray system

a preprimed water-filled piping system controlled by a squib-actuated deluge valve

The **squib-actuated ultra high-speed water spray system** is a preprimed water-filled piping system, controlled by a squib-actuated deluge valve. A squib is a small disc of munition that combusts rapidly in a chamber when a detection system detects a fire, which rapidly opens the deluge valve and transfers pressure to the preprimed water in the pipes. The water pressure increases to a point where blowoff caps are removed or rupture discs are broken, and water flows simultaneously on all system nozzles.

Design of squib-actuated ultra high-speed water spray systems should consider all of the following:

- Mount squib-actuated valves in their vertical position to avoid the formation of air pockets that may retard system response.
- Route pipe from the squib-actuated valve to the nozzles in the most direct manner possible.
- Slope all piping up to vent valves capable of removing trapped air.
- Group nozzles into counter-opposed groups of at least two nozzles when protecting a known point of hazard.
- Ensure a water supply of at least 50 psi (344.75 kPa), with greater pressures if possible, to enhance system effectiveness.
- Perform hydraulic calculations to ensure minimum densities required by NFPA 15.

Pilot-Actuated Ultra High-Speed Water Spray Systems

pilot-actuated ultra high-speed water spray system
a preprimed water discharge system with soledoid-operated pilot nozzles, supervised by a pilot piping system

A pilot-actuated ultra high-speed water spray system, shown on **Figure 7-9** and **Figure 7-10** consists of a preprimed water discharge system with soledoid-operated pilot nozzles, supervised by a pilot piping system.

Figure 7-9 *Pilot-operated control apparatus. (Courtesy of Superior Automatic Sprinkler Corp.)*

Figure 7-10
Discharge piping (large pipe) and pilot piping (smaller pipe) for a pilot-operated ultra high-speed water spray system. (Courtesy of Superior Automatic Sprinkler Corp.)

NFPA 15 Requirements for Ultra High-Speed Water Spray Systems

NFPA 15 permits the use of ultra high-speed water spray systems for extinguishment or control of deflagrations. The following are some of the minimum requirements for ultra high-speed water spray systems:

- Maximum response time: 100 milliseconds from the presentation of an energy source to the detector to flow of water from system nozzles is required.

- Local application: A minimum of 25 gpm (95 lpm) per nozzle is required.

- Area application: A minimum of 0.50 gpm/ft.^{2} (2 lpm/m^2) is required.

- Design pressure: A minimum of 50 psi (3.5 bar) is required.

- System volume limitation: A maximum of 500 gal (1893 l) system capacity is required; system capacity is the volume of water contained by the discharge system.

- System duration: A minimum water supply capable of providing water to the hazard for 15 minutes, or the minimum time required for safe egress of personnel, is required

- Piping is required to be sloped a minimum of 1 in per 10 ft. of pipe (25 mm per 3 m).

SUMMARY

Explosion suppression systems are designed in accordance with NFPA 69 for deflagrations in vessels where overpressurization is the primary concern, such as within vessels. For deflagrations where enhanced suppression time is required, but where overpressurization is not the primary concern, such as within rooms occupied by personnel, NFPA 15 is used for the design of ultra high-speed water spray systems.

An explosion suppression system consists of a suppression container with agent initiator, a control unit, and a detection system. Explosion pressure profiles are created for each commodity. Evaluation of the explosion profile with respect to the vessel yield strength is required to evaluate the need for an explosion suppression system. The suppressed explosion pressure profile must be below the vessel yield strength.

Alternatives or enhancements to explosion suppression systems include control of the oxidant concentration, control of the combustible concentration, deflagration pressure containment, pressure relief venting, and deflagration pressure isolation. Suppression agents work by absorbing the energy of the expanding flame front, by inhibiting combustion chain reactions, or by reducing the concentration of the combustible in an enclosed space.

REVIEW QUESTIONS

1. Compare and contrast the applications for an explosion suppression system and an ultra high-speed water spray system.

2. List and explain the available alternatives or enhancements to explosion suppression systems.

3. Using the information learned about each agent in the preceding chapters, list applica-tions for explosion suppression systems and suggest the best agent for each application.

4. Why is the sequence for an explosion unique for each combustible?

5. Explain each step in the sequence of an explosion in relation to an explosion pressure profile for a specific commodity.

DISCUSSION QUESTIONS

1. Obtain a copy of NFPA 15, NFPA 68, and NFPA 69. Compare and contrast the specific ways by which each NFPA document ad-dresses explosive hazards.

2. Using the SFPE *Handbook of Fire Protection Engineering*, determine how flammability diagrams are used to evaluate and mitigate explosive hazards.

ACTIVITIES

1. Visit a facility that has a functioning explosion suppression system. Inspect the system thoroughly, taking note of the kind of detection, the agent, volume of agent storage, and sophistication of the control unit apparatus. Are shutdown and alarm functions evident, and are personnel considerations designed into the system?

2. Visit a consulting or specifying engineer involved with an explosion suppression system. Determine the criteria used to specify an explosion suppression system.

3. How are personnel taken into account when an explosion suppression system is under consideration? What methodology is used to ensure that the correct explosion profile is employed for the commodity anticipated?

GASEOUS AND PARTICULATE AGENT SPECIAL HAZARD SYSTEMS

Chapter 8

CLEAN AGENT AND HALON REPLACEMENT EXTINGUISHING SYSTEM DESIGN

Learning Objectives

Upon completion of this chapter, you should be able to:

- Evaluate the Montreal Protocol relative to its effect on reducing depletion of the ozone layer.
- Describe the different types of halogenated hydrocarbons that have been used for fire protection, and identify the chemical symbol for each.
- Determine a halon using its halon identification number.
- List the uses for the types of halons encountered in fire protection.
- Discuss the conditions under which halon could be a hazard to personnel.
- List safety precautions for enclosures where personnel could become trapped or where egress time could be excessive.
- Discuss the reason for replacing halon systems.
- Evaluate the criteria that determine whether an agent can be classified as a clean agent.
- Compare and contrast clean agent systems with halon systems.

- List precautions that should be taken to protect personnel who may be affected by the discharge of a clean agent system.
- Discuss and compare the agents recognized by NFPA 2001 as clean agents.
- Calculate the quantity of clean agent required to protect a given occupancy.

Halon is a term shortened from *HALogenated hydrocarbON*, a gaseous agent used for fire protection. Halon systems were widely specified and installed from the 1940s until the 1980s, when it became clear that release of halon into the atmosphere was detrimental to the quality of the earth's protective ozone layer. Even though the reader will not likely become involved in the design of new halon systems, numerous halon systems continue to exist. Therefore, it is important to understand how they were originally conceived so replacement systems can be designed to meet the same performance objectives.

The action taken pursuant to the Montreal Protocol necessitated the formation of the NFPA Technical Committee on Alternative Protection Options to Halon in 1991. This committee researched the options available for Halon 1301 replacement design. The challenge was to find agents with the extinguishment capabilities of halon but without the environmental concerns associated with halon. In 1994, the first edition of NFPA 2001, *Standard on Clean Agent Fire Extinguishing Systems*, was published. This standard is used as the basis for clean agent design.

NFPA

NFPA 750, Standard on the Installation of Water Mist Fire Protection Systems

THE MONTREAL PROTOCOL

Two chemists at the University of California Irvine, Frank Sherwood Rowland and Mario Mocina, began a study of the effect of chlorofluorocarbons (CFCs) on the ozone layer. The results of their work won them the Nobel Prize and resulted in a landmark international agreement, the **Montreal Protocol**, signed by the United States and 24 other countries in 1987, with significant amendments in 1990 and 1992. Kofi Annan, former Secretary General of the United Nations, said "perhaps the single most successful international agreement to date has been the Montreal Protocol."[1]

The agreement is intended to sharply restrict the production of chemicals that had been identified as contributing to depletion of the stratospheric ozone layer. The **ozone layer** is a protective layer of our stratosphere that helps to filter the ultraviolet rays of the sun before they reach Earth. In the absence of the ozone layer, the incidence of skin cancer and melanoma increase. An ozone molecule consists of three oxygen atoms (O_3). Freon, released from air conditioners, and halogenated extinguishing agents rise to the stratosphere. Bromine and chlorine molecules from these agents break up the O_3 molecules and attach themselves to one of the free oxygen molecules. These gases, therefore, were included on the list of ozone depleters.

Montreal Protocol
an international agreement intended to sharply restrict the production of chemicals identified as contributing to depletion of the ozone layer

ozone layer
a protective layer of our stratosphere that helps to filter the ultraviolet rays of the sun before they reach Earth

In advance of the Montreal Protocol, the *Vienna Convention for the Protection of the Ozone Layer* provided the framework for negotiations in 1985. Immediately subsequent to the initial signing of the Montreal Protocol, evidence continued to mount that the ozone layer was continuing to shrink at a frightening rate. Numerous additional countries signed the Montreal Protocol, and the target date for ceasing production of halogenated hydrocarbons was advanced to January 1994. At present, 191 nations have signed the Montreal Protocol, making it one of the planet's most successful international agreements.

The cessation of halon production rapidly rendered existing halon systems the dinosaurs of the fire protection industry, placing owners of halon systems and the companies that insure the hazards protected by halon systems in an extremely uncomfortable position. Although the Montreal Protocol did not call for removing all existing halon systems, it prohibited the manufacture of new halon—making it impractical to legally purchase new halon. Owners and insurers of halon systems were faced with the prospect of a total loss of fire protection pursuant to an accidental or purposeful halon system discharge.

To protect against this eventuality, many system owners opted to replace their halon systems with either a substitute gaseous system replacement or a water-based system replacement. Although halon is covered in this chapter, the primary focus is on clean agents and halon replacements.

Unscrupulous individuals can import halon to replenish existing systems in defiance of the Montreal Protocol. The *Code of Ethics for Engineers* and the *Code of Ethics for Engineering Technicians*, referred to in Chapter 2, specifically prohibit engineers and technicians from entering into transactions that would place the public at risk.

Figure 8-1 is an aerial view of the largest extent of the Antarctic ozone hole. To illustrate, the town of Punta Arenas, at the Southern tip of Chile and at the southernmost tip of South America, periodically issues ozone "red alerts" warning citizens not to go out without protection during times of severe ozone depletion.

HALON FIRE SUPPRESSION SYSTEMS

To be able to evaluate and replace an existing halon system, some knowledge of halon systems is required. Halon was first used during World War II to extinguish fires on aircraft engines and fuel tanks caused by shrapnel shredding and exploding fuel in the tanks. The Army Corps of Engineers (COE) developed a numbering system for the identification of the various types of halon.

Halon Numbering System

The halon identification numbering system is a numerical way of avoiding having to say long chemical names. The first digit represents the number of carbon atoms in the chemical symbol, the second digit represents the number of fluorine

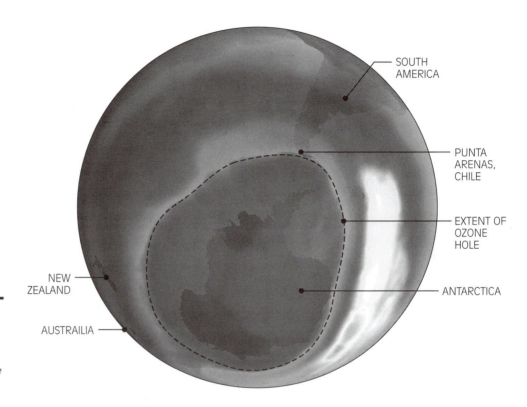

Figure 8-1 *Aerial view showing Antarctic ozone hole over the Earth from the South Pole in 2000.*

atoms, the third digit represents the number of chlorine atoms, the fourth digit represents the number of bromine atoms, and the fifth digit represents the number of iodine atoms.

Halon ID No.	Chemical Name	Chemical Symbol
Halon 1011	Bromochloromethane	$CBrCl$
Halon 1211	Bromochlorodifluoromethane	$CBrClF_2$
Halon 1202	Dibromodifluoromethane	CBr_2F_2
Halon 1301	Bromotrifluoromethane	$CBrF_3$
Halon 2402	Dibromotetrafluoroethane	$CBrF_2CBrF_2$

Even readers who have had very little chemistry will find it easy to describe the chemical composition of a halon by its identification number. For example, Halon 1301 can be described as follows:

Halon 1301	first digit = 1	1 carbon atom
	second digit = 3	3 fluorine atoms
	third digit = 0	0 chlorine atoms
	fourth digit = 1	1 bromine atom
	fifth digit = 0	0 iodine atoms

Using the prefixes "di" for two atoms, "tri" for three atoms, "tetra" for four atoms, and so forth, we can easily understand the chemical name for Halon 1301:

$$\text{Halon } 1301 = (\text{Bromo})(\text{tri-fluoro})(\text{Methane})$$

Chemical Name	*Number of Atoms*	*Chemical Symbol*
Bromo =	1 bromine atom	Br
Tri-fluoro =	3 fluorine atoms	F_3
Methane =	1 carbon atom	C

The chemical symbol for the halon agent simply uses the chemical symbol for each chemical element and assembles them for convenient identification:

$$\text{Halon } 1301 = CBrF_3$$

Note that if the fifth digit of the halon identification number is a zero, the digit is dropped.

Applying the halon numbering system to a substance with which you may be familiar, carbon tetrachloride:

Chemical Name	*Number of Atoms*	*Chemical Symbol*
Carbon	1 carbon atom	C
Tetrachloride	4 chlorine atoms	Cl_4

Carbon tetrachloride, once used as an extinguishing agent and commonly used as a cleaner, therefore can be referred to in three ways:

1. Carbon tetrachloride
2. CCl_4
3. Halon 104

Outdated Applications for Halon 1301

Note that of the five types of halon listed, Halon 1011 originally was used for aircraft engine fire protection. Halon 1301 was widely used as an extinguishing agent well into the 1980s, designed predominately for areas where personnel were expected to be present, with computer rooms and data processing centers the most prevalent applications. The Montreal Protocol had an immediate dramatic effect on reducing the number of new halon systems designed. Halon systems now are no longer designed, but designers may encounter existing Halon systems that require replacement by a clean agent system or some other halon system replacement.

The NFPA continues to publish standards for the design of Halon 1301 systems, NFPA 12A, *Standard on Halon 1301 Fire Extinguishing Systems*. Halon 1301 was commonly specified for use in areas occupied by personnel and where water damage by water-based systems was a concern.

NFPA

NFPA 12A, Standard on
Halon 1301 Fire
Extinguishing Systems

A fire protection designer likely will encounter an existing computer room, data processing facility, telecommunications equipment room, electrical room, or other facility where a Halon 1301 system is installed. Modifications to the room, changes in how the room is used, or concerns relative to an accidental discharge or refilling of the halon containers may force a decision to replace the system as a result of the Montreal Protocol prohibition of the manufacture of new halon. Because the majority of the fixed halon systems that were installed were Halon 1301 systems, this chapter focuses on this type of halon.

Properties of Halon 1301

Halon 1301 was an attractive agent for fire suppression in occupied areas involving delicate equipment, such as switchgear rooms and computer rooms, because of its physical properties that protected equipment without damage, with little effect on occupants of the room being protected.

In addition, Halon 1301 was highly desirable for protection of electronic equipment because it is noncorrosive and leaves no residue upon application. Also, because it is a gas, it can penetrate tightly packed electronic modules to provide adequate fire protection without concern for electrical conductivity.

Personnel Considerations with Halon 1301

Personnel considerations were largely responsible for the wide use of Halon 1301 in occupied areas. Experimentally, Halon 1301 in concentrations less than 4% for brief periods were found to have little or no effect on individuals exposed to the halon discharge, in the absence of flame.

Further, exposures between 5% and 7% up to 15 minutes were found to have few noticeable effects on exposed individuals in the absence of flame. Halon systems were designed predominately within this range.

For concentrations between 7% and 10%, some noticeable central nervous system (CNS) effects are apparent in the absence of flame—such as dizziness, tingling, and minor disorientation. The recommended human exposure for such concentrations is no longer than 1 minute. Theoretically, some systems possibly can be designed for a nominal 7% concentration to inject greater than 7%, if design conservatism is applied to the flow calculations or if the rate of halon leakage from the room is less than anticipated.

It was recommended that any halon system not be designed for concentrations greater than 10%. Persons exposed to concentrations in this range are subject to impending unconsciousness. The effects are reversible, and no long-term effects are anticipated if exposure is 30 seconds or less. A person's voice gets deeper and the person experiences some tingling from extremities when exposed to concentrations in this range. In the absence of any other indicators, a voice change should be a warning to exit immediately.

No studies of long-term human exposure to halon have been done, which helps to explain why the space shuttle must land immediately if the on-board halon fire suppression system discharges. For this reason, NASA decided to replace its existing halon systems with clean agent systems.

Products of Halon 1301 Decomposition

Although halon in its pure form is safe at low concentrations for short periods at room temperature, halon begins to decompose at about 900°F into halogen acids, such as HF and HBr. Breathing these products of halon decomposition is dangerous and potentially lethal, even at low halon concentrations. Immediately exiting a room that is discharging halon onto a flame is an absolute necessity.

Recommended Safety Features for Halon 1301

Because of the potential danger to occupants, rooms protected by Halon 1301 should be equipped with these safety features:

- A continuous alarm must sound upon or in advance of actuation of the halon system by a fire detection system.
- Breathing apparatus should be available to occupants.
- Consideration should be given to constructing more exits from the room than normally required by the applicable building or life safety code.
- Signs should be posted instructing occupants how to react in the event of a discharge.
- At regular intervals, training procedures, including a fire-escape plan and assignment of responsibilities in the event of a fire, should be conducted. New employees should be indoctrinated into the training system, followed by regular retraining.
- A time delay may be appropriate. For a system with a time delay, an alarm sounds upon detection. A voice alarm system may be desirable, to give instructions and the amount of time that occupants have to escape before the system activates. The system then activates automatically after a preset time.
- Manual, as opposed to automatic, activation of the system may be appropriate in rare cases of severe concern for personnel or where personnel are always present, highly trained, and responsible. In other cases, an override switch, delay switch, or abort switch may be considered.

Manual means of activation or halting discharge is not recommended for most applications because lack of training, panic, or personal considerations may interfere with the proper operation of these crucial devices. Security guards, janitors, or desk clerks may not be trained or capable of taking swift action to actuate the system. Whenever a decision is involved in activating a system, fears or doubts related to the effect of the system on any personnel who may be trapped in the enclosure, lack of knowledge, fear of reprimand or firing in case

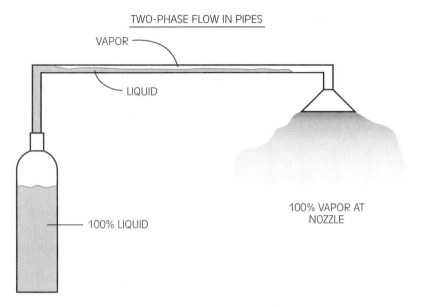

Figure 8-2
Two-phase flow.

of false activation or many other factors may delay system activation, or may result in no action to activate the system.

Halon 1301 Storage

two-phase flow
liquid and vapor halon phases flowing in pipes simultaneously

Halon is stored in compressed liquid form, superpressurized with dry nitrogen in cylinders capable of holding 360 psi or 600 psi, as shown on **Figure 8-2**. Halon flows in pipes as a **two-phase flow**, both liquid and vapor, with 100% vapor being discharged at the nozzles.

The cylinders must be installed in an area protected from mechanical injury, protected from the weather, and where the temperature is between −20°F and 130°F, with 70°F being the ideal temperature. The humidity of the room storing the cylinders should be controlled to minimize corrosion of the cylinders and components. The cylinders are installed as close to the hazard as possible without compromising the safety of the cylinders from a fire in the room being protected.

Types of Halon 1301 Systems

Halon 1301 systems were designed for total flooding. Because of the relatively low discharge percentages with which the agent is being applied, typically between 5% and 7%, Halon 1301 usually makes a poor local application system.

Total flooding systems discharge halon into an enclosure and flood the entire volume of the enclosure with halon at a specified percentage. Total flooding Halon 1301 systems were designed to protect ordinary combustibles, for data-processing equipment, control rooms, computer rooms, electrical hazards, and hazards involving gaseous or liquid flammable materials.

Halon 1301 System Decommissioning

Halon 1301 systems are decommissioned in accordance with *The Safety Guide for Decommissioning of Halon Systems*, published by the Environmental Protection Agency (EPA). The EPA defines **decommissioning** as "removing a halon system from service," and is performed in two steps:

1. *System decommissioning.* Disconnection and removal of halon cylinders from the system manifolds and distribution hardware by a qualified fire suppression systems service company. Once the halon cylinders are safely removed from the site, the piping and nozzles may be removed and an alternative fire suppression system installed. Steps in the system decommissioning are as follows:

   a. Secure cylinders.

   b. Disable actuation devices.

   c. Install anti-recoil devices.

   d. Pack cylinders for shipment.

   e. Receive shipped cylinders.

2. *Reclaiming of halon agent from the storage cylinders.* Usually performed offsite, and must be done carefully and professionally by properly trained personnel to ensure that all halon is reclaimed and that none is permitted to discharge into the atmosphere.

Improper decommissioning can result in releasing the agent and also in serious injury to decommissioning personnel, as pressurized halon cylinders can inadvertently discharge violently and become projectiles if emptied improperly. Injury and death can result. The EPA has established strict guidelines for training of decommissioning personnel. A number of recorded injuries emphasize the need for extreme caution throughout the decommissioning process.

CLEAN AGENTS

NFPA 2001 defines a **clean agent** as an electrically nonconducting, volatile, or gaseous fire extinguishant that does not leave a residue upon evaporation. For an agent to qualify as a clean agent, it must have no known effect on the ozone layer, it must have no effect on human survival within an enclosure protected by a clean agent, and in normally occupied areas must be used in a concentration that is less than the **NOAEL**—an abbreviation for "no observed adverse effect level." NOAEL is a measure of clean agent toxicity to humans, under test conditions.

At the present time, no **drop-in agent** is available that would allow Halon 1301 to be removed and an equivalent amount of replacement agent inserted. Systems with gaseous halon replacement agents require that more gas than halon

decommissioning
removing a halon system from service

clean agent
an electrically nonconducting, volatile, or gaseous fire extinguishant that does not leave a residue upon evaporation

NOAEL
abbreviation for "no observed adverse effect level," a measure of clean agent toxicity to humans, under test conditions

drop-in agent
an agent that allows Halon 1301 to be removed and an equivalent amount of replacement agent inserted

Table 8-1 *NOAEL and LOAEL percentages for halocarbon clean agents.*

Agent	NOAEL (%)	LOAEL (%)
FC-3-1-10	40	>40
FK-5-1-12	10.0	>10.0
HCFC Blend A	10.0	>10.0
HCFC-124	1.0	2.5
HFC-125	7.5	10.0
HFC-227ea	9.0	>10.5
HFC-23	30	>50
HFC-236fa	10	15

Source: NFPA 2001 (2004), Table 1.5.1.2.1(a). Reprinted with permission from NFPA 2001, *Clean Agent Fire Extinguishing Systems,* Copyright © 2004, National Fire Protection Association, Quincy, MA 02169. This reprinted material is not the complete and official position of the National Fire Protection Association on the referenced subject which is represented only by the standard in its entirety.

be stored on a volumetric basis, with differing devices and appurtenances required.

Clean agents have been found to be effective for electrical or electronic applications, telecommunication facilities, flammable liquids and gases, and high-value assets. They also may be considered for explosion suppression systems, as discussed in Chapter 7.

NOAEL and **LOAEL**—lowest observable adverse effect level—percentages for halocarbon suppression agents are shown on **Table 8-1**.

Clean Agent Classification

The two basic classifications of clean agents are halocarbon agents and inert gas agents. Agents addressed by NFPA 2001 are listed on **Table 8-2**.

Halocarbon agents consist of hydrofluorocarbons (HFCs), hydrochlorofluorocarbons (HCFCs), and perfluorocarbons (PFCs) and are given numerical descriptions as shown in Table 8-2 in accordance with ANSI (American National Standards Institute) and ASHRAE (American Society of Heating, Refrigerating and Air-Conditioning Engineers) standards. Halocarbons are stored as a liquid and distributed to the hazard as a gas, and extinguish fires by chemical and physical mechanisms, as opposed to oxygen deprivation. The extinguishment mechanism is breaking the combustion chain, as discussed in the fundamentals unit 1.

Inert gas agents contain one or more non-reactive gases, such as helium, neon, and argon, mixed with nitrogen or carbon dioxide. They extinguish fires by lowering the oxygen concentration within a room from the normal or ambient condition of 21% to a level below 15%—usually 12% to 13%, less than the level required to

LOAEL
abbreviation for lowest observable adverse effect level

halocarbon agents
hydrofluorocarbons (HFCs), hydrochlorofluorocarbons (HCFCs), and perfluorocarbons (PFCs)

inert gas agents
agents that contain one or more non-reactive gases, such as helium, neon, argon, nitrogen, and carbon dioxide

Table 8-2   *Agents addressed in NFPA 2001.*

FC-3-1-10	Perfluorobutane	C_4F_{10}
FK-5-1-12	Dodecafluoro-2-methylpentan-3-one	$CF_2CF_2C(O)CF(CF_3)_2$
HCFC Blend A	Dichlorotrifluoroethane HCFC-123 (4.75%)	$CHCl_2CF_3$
	Chlorodifluoromethane HCFC-22 (82%)	$CHClF_2$
	Chlorotetrafluoroethane HCFC-124 (9.5%)	$CHClFCF_3$
	Isopropenyl-1-methylcyclohexene (3.75%)	
HCFC-124	Chlorotetrafluoroethane	$CHClFCF_3$
HFC-125	Pentafluoroethane	CHF_2CF_3
HFC-227ea	Heptafluoropropane	CF_3CHFCF_3
HFC-23	Trifluoromethane	CHF_3
HFC-236fa	Hexafluoropropane	$CF_3CH_2CF_3$
FIC-13I1	Trifluoroiodide	CF_3I
IG-01	Argon	Ar
IG-100	Nitrogen	N_2
IG-541	Nitrogen (52%)	N_2
	Argon (40%)	Ar
	Carbon dioxide (8%)	CO_2
IG-55	Nitrogen (50%)	N_2
	Argon (50%)	Ar

Notes:

1. Other agents could become available at later dates. They could be added via the NFPA process in future editions or amendments of the standard.

2. Composition of inert gas agents are given in percent by volume. Composition of HCFC Blend A is given in percent by weight.

3. The full analogous ASHRAE nomenclature for FK-5-1-12 is FK-5-1-12mmy2.

Source: NFPA 2001 (2004), Table 1.4.1.2. Reprinted with permission from NFPA 2001, *Clean Agent Fire Extinguishing Systems,* Copyright © 2004, National Fire Protection Association, Quincy, MA 02169. This reprinted material is not the complete and official position of the National Fire Protection Association on the referenced subject which is represented only by the standard in its entirety.

sustain combustion for most combustibles. Inert gas agents are approximately the same density as air and, therefore, mix better and display less settling to the floor than other gaseous agents such as carbon dioxide.

Personnel Concerns with Clean Agents

NFPA 2001 does not recommend exposure to halocarbon clean agents for more than 5 minutes with less exposure in higher concentrations, as shown on **Table 8-3**.

Designers of fire protection systems must exercise care in the design of clean agent systems for enclosures where human exposure to the agent is possible.

Table 8-3   *Time for safe egress for halocarbon clean agents.*

Time for Safe Human Exposure at Stated Concentrations for HFC-125

HFC-125 Concentration		Maximum Permitted Human Exposure Time (minutes)
% v/v	ppm	
7.5	75,000	5.00
8.0	80,000	5.00
8.5	85,000	5.00
9.0	90,000	5.00
9.5	95,000	5.00
10.0	100,000	5.00
10.5	105,000	5.00
11.0	110,000	5.00
11.5	115,000	5.00
12.0	120,000	1.67
12.5	125,000	0.59
13.0	130,000	0.54
13.5	135,000	0.49

Notes:
1. Data derived from the EPA-approved and peer-reviewed physiologically based pharmacokinetic (PBPK) model or its equivalent.
2. Based on LOAEL of 10.0 percent in dogs.

Time for Safe Human Exposure at Stated Concentrations for HFC-236fa

HFC-236fa Concentration		Maximum Permitted Human Exposure Time (minutes)
% v/v	ppm	
10.0	100,000	5.00
10.5	105,000	5.00
11.0	110,000	5.00
11.5	115,000	5.00
12.0	120,000	5.00
12.5	125,000	5.00
13.0	130,000	1.65
13.5	135,000	0.92
14.0	140,000	0.79
14.5	145,000	0.64
15.0	150,000	0.49

Notes:
1. Data derived from the EPA-approved and peer-reviewed PBPK model or its equivalent.
2. Based on LOAEL of 15.0 percent in dogs.

Time for Safe Human Exposure at Stated Concentrations for HFC-227ea

HFC-227ea Concentration		Maximum Permitted Human Exposure Time (minutes)
% v/v	ppm	
9.0	90,000	5.00
9.5	95,000	5.00
10.0	100,000	5.00
10.5	105,000	5.00
11.0	110,000	1.13
11.5	115,000	0.60
12.0	120,000	0.49

Notes:
1. Data derived from the EPA-approved and peer-reviewed PBPK model or its equivalent.
2. Based on LOAEL of 10.5 percent in dogs.

Time for Safe Human Exposure at Stated Concentrations for FIC-13I1

FIC-13I1 Concentration		Maximum Permitted Human Exposure Time (minutes)
% v/v	ppm	
0.20	2000	5.00
0.25	2500	5.00
0.30	3000	5.00
0.35	3500	4.30
0.40	4000	0.85
0.45	4500	0.49
0.50	5000	0.35

Notes:
1. Data derived from the EPA-approved and peer-reviewed PBPK model or its equivalent.
2. Based on LOAEL of 0.4 percent in dogs.

Table 8-4   *Toxicity information for halocarbon clean agents.*

Agent	LC_{50} or ALC (%)	NOAEL (%)	LOAEL (%)
FC-3-1-10	>80	40	>40
FIC-13I1	>12.8	0.2	0.4
FK-5-1-12	>10.0	10	>10.0
HCFC Blend A	64	10	>10.0
HCFC-124	23–29	1	2.5
HFC-125	>70	7.5	10
HFC-227ea	>80	9	10.5
HFC-23	>65	50	>50
HFC-236fa	>18.9	10	15

Notes:

1. LC_{50} is the concentration lethal to 50 percent of a rat population during a 4-hour exposure. The ALC is the approximate lethal concentration.
2. The cardiac sensitization levels are based on the observance or nonobservance of serious heart arrhythmias in a dog. The usual protocol is a 5-minute exposure followed by a challenge with epinephrine.
3. High concentration values are determined with the addition of oxygen to prevent asphyxiation.

Source: NFPA 2001 (2004), Table A.1.5.1.2(a). Reprinted with permission from NFPA 2001, *Clean Agent Fire Extinguishing Systems,* Copyright © 2004, National Fire Protection Association, Quincy, MA 02169. This reprinted material is not the complete and official position of the National Fire Protection Association on the referenced subject which is represented only by the standard in its entirety.

Of particular concern is human exposure to the decomposition byproducts formed by breakdown of the extinguishant when exposed to high temperatures or an open flame. For example, halocarbon agents containing fluorine have the potential to form toxic hydrogen fluoride. Inert gas agents do not create decomposition products, but care must be taken to avoid high application concentrations. **Table 8-4** lists the toxicity of clean agents. Inert gas agents contain about 8% carbon dioxide, but the CO_2 is not a concern at normal inert gas concentrations. Care must be taken to avoid over-design which could result in excessive inert gas concentrations and reduce oxygen concentrations below 10%.

NFPA 2001 prohibits the application of halocarbon agents into occupied rooms at concentrations greater than 24% and requires that the NOAEL limits listed in Table 8-1 not be exceeded for any clean agent.

Table 8-4 provides information for designers relative to NOAEL and LOAEL percentages for halocarbon clear agents and time for safe exposure for HFC-125, HFC-236fa, HFC-277ea, and FIC-1311. These tables enable designers of clean agent systems to consider methodologies for keeping human exposure to clean agents to a minimum. Methods to protect personnel exposed to clean agents include

- ensuring that exits are well situated, well marked, and well lighted, are of adequate number and width to allow rapid egress of all occupants, and are readily accessible with clear and unobstructed aisles or passageways to the exits

- considering the provision of extra egress doors
- specifying that doors are required to swing in the direction of egress travel and to reclose automatically
- providing adequate alarm notification before clean agent discharge
- providing training of personnel to ensure proper identification and response to an alarm
- providing continuous alarms during discharge and agent containment
- providing alarms, locks, signs, and other methods to prevent reentry to the room during agent containment
- specifying placement of breathing apparatus
- specifying room ventilation
- provision of abort switches
- specifying a plan for rescue of anyone who may become trapped within the room or otherwise overexposed to the suppressing agent and its combustion byproducts

Consideration also must be given to the possibility of confusion or disorientation of occupants during discharge. Clean agent discharge may be noisy, and the force of discharge may create reduced visibility, may produce a swirl of dislodged papers or other loose materials, and the low temperature of discharging gas may be a shock to personnel. Training of responsible personnel is a necessity, and the specification of clean agent systems should not be considered for "at risk" persons, such as in public or patient areas in hospitals and nursing homes.

Clean Agent System Design

Because local application had not been found effective by the committee responsible for the technical content of NFPA 2001, clean agent systems are to be specified and designed for total flooding of enclosures. The enclosure protected by a clean agent must be rendered amenable to the application and retention of agent by

- arranging for the automatic closing of doors
- sealing openings and cracks around doors and windows
- clipping down and restraining ceiling tiles and sealing them where necessary
- shutting down supply and return air to the room with dampers in the ducts to prevent loss of clean agent
- attempting to limit loss of clean agent through floor drains, trenches, pipe penetrations through walls, and other wall and floor penetrations
- shutting down gas or other flammable supplies
- shutting down electrical power to energized electrical components where necessary

Systems must be designed such that the agent containers are not in the hazard area, and are in a protected location as close as possible to the hazard. Piping and fittings must be of a pressure rating commensurate with expected system pressures, and must be corrosion-resistant. Piping and fittings must be metallic, and fittings cannot be cast iron. They can be welded, brazed, or malleable iron. Detection shall be selected to be appropriate for the anticipated fire, as discussed in Chapters 11 through 15. An existing detection system possibly may be reused when designing a clean agent system for a room currently protected by halon, provided that the characteristics of the anticipated fire have not changed.

Determining Halocarbon Total Flooding Quantity

Inerting
reducing the flammable concentration in an atmosphere to below one-half of its lower flammable limit

Clean agent systems are designed to extinguish fires either by flame extinguishment or by inerting. **Inerting** is intended to reduce the flammable concentration in an atmosphere to below one-half its lower flammable limit. Flame extinguishment is designed to cease combustion of a combustible solid or a flammable or combustible liquid.

Halogenated clean agents are required to possess the properties listed in **Table 8-5**, and halogenated clean agent systems are required to be designed to operate within the working pressure shown on **Table 8-6**.

Halocarbon agent total flooding quantity, assuming normal leakage from a tight enclosure, is calculated by using the same formula used for halon:

$$W = \frac{(V) \times (C) \times (A)}{(s) \times (100 - C)}$$

$$s = (k_1) + (k_2 \times T)$$

where W = weight of halocarbon clean agent (lbs) (kg)
V = net volume of protected enclosure (ft^3) (m^3)
S = specific volume (ft^3/lb) (m^3/kg)
A = altitude correction factor, per **Table 8-7**

Table 8-5 *Halogenated agent quality requirements.*

Property	Specification
Agent purity, mole %, minimum	99.0
Acidity, ppm (by weight HCl equivalent), maximum	3.0
Water content, % by weight, maximum	0.001
Nonvolatile residues, g/100 ml maximum	0.05

Source: NFPA 2001 (2004), Table 4.1.2(a). Reprinted with permission from NFPA 2001, *Clean Agent Fire Extinguishing Systems*, National Fire Protection Association, Quincy, MA 02169. This reprinted material is not the complete and official position of the National Fire Protection Association on the referenced subject which is represented only by the standard in its entirety.

Table 8-6 *Minimum design working pressure for halocarbon clean agent system piping.*

Agent	Agent Container Maximum Fill Density (lb/ft³)	Agent Container Charging Pressure at 70°F (21°C) (psig)	Agent Container Pressure at 130°F (55°C) (psig)	Minimum Piping Design Pressure at 70°F (21°C) (psig)
HFC-227ea	75	150	249	200
	72	360	520	416
	72	600	1025	820
FC-3-1-10	80	360	450	360
HCFC Blend A	56.2	600	850	680
	56.2	360	540	432
HFC 23	48	608.9*	1713	1371
	45	608.9*	1560	1248
	40	608.9*	1382	1106
	35	608.9*	1258	1007
	30	608.9*	1158	927
HCFC-124	74	240	354	283
HCFC-124	74	360	580	464
HFC-125	54	360	615	492
HFC 125	56	600	1045	836
HFC-236fa	74	240	360	280
HFC-236fa	75	360	600	480
HFC-236fa	74	600	1100	880
FK-5-1-12	90	360*	413	360

*Not superpressurized with nitrogen.

Source: NFPA 2001 (2004), Table 4.2.1.1(b). Reprinted with permission from NFPA 2001, *Clean Agent Fire Extinguishing Systems,* Copyright © 2004, National Fire Protection Association, Quincy, MA 02169. This reprinted material is not the complete and official position of the National Fire Protection Association on the referenced subject which is represented only by the standard in its entirety.

Table 8-7 *Atmospheric correction factors.*

Equivalent Altitude		Enclosure Pressure		Atmospheric Correction Factor
ft	km	psia	mm Hg	
−3,000	−0.92	16.25	840	1.11
−2,000	−0.61	15.71	812	1.07
−1,000	−0.30	15.23	787	1.04
0	0.00	14.70	760	1.00
1,000	0.30	14.18	733	0.96
2,000	0.61	13.64	705	0.93
3,000	0.91	13.12	678	0.89
4,000	1.22	12.58	650	0.86
5,000	1.52	12.04	622	0.82
6,000	1.83	11.53	596	0.78
7,000	2.13	11.03	570	0.75
8,000	2.45	10.64	550	0.72
9,000	2.74	10.22	528	0.69
10,000	3.05	9.77	505	0.66

Source: NFPA 2001 (2004), Table 5.5.3.3. Reprinted with permission from NFPA 2001, *Clean Agent Fire Extinguishing Systems,* Copyright © 2004, National Fire Protection Association, Quincy, MA 02169. This reprinted material is not the complete and official position of the National Fire Protection Association on the referenced subject which is represented only by the standard in its entirety.

C = halocarbon clean agent design concentration that represents percentage of clean agent per volume. For example, if the halocarbon clean agent concentration is 6%, $C = 6$, not 0.06, and if the concentration is 6.5%, C = 6.5, not 0.065.

k_1 and k_2 = constants that relate to the specific volume of halocarbon agent used; these constants are listed in Table 8-8.

Alternatively, the required agent quantity can be can be determined using the flooding factors found in Annex A of NFPA 2001. For each agent, the flooding factor multiplied by the room volume gives the agent quantity, which is multiplied by the altitude correction factor.

EXAMPLE 8-1

Calculation of HFC-227ea Halocarbon Clean Agent Quantity

A 10-ft. wide, 20-ft. long, 10-ft. high room with an ambient temperature of 70°F is protected by a halocarbon clean agent, HFC-227ea, at a 6% design concentration. Determine the weight of HFC-227ea required to protect the room, assuming an elevation at sea level.

Solution

$$s = (k_1) + (k_2 \times T)$$

From **Table 8-8**, $k_1 = 1.885$ and $k_2 = 0.0046$ for HFC-227ea and T is given as 70°F.

$$s = k_1 + (k_2 \times T)$$

$$s = 1.885 + (0.0046 \times 70)$$

$$= 2.207$$

$$V = (10 \text{ ft.}) \times (20 \text{ ft.}) \times (10 \text{ ft.}) = 2000 \text{ ft.}^3$$

$$C = 6 \text{ (given)}$$

A = sea level, or 0 feet elevation, and the correction factor is therefore 1, per Table 8-7.

$$W = \frac{(V) \times (C) \times (A)}{s \times (100 - C)}$$

$$= \frac{(2000) \times (6) \times (1)}{(2.207) \times (100 - 6)}$$

$$= 57.8 \text{ lb}$$

An alternative method of determining total flooding quantity is to use the tables contained in the annex of NFPA 2001. The table for HFC-227ea is included as Table 8-8. The weight requirement corresponding to 70°F and 6% is shown as 0.0289. Multiplying this factor times the room volume gives:

$$W = \left(0.0289 \frac{\text{lb}}{\text{ft.}^3}\right) \times (2000 \text{ ft.}^3)$$

$$W = 57.8 \text{ lb}$$

Note that for this example the results are identical whether using the formula or the table. NFPA 2001 includes a table for each clean agent recognized by the standard. The value 0.0289 lb/ft.[3] is a flooding factor, representing the quantity of halocarbon clean agent required to achieve a selected design concentration (6%) at a specified temperature (70°F).

Table 8-8 *HFC-227ea total flooding quantity (English units*).*

Temp. t (°F)[c]	Specific Vapor Volume s (ft³/lb)[d]	Weight Requirements of Hazard Volume, W/V (lb/ft³)[b] Design Concentration (% by Volume)[e]									
		6	7	8	9	10	11	12	13	14	15
10	1.9264	0.0331	0.0391	0.0451	0.0513	0.057	0.0642	0.0708	0.0776	0.0845	0.0916
20	1.9736	0.0323	0.0381	0.0441	0.0501	0.0563	0.0626	0.0691	0.0757	0.0825	0.0894
30	2.0210	0.0316	0.0372	0.0430	0.0489	0.0550	0.0612	0.0675	0.0739	0.0805	0.0873
40	2.0678	0.0309	0.0364	0.0421	0.0478	0.0537	0.0598	0.0659	0.0723	0.0787	0.0853
50	2.1146	0.0302	0.0356	0.0411	0.0468	0.0525	0.0584	0.0645	0.0707	0.0770	0.0835
60	2.1612	0.0295	0.0348	0.0402	0.0458	0.0514	0.0572	0.0631	0.0691	0.0753	0.0817
70	2.2075	0.0289	0.0341	0.0394	0.0448	0.0503	0.0560	0.0618	0.0677	0.0737	0.0799
80	2.2538	0.0283	0.0334	0.0386	0.0439	0.0493	0.0548	0.0605	0.0663	0.0722	0.0783
90	2.2994	0.0278	0.0327	0.0378	0.0430	0.0483	0.0538	0.0593	0.0650	0.0708	0.0767
100	2.3452	0.0272	0.0321	0.0371	0.0422	0.0474	0.0527	0.0581	0.0637	0.0694	0.0752
110	2.3912	0.0267	0.0315	0.0364	0.0414	0.0465	0.0517	0.0570	0.0625	0.0681	0.0738
120	2.4366	0.0262	0.0309	0.0357	0.0406	0.0456	0.0507	0.0560	0.0613	0.0668	0.0724
130	2.4820	0.0257	0.0303	0.0350	0.0398	0.0448	0.0498	0.0549	0.0602	0.0656	0.0711
140	2.5272	0.0253	0.0298	0.0344	0.0391	0.0440	0.0489	0.0540	0.0591	0.0644	0.0698
150	2.5727	0.0248	0.0293	0.0338	0.0384	0.0432	0.0480	0.0530	0.0581	0.0633	0.0686
160	2.6171	0.0244	0.0288	0.0332	0.0378	0.0425	0.0472	0.0521	0.0571	0.0622	0.0674
170	2.6624	0.0240	0.0283	0.0327	0.0371	0.0417	0.0464	0.0512	0.0561	0.0611	0.0663
180	2.7071	0.0236	0.0278	0.0321	0.0365	0.0410	0.0457	0.0504	0.0552	0.0601	0.0652
190	2.7518	0.0232	0.0274	0.0316	0.0359	0.0404	0.0449	0.0496	0.0543	0.0592	0.0641
200	2.7954	0.0228	0.0269	0.0311	0.0354	0.0397	0.0442	0.0488	0.0535	0.0582	0.0631

[a] The manufacturer's listing specifies the temperature range for operation.
[b] W/V [agent weight requirements (lb/ft³)] = pounds of agent required per cubic foot of protected volume to produce indicated concentration at temperature specified.

$$W = \frac{V}{s}\left(\frac{C}{100-C}\right)$$

[c] t [temperature (°F)] = the design temperature in the hazard area.
[d] s [specific volume (ft³/lb)] = specific volume of superheated HFC-227ea vapor can be approximated by the formula:

$$s = 1.885 + 0.0046t$$

where t = temperature (°F).
[e] C [concentration (%)] = volumetric concentration of HFC-227ea in air at the temperature indicated.

*For metric units, please refer to NFPA 2001. For other halocarbon clean agents, refer to NFPA 2001.

Source: NFPA 2001 (2004), Table A.5.5.1(k). Reprinted with permission from NFPA 2001, *Clean Agent Fire Extinguishing Systems*, Copyright © 2004, National Fire Protection Association, Quincy, MA 02169. This reprinted material is not the complete and official position of the National Fire Protection Association on the referenced subject which is represented only by the standard in its entirety.

Inert Gas Clean Agent Total Flooding Quantity

Inert gas agents are required to be used at the minimum working pressures shown in **Table 8-9** and must possess the quality shown in **Table 8-10**.

Inert gas quantity is based on finding the volume of gas needed, as opposed to finding the weight, as we did with halocarbons. The formula for determining

Table 8-9 *Minimum design working pressure for inert gas clean agent system piping.*

Agent	Agent Container Pressure at 70°F (21°C)		Agent Container Pressure at 130°F (55°C)		Minimum Design Pressure at 70°F (21°C) of Piping Upstream of Pressure Reducer	
	psig	kPa	psig	kPa	psig	kPa
IG-01	2,370	16,341	2,650	18,271	2,370	16,341
	2,964	20,436	3,304	22,781	2,964	20,436
IG-541	2,175	14,997	2,575	17,755	2,175	14,997
	2,900	19,996	3,433	23,671	2,900	19,996
IG-55	2,222	15,320	2,475	17,065	2,222	15,320
	2,962	20,423	3,300	22,753	2,962	20,423
	4,443	30,634	4,950	34,130	4,443	30,634
IG-100	2,404	16,575	2,799	19,299	2,404	16,575
	3,236	22,312	3,773	26,015	3,236	22,312
	4,061	28,000	4,754	32,778	4,061	28,000

Source: NFPA 2001 (2004), Table 4.2.1.1(a). Reprinted with permission from NFPA 2001, *Clean Agent Fire Extinguishing Systems,* Copyright © 2004, National Fire Protection Association, Quincy, MA 02169. This reprinted material is not the complete and official position of the National Fire Protection Association on the referenced subject which is represented only by the standard in its entirety.

Table 8-10 *Inert gas agent quality requirements.*

		IG-01	IG-100	IG-541	IG-55
Composition, % by volume	N_2		Minimum 99.9%	52% ± 4%	50% ± 5%
	Ar	Minimum 99.9%		40% ± 4%	50% ± 5%
	CO_2			8% + 1% − 0.0%	
Water content, % by weight		Maximum 0.005%	Maximum 0.005%	Maximum 0.005%	Maximum 0.005%

Source: NFPA 2001 (2004), Table 4.1.2(b). Reprinted with permission from NFPA 2001, *Clean Agent Fire Extinguishing Systems,* Copyright © 2004, National Fire Protection Association, Quincy, MA 02169. This reprinted material is not the complete and official position of the National Fire Protection Association on the referenced subject which is represented only by the standard in its entirety.

the volume of gas required is:

$$X = (2.303) \times (V) \times \left(\frac{V_s}{s}\right) \times \log_{10}\left(\frac{100}{100 - C}\right) \times (A)$$

where $s = (k_1) + (k_2 \times T)$
V = net volume of protected enclosure (ft.^{3}) (m^3)
s = specific volume of inert gas (ft.^{3}/lb) (m^3/kg) at specified temperature
C = inert gas clean agent design concentration (volume percent)
V_s = specific volume of inert gas at 70°F (21°C)
A = altitude correction factor, per Table 8-7.
$k_1 + k_2$ = constants that relate to the specific volume of inert gas clean agent used, as listed on Table 8-10. Note that Table 8-10 provides constants for IG-541 only, and that NFPA 2001 should be consulted for other inert gas agents. Also note that Table 8-10 provides English units only and NFPA 2001 should be consulted for the corresponding metric units.
X = volume of inert gas added at standard conditions per volume of hazard space

EXAMPLE 8-2

Calculation of IG-541 Inert Gas Clean Agent Quantity

A room at sea level with an ambient temperature of 70°F, 20 ft. wide by 50 ft. long by 8 ft. high, is protected by inert gas clean agent IG-541, with a concentration of 34%. Determine the volume of IG-541 required to protect the room. V_s for IG-541 at 70°F is 11.358.

Solution

$$s = (k_1) + (k_2 \times T)$$

From **Table 8-11**, $k_1 = 9.8579$ and $k_2 = 0.02143$ for IG-541

$$s = (9.8579) + (0.02143 \times 70)$$
$$= 11.358$$
$$V = (20') \times (50') \times (8')$$
$$= 8000 \text{ ft.}^3$$
$$C = 34 \text{ (given)}$$
$$V_s = 11.358 \text{ ft.}^3/\text{lb per NFPA 2001 (given)}$$

A = sea level, or 0 feet elevation, and the correction factor is therefore 1, per Table 8-7.

$$W = (2.303) \times (8000) \times \left(\frac{11.358}{11.358}\right) \times \log_{10}\left(\frac{100}{100 - 34}\right) \times (1)$$
$$= (2.303) \times (8000) \times (1) \times (0.1804) \times (1)$$
$$= (18424) \times (1) \times (0.1804) \times (1)$$
$$= 3323.689 \text{ ft.}^3$$

In a manner analogous to the tabular solution presented in Example 8-1, designers can use Table 8-11 to obtain a solution. Note that the flooding factor corresponding to 70°F and 34% concentration is $\left(0.416 \dfrac{\text{ft.}^3}{\text{ft.}^3}\right) \times (8000 \text{ ft.}^3) = 3328 \text{ ft.}^3$ of IG-541, which is slightly more than was obtained using the calculation method, which could be attributed to rounding of logarithmic functions.

Table 8-11 *IG-541 Total flooding quantity (English units*).*

| Temp. *t* (°F)[c] | Specific Vapor Volume *s* (ft³/lb)[d] | Volume Requirements of Agent per Unit Volume of Hazard, $V_{agent}/V_{enclosure}$[b] | | | | | | | |
| | | Design Concentration (% by Volume)[e] | | | | | | | |
		34	38	42	46	50	54	58	62
−40	9.001	0.524	0.603	0.686	0.802	0.873	0.977	1.096	1.218
−30	9.215	0.513	0.590	0.672	0.760	0.855	0.958	1.070	1.194
−20	9.429	0.501	0.576	0.657	0.743	0.836	0.936	1.046	1.166
−10	9.644	0.490	0.563	0.642	0.726	0.817	0.915	1.022	1.140
0	9.858	0.479	0.551	0.628	0.710	0.799	0.895	1.000	1.116
10	10.072	0.469	0.539	0.615	0.695	0.782	0.876	0.979	1.092
20	10.286	0.459	0.528	0.602	0.681	0.766	0.858	0.958	1.069
30	10.501	0.450	0.517	0.590	0.667	0.750	0.840	0.939	1.047
40	10.715	0.441	0.507	0.578	0.653	0.735	0.824	0.920	1.026
50	10.929	0.432	0.497	0.566	0.641	0.721	0.807	0.902	1.006
60	11.144	0.424	0.487	0.555	0.628	0.707	0.792	0.885	0.987
70	11.358	0.416	0.478	0.545	0.616	0.693	0.777	0.868	0.968
80	11.572	0.408	0.469	0.535	0.605	0.681	0.762	0.852	0.950
90	11.787	0.401	0.461	0.525	0.594	0.668	0.749	0.836	0.933
100	12.001	0.393	0.453	0.516	0.583	0.656	0.735	0.821	0.916
110	12.215	0.386	0.445	0.507	0.573	0.645	0.722	0.807	0.900
120	12.429	0.380	0.437	0.498	0.563	0.634	0.710	0.793	0.884
130	12.644	0.373	0.430	0.489	0.554	0.623	0.698	0.779	0.869
140	12.858	0.367	0.422	0.481	0.544	0.612	0.686	0.766	0.855
150	13.072	0.361	0.415	0.473	0.535	0.602	0.675	0.754	0.841
160	13.287	0.355	0.409	0.466	0.527	0.593	0.664	0.742	0.827
170	13.501	0.350	0.402	0.458	0.518	0.583	0.653	0.730	0.814
180	13.715	0.344	0.396	0.451	0.510	0.574	0.643	0.718	0.801
190	13.930	0.339	0.390	0.444	0.502	0.565	0.633	0.707	0.789
200	14.144	0.334	0.384	0.437	0.495	0.557	0.624	0.697	0.777

[a] The manufacturer's listing specifies the temperature range for operation.

[b] *X* [agent volume requirements (lb/ft³)] = volume of agent required per cubic foot of protected volume to produce indicated concentration at temperature specified.

$$X = 2.303 \times \left(\frac{V_S}{s}\right) \times Log_{10}\left(\frac{100}{100-C}\right) = \left(\frac{V_S}{s}\right) \times \ln\left(\frac{100}{100-C}\right)$$

[c] *t* [temperature (°F)] = the design temperature in the hazard area.

[d] *s* [specific volume (ft³/lb)] = specific volume of superheated IG-541 vapor can be approximated by the formula:

$$s = 9.8579 + 0.02143t$$

where *t* = temperature (°F).

[e] *C* [concentration (%)] = volumetric concentration of IG-541 in air at the temperature indicated.

Note: V_s = the term $X = \ln[100/(100-C)]$ gives the volume at a rated concentration (%) and temperature to reach an air–agent mixture at the end of flooding time in a volume of 1 ft³.

*For metric units, please refer to NFPA 2001.

Source: NFPA 2001 (2004), Table A.5.5.2(e) (For other inert gas agents, refer to NFPA 2001). Reprinted with permission from NFPA 2001, *Clean Agent Fire Extinguishing Systems*, Copyright © 2004, National Fire Protection Association, Quincy, MA 02169. This reprinted material is not the complete and official position of the National Fire Protection Association on the referenced subject which is represented only by the standard in its entirety.

Halocarbon and Inert Gas Discharge Time

Halocarbon clean agents must be discharged within 10 sec. Inert gas agents that do not create decomposition products may be discharged within 1 minute. The room must hold the gas for a time sufficient to extinguish a deep-seated fire without reignition.

Clean Agent Storage Arrangement

A clean agent storage arrangement is shown on **Figure 8-3**. A clean agent nozzle and a releasing system detector is shown on **Figure 8-4**. Clean agent discharge is shown on **Figure 8-5**.

Pressure Relief Venting for Clean Agent Systems

NFPA 2001 requires that where clean agent valving arrangements on the pilot piping or on the discharge piping create closed piping arrangements where pressure could increase beyond the pressure rating of the piping, fittings, and nozzles, pressure relief devices are to be installed. The pressure relief devices are

Figure 8-3 *Clean agent storage cylinder arrangement. (Courtesy of Ansul, Inc.)*

Figure 8-4 *Clean agent nozzle and detector for clean agent releasing system. (Courtesy of Ansul, Inc.)*

Figure 8-5 *Clean agent nozzle discharge. (Courtesy of Ansul, Inc.)*

required to discharge in such a manner as not to be hazardous to personnel. The NFPA 2001 Annex describes pressure relief isometric diagrams for clean agent cylinders, showing pressure compatibilities for a variety of clean agent storage conditions. The Annex further recommends that pressure relief venting for closed piping sections follow the *FSSA Pipe Design Handbook*.

OTHER HALON REPLACEMENT OPTIONS

Three water-based options for Halon replacement are:
1. Water mist system (discussed in Chapter 6)
2. Double-interlocked preaction automatic sprinkler systems
3. Standard automatic sprinkler systems

Sprinkler Systems as a Halon Replacement

Some system designers are considering a standard sprinkler system as a replacement for a halon system. Sprinkler systems present no personnel safety concerns, nor do they have any effect on the ozone layer. The only concern in specifying a well-designed sprinkler system is the possibility of property damage where electronic devices or irreplaceable objects are protected.

NFPA

NFPA 13, Standard for the Installation of Sprinkler Systems

These concerns may be addressed by installing a double-interlocked preaction system in accordance with NFPA 13, *Standard for the Installation of Sprinkler Systems*. A double-interlocked preaction system is filled with air instead of water, which requires a detector as well as a sprinkler to actuate before water flows to the actuated sprinkler. For additional information on preaction sprinkler systems, please refer to *Design of Water-Based Fire Protection Systems*, also published by Delmar Learning.

SUMMARY

The design of new halon systems has been essentially halted as the result of cessation of production of halon in accordance with the Montreal Protocol, which prohibits the manufacture of halogenated agents in countries participating in the agreement. Although pure halon in concentrations between 5% and 10% is considered nontoxic to humans during brief exposure, the products of decomposition can be dangerous if breathed.

Clean agent systems may be considered as halon system replacements when designed in accordance with NFPA 2001. Clean agents include halocarbon and inert gas agents that are in conformance with NOAEL and EPA guidelines. Halocarbon agents develop products of decomposition that may be harmful to personnel.

REVIEW QUESTIONS

1. Why is halon no longer manufactured in the United States? What problem does this pose for owners of halon systems, and what ethical dilemma does it pose for designers of fire protection systems?

2. Why is halon considered a poor local application system?

3. List systems that may be considered as replacements for a halon system, and give advantages, disadvantages, personnel concerns, and design guidelines for each. State the NFPA standard that applies to each system.

4. What criteria are used for classifying an agent as a clean agent?

5. Determine the weight of HFC-227ea required to protect an enclosure 20 ft. wide, 40 ft. long, and 8 ft. high, at an ambient temperature of 60°F and an elevation of 6000 ft., using a 9% design concentration.

6. Determine the volume of IG-541 required to protect an enclosure 12 ft. wide, 12 ft. long, and 10 ft. high, at an ambient temperature of 75°F and an elevation of −3000 ft., using a 40% design concentration.

DISCUSSION QUESTIONS

1. Discuss the potential harm to personnel exposed to halon in a fire.

2. List and discuss safety features intended to protect personnel from accidental or purposeful discharge.

ACTIVITIES

1. Conduct research related to the first Montreal Protocol agreement, and provide details on how the agreement has been modified since 1987.

2. Visit a company that maintains or inspects halon systems, and determine how halon systems are maintained. What will happen if halon accidentally discharges or leaks? Explain.

3. Conduct research on halon safety. Has anyone been killed or seriously injured in a fire involving a halon system? Are there any recorded cases of persons being injured by breathing products of halon decontamination?

4. Interview a fire service officer or an official responsible for inspecting and approving fire protection systems. Do any local or state regulations deal with the removal of existing halon systems?

NOTE

1. http://www.theozonehole.com/montreal.htm

Chapter 9

CARBON DIOXIDE SYSTEM DESIGN

Learning Objectives

Upon completion of this chapter, you should be able to:

- Demonstrate understanding of the carbon dioxide phase diagram.
- Explain why storing carbon dioxide in its liquid form is desirable.
- Describe two methods for maintaining carbon dioxide in its liquid form, using the carbon dioxide phase diagram as a basis.
- List potential uses for a carbon dioxide fire protection system.
- Detail the limitations and personnel concerns that must be considered when specifying or designing a carbon dioxide system.
- Compare and contrast the types of carbon dioxide systems.
- Calculate the carbon dioxide required for a rate-by-volume or rate-by-area local application fire protection system.
- Calculate the carbon dioxide required for a total flooding application fire protection system.

NFPA

NFPA 12, Standard on Carbon Dioxide Extinguishing Systems

NFPA 12, *Standard on Carbon Dioxide Extinguishing Systems*, is used for design of carbon dioxide systems. Although this chapter provides the minimum information needed to perform some basic carbon dioxide system design and calculation, it is recommended that NFPA 12 be obtained and used in conjunction with this chapter when designing or specifying carbon dioxide systems.

CARBON DIOXIDE

carbon dioxide
a gaseous fire protection agent, also known by its chemical designation CO_2

Carbon dioxide is a gaseous fire protection agent, also known by its chemical designation CO_2, one molecule of carbon and two molecules of oxygen. Normally, the air we breathe contains 21% oxygen, 79% nitrogen, and only a trace amount of carbon dioxide, 0.03%. The presence of significantly higher percentages of carbon dioxide in a room cannot be detected by human senses because it is colorless and odorless.

These properties of carbon dioxide, combined with its property of being incapable of conducting or transmitting electrical charges provided that a proper clearance is maintained between system components and live uninsulated electrical components, make it useful for protecting many items of equipment. These properties also create a scenario for a potential hazard to human life when high volumes of carbon dioxide are injected into a room.

Carbon dioxide is 1.5 times heavier than air, so it forces oxygen out of a room or significantly reduces the concentration of oxygen at breathing level. The primary mode of extinguishment of a carbon dioxide system, therefore, is to reduce the concentration of oxygen in a room to below the concentration required for sustained combustion.

When applied in high concentrations, carbon dioxide can be dangerous to personnel, as described later in this chapter. NFPA 12 requires that signs be provided in every protected space, as shown in **Figure 9-1**, to warn personnel of the potential for impending carbon dioxide discharge.

Carbon Dioxide Phase Diagram

phase diagram
a graph that represents the physical state of a specific substance at varying pressures and temperatures

Gases can be represented pictorially by a **phase diagram**, a graph of the physical state of a specific substance at varying pressures and temperatures. We know, for instance, that water (H_2O) exists at normal atmospheric pressure as a solid below 32°F (0°C), as a liquid between 32°F (0°C) and 212°F (100°C), and as a vapor above 212°F (100°C). Carbon dioxide also can exist in three physical states, or phases, as shown in **Figure 9-2**. Figure 9-2 depicts the three physical states as the liquid region, solid region, and vapor region, depending on the pressure and temperature at which carbon dioxide is maintained.

1. *Solid phase.* The solid region of carbon dioxide is shown on the left side of Figure 9-2. This region exists at temperatures less than −70°F (−56.6°C) and at pressures higher than the dark-lined curve shown. As an example,

Figure 9-1 *A warning sign—placed in every protected space; similar signs are required at every entrance to a protected space, at every entrance to a protected space for systems provided with a wintergreen odorizer, in every nearby space where carbon dioxide could accumulate to hazardous levels, and outside each entrance to carbon dioxide storage rooms.*

Source: NFPA 12 (2005), Figure 4.3.2.3.1. Reprinted with permission from NFPA 12, *Carbon Dioxide Extinguishing Systems*, Copyright © 2005, National Fire Protection Association, Quincy, MA 02169. This reprinted material is not the complete and official position of the National Fire Protection Association on the referenced subject which is represented only by the standard in its entirety.

carbon dioxide stored at −100°F (−73.33°C) and 40 lb/sq. in. absolute (or psia) (275.79 kPa) would exist in its solid state, but if the pressure were to drop to atmospheric pressure (14.7 psi or 101.283 kPa), it would transition to a vapor. Carbon dioxide in its solid state, also known as *dry ice*, exudes a spooky vapor that drops to the floor as its temperature begins to decrease, making it a popular Halloween special effect or science fair exhibit. The visual effect of dry ice exemplifies the vaporization of solid carbon dioxide as its temperature decreases.

2. *Vapor phase.* When the temperature exceeds −120°F and pressures are less than the dark-lined graph, carbon dioxide exists as a vapor. For instance, carbon dioxide stored at 0°F (−17.7°C) and 40 psia (275.79 kPa) exists in its pure vapor form. Vaporized carbon dioxide is used to add the fizzy bubbles to our soft drinks and carbonated beverages. Note that at **atmospheric pressure**, the sea-level pressure at which we consume carbonated beverages, represented by the dashed line shown at 14.7 psi (101.283 kPa), carbon dioxide exists in its vapor phase at all temperatures above about −110°F (−78.8°C). To illustrate atmospheric pressure, consider a 1-inch square area of the top of your hand, with the weight of the

atmospheric pressure
sea-level pressure, 14.7 psia (101.283 kPa) or one "G"

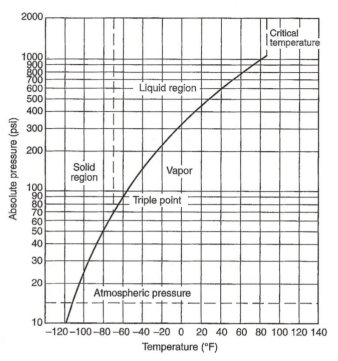

Figure 9-2 *Carbon dioxide phase diagram, variation of pressure of carbon dioxide with change in temperature (constant volume).*

For SI units, 1 psi = 6.89 kPa; °C = 5⁄9 (°F − 32).

earth's atmosphere (14.7 psia or 1 G) pressing down on it, then consider the increased pressure as you accelerate in a car or an elevator, as pressure in excess of one "G" is applied to that one-square inch area. As can be seen in Figure 9-2, at room temperature (70°F) (21.1°C) and atmospheric pressure, special steps are required to cool or pressurize carbon dioxide if we want it to exist in a phase other than its vaporous state.

3. *Liquid phase.* Between −70°F and 87.8°F (−56.6 and 31°C) and at pressures higher than the dark-lined graph, carbon dioxide exists in its liquid state. Storing carbon dioxide for fire protection systems in its liquid state is desirable because the container volume required to store it as a vapor is

extremely large and inefficient, and because carbon dioxide in its solid state involves a significant time delay for a change of phase before it can be injected into a room for fire protection. To store carbon dioxide in its liquid state, we have to maintain the liquid at the temperatures and pressures shown in Figure 9-2.

Carbon Dioxide Triple Point and Critical Temperature

triple point

a point where carbon dioxide exists in all three states simultaneously

critical temperature

a temperature beyond which carbon dioxide can exist only in its vapor phase

As with other gases, carbon dioxide has a point, called its **triple point**, where it exists in all three phases simultaneously. The triple point for carbon dioxide is 75.35 psia at $-69.88°F$ (519.52 kPa at $-56.6°C$). The reader should be able to identify this point on Figure 9-2. The **critical temperature** for carbon dioxide is 87.8°F (31°C), a temperature above which carbon dioxide can exist only in its vapor phase.

CARBON DIOXIDE STORAGE

To maintain carbon dioxide in its liquid state, it must be kept in pressurized containers at a controlled temperature. The two distinct types of containers for carbon dioxide fire protection systems are high-pressure cylinders and low-pressure containers.

High-Pressure Cylinders

High-pressure cylinders, shown in **Figures 9-3** and **9-4**, contain liquid carbon dioxide at an ambient room temperature of 70°F (21.1°C), and at a pressure above 850 psi (5860.5 kPa). The reader is urged to verify that this point falls within the liquid phase in Figure 9-2. Although the range of room storage temperatures is permitted to vary between 32°F and 120°F (0°C and 48.88°C), special arrangements, including automated pressure adjustment, must be performed to ensure that the carbon dioxide is maintained in its liquid state in accordance with the phase diagram. Electrical supervision of the room temperature in which the cylinders are stored is also needed to ensure proper storage of the carbon dioxide in its liquid phase.

High-pressure cylinders can be as small as 5 pounds (2.267 kg) liquid capacity or up to 120 pounds (54.43 kg) liquid capacity. This range of weights is intended to ensure that liquid-filled cylinders can be removed readily for testing and refilling. A relief valve must be installed to maintain pressures below the failure pressure of the container. Relief valves usually are set at about 2500 to 3000 psi (17236.89 to 20684.27 kPa).

The amount of carbon dioxide required to protect a space is determined in accordance with the calculation method outlined in NFPA 12 and as illustrated in this chapter, and the number of cylinders is determined by dividing the maximum cylinder capacity into the total carbon dioxide demand calculated.

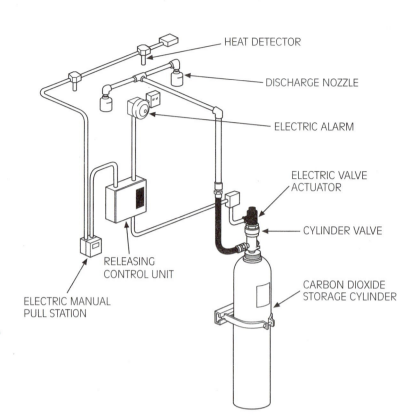

HEAT DETECTOR

DISCHARGE NOZZLE

ELECTRIC ALARM

ELECTRIC VALVE ACTUATOR

CYLINDER VALVE

CARBON DIOXIDE STORAGE CYLINDER

RELEASING CONTROL UNIT

ELECTRIC MANUAL PULL STATION

Figure 9-3 *Carbon dioxide high-pressure storage cylinders. (Courtesy of Ansul, Inc.)*

Figure 9-4 *High-pressure carbon dioxide cylinder installation arrangement. (Courtesy of Ansul, Inc.)*

NFPA 12 requires that a reserve supply with a weight equal to the main supply of carbon dioxide be installed adjacent to the main supply calculated.

To minimize the probability for overpressurization, the filling density of a cylinder usually is held to no more than 68% of its water-holding capacity. A carbon dioxide cylinder listed as a 100-pound (45.359 kg) cylinder holds 100 pounds (45.359 kg) of carbon dioxide, with additional volume to account for a filling density of 68%.

EXAMPLE 9-1

Calculating Carbon Dioxide Cylinder Quantity

Determine the number of high-pressure carbon dioxide cylinders required for a total system demand of 1600 lbs (725.74 kg) of carbon dioxide.

Solution

Using 100-lb cylinders, the minimum number of cylinders required would be 16 for the main supply, with 16 cylinders required to contain the reserve supply, for a total of 32 cylinders. Space must be allocated for a minimum of 32 high-pressure cylinders in a room with an ambient temperature controlled at 70°F at atmospheric pressure. If 120-lb cylinders are used, 14 main cylinders and 14 reserve cylinders are required.

Low-Pressure Storage Containers

A low-pressure storage container, as shown in **Figures 9-5** and **9-6**, contains an insulated refrigeration unit capable of maintaining the carbon dioxide in its liquid state at 0°F and at a low pressure of above 300 psi. Electrical supervision of the pressure and temperature is required to ensure proper storage of the carbon dioxide in its liquid state.

Low-pressure containers are used predominately for hazards where the calculated carbon dioxide storage requirement is large and when the number of high-pressure cylinders and the space required to house the cylinders becomes excessive. Low-pressure containers can be ordered in sizes as small as 500 pounds of storage capacity, and can be obtained in sizes containing tens of thousands of pounds of liquid carbon dioxide. Low-pressure storage containers can be installed within a building, space permitting, or outside a building, either at ground level or on the roof.

Determination of High Pressure Versus Low Pressure

A rule of thumb that can be applied for selecting the storage arrangement for carbon dioxide systems is to use high-pressure cylinders for hazards requiring up to 4000 pounds (1814.369 kg) of carbon dioxide, interior cylinder storage space permitting, and to use low-pressure systems for hazards requiring more than

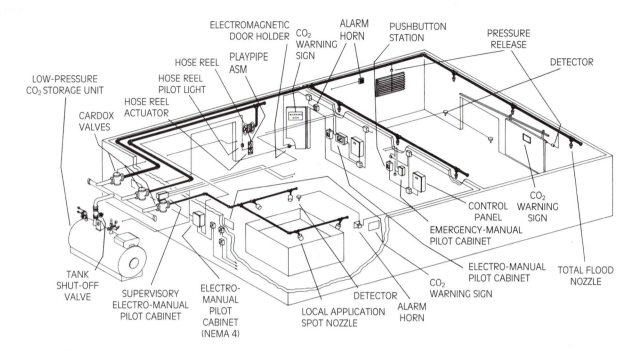

Figure 9-5 *Carbon dioxide low-pressure system arrangement. (Courtesy of Chemetron, Inc.)*

Figure 9-6 *Low-pressure carbon dioxide storage container. (Courtesy of Ansul, Inc.)*

4000 pounds (1814.369 kg) of carbon dioxide. A comparative cost analysis, combined with an evaluation of available interior storage space, is required to determine the optimal selection of a system.

USES FOR CARBON DIOXIDE SYSTEMS

Carbon dioxide is an effective extinguishant for

- ordinary combustibles—Class A commodities
- flammable liquids—Class B commodities
- electrical hazards—Class C commodities

Class A Commodities

In some isolated circumstances, a carbon dioxide system may be appropriate for ordinary combustibles, but in most cases a standard sprinkler system is more effective and less hazardous to personnel for this application. Clean-agent suppression systems should be considered in areas such as record storage rooms containing rare or irreplaceable documents, to avoid the potential for water damage. Normally-unoccupied vessels such as dust collectors, grain elevators, and food-milling machines may be amenable to protection by a carbon dioxide system.

Class B Commodities

Some flammable liquid hazards are typically protected by carbon dioxide systems. Examples of these applications include:

- open-top lube oil pits—open pits for storage of lube oil for lubrication of power plant turbine generators
- industrial fryers, range hoods, and grease ducts
- diptanks—containing flammable liquids for plating or stripping of metal components
- marine engine protection
- solvent storage rooms
- printing presses and folders—flammable ink, lubricating oil, and paper dust; carbon dioxide is a common application for this hazard
- paint spray booths and paint storage rooms
- chemical storage labs
- flammable liquid storage
- test cells for aircraft engine testing

Class C Commodities

A carbon dioxide system may be ideal for protecting normally-unoccupied electrical hazards because carbon dioxide does not conduct electrical charges. The piping system and the nozzles must be grounded to avoid the transfer of electrostatic charges resulting from frictional resistance at the nozzles, and electrical components protected by the system should be automatically shut down by the carbon dioxide fire alarm control and releasing unit. Electrical applications include:

- electronic data processing—computer room or protection of subfloor below computer room. This hazard is predominately protected by standard automatic sprinkler systems or preaction systems.
- Internet switching facilities—subfloor or internal protection for each unit
- turbine generator enclosures
- electric switchgear rooms
- battery storage rooms
- electrical cabinet interiors
- electric transformer vaults
- wave solder machines—enclosures where tin cans are soldered shut or where electronic components are manufactured

CARBON DIOXIDE SYSTEM LIMITATIONS

Carbon dioxide systems are not to be used for materials containing their own oxygen supply, for hazards involving reactive metals such as magnesium, and for metal hydrides. Carbon dioxide systems also should not be considered for enclosures where personnel may be present, or for enclosures where a person might become trapped or where extensive time is required to exit the room, as detailed next. Even in normally-unoccupied spaces, carbon dioxide may not be a good choice if inspection or maintenance personnel may be present intermittently.

Personnel Hazards Related to Carbon Dioxide

Carbon dioxide systems should be considered only for enclosures or volumes where personnel are not present. CO_2 has an immediate disabling effect at 34% concentration and is lethal in a short amount of time at this concentration. CO_2 paralyzes the respiratory system at elevated injection percentages. The injection of CO_2 at high concentrations reduces the oxygen level in a room, but the CO_2, not the low oxygen level, is the primary danger. **Table 9-1** shows the acute health effects of high concentrations of CO_2.

Table 9-1 *Acute health effects of high concentrations of carbon dioxide.*

Concentration of Carbon Dioxide in Air (%)	Time	Effects
2	Several hours	Headache, dyspnea upon mild exertion
3	1 hour	Dilation of cerebral blood vessels, increased pulmonary ventilation, and increased oxygen delivery to the tissues
4–5	Within a few minutes	Mild headache, sweating, and dyspnea at rest
6	1–2 minutes	Hearing and visual disturbances
	<16 minutes Several hours	Headache and dyspnea Tremors
7–10	Few minutes	Unconsciousness or near unconsciousness
	1.5 minutes– 1 hour	Headache, increased heart rate, shortness of breath, dizziness, sweating, rapid breathing
10–15	1+ minute	Dizziness, drowsiness, severe muscle twitching, and unconsciousness
17–30	<1 minute	Loss of controlled and purposeful activity, unconsciousness, convulsions, coma, and death

Source: EPA 430-R-00-002, "Carbon Dioxide as a Fire Suppressant: Examining the Risks," February 2000.

A designer of a fire protection system for an enclosure must be vigilant in protecting personnel within an enclosure and personnel in adjacent spaces, because carbon dioxide could leak from the protected room and migrate to other areas. Even in normally-unoccupied spaces, a maintenance person could be in a low-lying enclosure adjacent or below the hazard area at the time of system discharge, and even under ideal conditions with trained personnel, death or injury could result. In the room where the carbon dioxide cylinders are stored, some gas may be released through the relief valves, producing a dangerous atmosphere. Because carbon dioxide is 1.5 times heavier than air, the gas could settle into spaces below the protected area and endanger personnel. A designer must specify room containment methods or specify and protect a route expected to be followed by migrating carbon dioxide.

The Environmental Protection Agency (EPA) issued a report in 2000, entitled "Carbon Dioxide as a Fire Suppressant: Examining the Risks," available for download on epa.gov/ozone/snap/fire/co2/co2report.pdf. It evaluates CO_2 as an extinguishing agent and as a halon replacement and provides advice relative to the hazards of CO_2. A number of accidental deaths involving CO_2 have occurred during maintenance of carbon dioxide systems.

Some actions that may be taken to protect personnel, prevent entry or reentry to the hazard area, and enhance rescue for an enclosure protected by carbon dioxide include the following:

- *Continuous predischarge alarms.* In advance of any automatic carbon dioxide discharge, a distinctive alarm is required to be sounded to warn occupants in the space. A voice alarm that gives instructions to personnel is an excellent idea and may be more effective than a horn or bell, which may be ignored or misinterpreted.

- *Breathing apparatus.* Placing such apparatus in an enclosure is no guarantee that personnel will know how to find it, use it, or be capable of accessing it in a timely fashion. In some extreme cases trained personnel were found dead within feet of such apparatus.

- *Voice alarm systems.* Instructions given on a voice alarm system and a flashing sign next to breathing apparatus may increase the probability that the apparatus will be located and used. NFPA 12 refers to breathing apparatus in an enclosure as equipment for rescue personnel, but a designer should make every effort to ensure that trapped personnel have access to the equipment in extreme cases.

- *Exits.* If an enclosure is being considered for a carbon dioxide system, all options for rapid egress from the space must be considered. As an example, a maintenance person may be required to enter a cable spread room—the room directly below the control room in a power plant, containing a maze of cable trays, ducts, and other ceiling-mounted equipment. The time to egress remote portions of these confined ceiling spaces could greatly exceed 10 minutes. An analysis of the maximum amount of time to egress a space is fundamental to the design of a carbon dioxide system. If the egress time can be reduced by the addition of more exit doors than normally would be necessary, the extra doors should be installed, with locks that prohibit reentry by unauthorized personnel. If the egress time cannot be reduced to complete the egress before the initiation of carbon dioxide, a carbon dioxide system should not be considered.

- *Signs.* Warning signs, required by NFPA 12, can be an effective means of transmitting information. Signs within the room can warn an occupant to egress a room upon hearing an alarm. Such a sign is most effective if it is in the language spoken by the occupant, is seen, is read, and is understood, and if the alarm is taken seriously and not considered as a false alarm. A sign outside a space could warn people from entering the space, but some may be unaware of a carbon dioxide discharge or may knowingly risk danger to search for persons who may have become trapped in the room.

  Rooms adjacent to the room of discharge may develop lower oxygen concentrations resulting from carbon dioxide leaking from the room of discharge. Signs may be effective in warning personnel in such adjacent spaces. Installing signs may make the designer feel better, but this is not by itself an effective means of informing personnel. Extra measures are needed to ensure that personnel are aware of the sign, understand what it means, and are trained rigorously to know what to do to comply.

- *Training procedures.* Personnel training is required for rooms protected by a carbon dioxide system, in accordance with NFPA 12. The training must be regular and continuous, because personnel in or near a given space change with time. Any person entering or occupying a space must be trained to react properly, but a trained person may react inappropriately or forget any training that was conducted.

- *Time delay.* Under no circumstance should carbon dioxide discharge commence immediately or without warning into a space. Even in a normally unoccupied space, the discharge must be delayed to allow any transient personnel that may be present to exit before the initiation of discharge. The designer must recognize that during this time delay the fire is growing and spreading. If personnel are endangered by a rapidly spreading fire during the predischarge time delay, or if the fire may grow to an extent that a carbon dioxide system is ineffective at the time of discharge, another type of system should be considered.

- *Manual activation.* An alternative to an automatic system is one that requires human intervention to release gas. Although manual activation may alleviate concerns about protecting personnel, it does not eliminate them. A system can be activated manually while a person is trapped in a room unknowingly. Also the manual activation may not be performed, or it could be delayed unnecessarily because of lack of training or fear of liability resulting from a person being injured or killed by the discharge. Designers considering manual activation of a carbon dioxide system because of personnel concerns should consider another type of system.

abort switch
a type of switch that cancels discharge

- *Manual override.* NFPA 12 does not permit **abort switches**—switches that cancel discharge. Some manufacturers may be able to supply a "dead man's abort"—a switch that delays discharge as long as the switch is held down continuously.

- *Scented gas.* Because carbon dioxide is colorless and odorless, a scent, such as wintergreen, could be added to the gas upon discharge. Warning signs would warn occupants who detect this scent, and advise them to exit.

TYPES OF CARBON DIOXIDE SYSTEMS

Four types of carbon dioxide systems are recognized by NFPA 12:

1. Total flooding carbon dioxide systems
2. Local application carbon dioxide systems
3. Hand hose line carbon dioxide systems
4. Standpipe systems with mobile supply

Total Flooding Carbon Dioxide Systems

A total flooding system, shown on **Figure 9-7**, is designed to completely fill a volume with a predetermined percentage of gas discharged by a fixed piping system and supply. For an enclosure to qualify for carbon dioxide protection, it

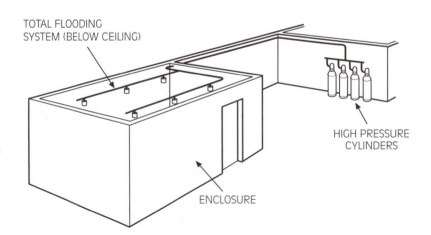

Figure 9-7 *A high-pressure total flooding system. (Courtesy of Ansul, Inc.)*

must have walls, a floor, and a ceiling that are reasonably tight to contain the gas until the fire is out, and openings must either be capable of being automatically closed or small enough for the effective injection of extra gas as a compensatory measure.

Enclosures can be made amenable for holding carbon dioxide by the following means:

- *Sealing wall joints.* Applying plaster or putty to joints that may leak gas could reduce the leakage rate.

- *Restraining ceiling tiles.* Suspended ceiling drop-in tiles may become disengaged by the considerable force of a carbon dioxide system discharge. Clips that retain the tiles in their runners will minimize this, and sealing the spaces between the tiles and the tile supports may be necessary.

- *Door and window closers.* Electromagnetic releases can hold a door open in advance of activation. An electric signal emitted by a fire alarm control unit upon detection of a fire reverses the polarity of the release and allows the door or window to be closed.

- *Duct dampers.* Ventilation and return ducts are shut down in advance of discharge, but these ducts will fill with gas without the installation of duct dampers, which are louvers that are actuated electrically to close off the duct upon activation of the system.

- *Sealing floor openings.* Sealing floor openings such as pipe cable penetration is important because the weight of carbon dioxide causes the gas to collect at the floor and seek an exit.

- *Wall reinforcement.* Carbon dioxide exerts significant pressure on the lower elevations of walls in tight rooms where high concentrations of carbon dioxide are injected. Even cinder block walls have been damaged by pressure exerted by carbon dioxide discharges at high concentrations. If the wall cannot be reinforced sufficiently, another type of system should be considered.

- *Fire spread prevention.* In cases where fire could spread through unclosable openings, extra discharge nozzles at those openings should be installed.

surface fire
a fire on the exterior of an object

deep-seated fire
a fire within an object or enclosure

Total flooding systems protect enclosures where either surface fires or deep-seated fires are present A **surface fire** is a fire on the exterior surface of an object, exemplified by a burning flammable liquid, gas, or exposed solid. A **deep-seated fire** occurs within an object or enclosure, such as a fire deep in a haystack or bag of fertilizer, or dry electrical hazards within a tightly sealed mechanical enclosure, where the presence of flame is preceded by a significant period of smoldering. Measures should be taken to make tightly sealed enclosures amenable to the application of carbon dioxide, either by providing louvers or other avenues for the gas to enter, or by installing carbon dioxide discharge nozzles within a mechanical or electrical enclosure.

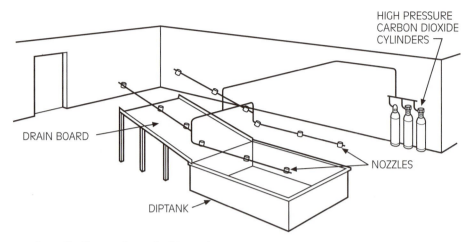

HIGH PRESSURE
CARBON DIOXIDE
CYLINDERS

DRAIN BOARD

NOZZLES

DIPTANK

Figure 9-8 *A high-pressure local application system. (Courtesy of Ansul, Inc.)*

Local Application Carbon Dioxide Systems

In spaces that are too large for a total flooding system, or where a total flooding system could endanger personnel, or in enclosures incapable of holding gas, a local application system, shown in **Figure 9-8**, can be considered. A local application system is a fixed system of piping, nozzles, and carbon dioxide storage that provides a discharge of carbon dioxide directly onto an object where combustion is likely. Local application systems are designed for surface fires only. Designers should be cautioned that local application systems could create concentrations of carbon dioxide that may fall within the critical ranges shown on Table 9-1.

A local application system can be effective if the object of protection is discrete and identifiable, if the combustible is compatible for extinguishment by a carbon dioxide system, and if the hazard is separated from other combustibles by a distance that would prevent ignition of adjacent hazards. A local application system can be designed for numerous hazards, with discharge being provided for only one hazard undergoing combustion, provided that acceptable separations exist between the hazards. Selector valves can permit the application of carbon dioxide onto each object individually, as shown on Figure 9-5.

Local application systems are ordinarily installed indoors, but are permitted to be installed outdoors, provided that adequate shielding is installed to avoid disruption of the discharge by wind.

A local application low-pressure carbon dioxide system protecting a printing press is shown on **Figure 9-9**. The printing presses are protected predominately with local application carbon dioxide systems. A high-pressure local application carbon dioxide system protecting a diptank is shown on **Figure 9-10**.

Hand Hose Line Systems

Carbon dioxide hosereels may be installed in the vicinity of a perceived hazard, either as a supplement to a total flooding or local application system, or as a

Figure 9-9 *Low-pressure carbon dioxide protecting a printing press using the local application method. (Courtesy of Ansul, Inc.)*

Figure 9-10 *A high-pressure carbon dioxide system protecting a diptank surrounded by a dike, using the local application method. (Courtesy of Ansul, Inc.)*

stand-alone system. The presence of a hand hose line system assumes that trained personnel are present at all times, are equipped with the necessary life-safety equipment, and are willing and capable of using the hand hose line system to approach and extinguish a fire.

If the hazard area is unoccupied or is occupied sporadically, a hand hose line system may be of little value. If the area normally is occupied by transient personnel, the system may never be used. If untrained personnel are likely to use the hose apparatus, extra carbon dioxide is needed to account for the lack of extinguishment efficiency and additional personnel protective measures may be necessary. The number of hosereels expected to be discharged simultaneously must be considered conservatively in determining the amount of carbon dioxide required per discharge incident, and this value must be added to any total flooding or local application quantities that are likely to occur simultaneously with any hose line discharge.

Standpipe Systems and Mobile Supply

In a hazard serviced by highly trained, immediately available fire service personnel having the proper equipment, it may be permissible to install a fixed total flooding, local application, or hand hose line system without a permanent high-pressure or low-pressure supply. An example of such a situation is a chemical plant with a full-time fire service brigade located close to the hazard area.

Hose couplings are provided at locations convenient for attachment to a mobile carbon dioxide tank or truck. When designing such systems, the speed of response, attachment of mobile supply, and initiation of discharge must be considered. Hazards involving rapidly spreading fires, hazards that are occupied sporadically or are normally not occupied by trained personnel, or hazards where the response time is too slow, are not compatible with a mobile supply system.

LOCAL APPLICATION CARBON DIOXIDE SYSTEM DESIGN PROCEDURE

Two basic design methods exist for local application design—the rate-by-volume method and the rate-by-area method. Each method applies carbon dioxide directly on an object without the intent of filling a volume with carbon dioxide.

rate-by-volume method
a method of local application of carbon dioxide where an imaginary volume larger than the hazard is created to account for the dissipation and loss of carbon dioxide during discharge

RATE-BY-VOLUME CARBON DIOXIDE LOCAL APPLICATION: DESIGN PROCEDURE

A three-dimensional object with irregularly dimensioned sides, such as a printing press, may be a candidate for local application of carbon dioxide using the **rate-by-volume method**. In performing this calculation, an imaginary volume

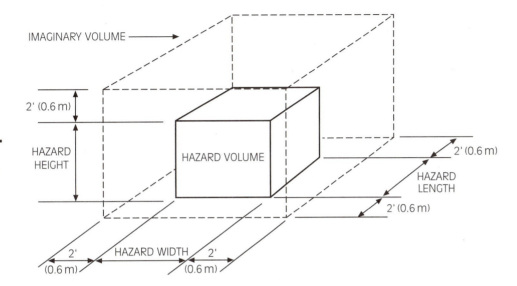

IMAGINARY VOLUME

2' (0.6 m)

HAZARD HEIGHT

HAZARD VOLUME

2' (0.6 m)

HAZARD LENGTH

2' (0.6 m)

2' (0.6 m)

HAZARD WIDTH   2' (0.6 m)

Figure 9-11   *Local application rate-by-volume method; assuming that the hazard is mounted to a solid floor, no distance is added below the hazard height.*

larger than the hazard is created to account for the dissipation and loss of carbon dioxide during discharge. **Figure 9-11** illustrates this imaginary volume.

The imaginary volume assumes that the object is flat on the floor and that the floor is solid. If a hazard is raised above a solid floor, the imaginary volume must extend 2 ft. (0.6 m) below the hazard, or to the solid floor if the hazard is raised to an elevation less than 2 ft. (0.6 m) above the floor. The imaginary volume is determined by adding a total of 4 ft. (1.2 m) to the width of the hazard, by adding 4 ft. (1.2 m) to the length of the hazard, and adding 2 ft. (0.6 m) to the height of the hazard, assuming that the bottom of the hazard is mounted to a solid floor.

Local Application Imaginary Volume Calculation—Mounted to Solid Floor

- The design volume of an object mounted to a solid floor is the product of the length, width, and height of the imaginary volume, as shown in Figure 9-11.

$V_{\text{imaginary}} = (\text{length} + 4\text{ ft.}) \times (\text{width} + 4\text{ ft.}) \times (\text{height} + 2\text{ ft.})$
$V_{\text{imaginary}} = (\text{length} + 1.2\text{ m}) \times (\text{width} + 1.2\text{ m}) \times (\text{height} + 0.6\text{ m})$

- No deduction is permitted for any solid objects within the imaginary volume.

Local Application Imaginary Volume Calculation—Raised 2 Feet (0.6 m) Above Solid Floor

- If a hazard is raised 2 ft. (0.6 m) or more above a floor or is mounted to a floor that is not solid, the imaginary volume is:

$V_{\text{imaginary}} = (\text{length} + 4\text{ ft.}) \times (\text{width} + 4\text{ ft.}) \times (\text{height} + 4\text{ ft.})$
$V_{\text{imaginary}} = (\text{length} + 1.2\text{ m}) \times (\text{width} + 1.2\text{ m}) \times (\text{height} + 1.2\text{ m})$

- No deduction is permitted for any solid objects within the imaginary volume.

Local Application Imaginary Volume Calculation—Raised Less Than 2 Feet Above Solid Floor

- If a hazard is raised less than 2 ft. (0.6 m) above a solid floor, the imaginary volume is:

$$V_{imaginary} = (length + 4\ ft.) \times (width + 4\ ft.) \times [(height + 2\ ft.) \\ + (Distance\ from\ floor\ to\ bottom\ of\ hazard)]$$

$$V_{imaginary} = (length + 1.2\ m) \times (width + 1.2\ m) \times [(height + 0.6\ m) \\ + (Distance\ from\ floor\ to\ bottom\ of\ hazard)]$$

- No deduction is permitted for any solid objects within the imaginary volume.

Determination of Local Application Rate-By-Area Carbon Dioxide Quantity—Walls Remote From Hazard

To determine the minimum rate of carbon dioxide required, multiply the imaginary volume by a factor of 1 lb/min/ft³ (16 kg/min·m³). NFPA 12 requires that high-pressure local application systems have quantities increased by 40%.

Low-pressure systems:
$$R = (V_{imaginary}) \times (1\ lb/min/ft.^3)$$
$$R = (V_{imaginary}) \times (16\ kg/min \cdot m^3)$$

High-pressure systems:
$$R = (V_{imaginary}) \times (1\ lb/min/ft.^3) \times (1.4)$$
$$R = (V_{imaginary}) \times (16\ kg/min \cdot m^3) \times (1.4)$$

Determination of Local Application Rate-By-Area Carbon Dioxide Quantity—Walls Very Close to Hazard

If the hazard is mounted to a solid floor and is surrounded by walls located "very close" to the hazard, extending at least 2 ft. above the hazard, the minimum application rate is permitted to be 0.25 lb/min/ft³ (4 kg/min · m³), applied to the actual enclosed area. NFPA 12 requires that high-pressure local application systems have quantities increased by 40%. NFPA 12 does not define the term "very close."

Low-pressure systems:
$$R = (V_{enclosed}) \times (0.25\ lb/min/ft.^3)$$
$$R = (V_{enclosed}) \times (4\ kg/min \cdot m^3)$$

High-pressure systems:
$$R = (V_{enclosed}) \times (0.25\ lb/min/ft.^3) \times (1.4)$$
$$R = (V_{enclosed}) \times (4\ kg/min \cdot m^3) \times (1.4)$$

Although the 40% factor does not apply to low-pressure systems, most local application systems are of relatively low volumes and are protected predominately by high-pressure systems.

Determination of Local Application Carbon Dioxide Weight

The total weight of carbon dioxide required is determined by multiplying the rate (R) by the required duration (D) in minutes.

$$W = (R) \times (D)$$

The minimum acceptable duration is 30 sec.

EXAMPLE 9-2

Calculation of Local Application Rate-By-Volume Quantity

Use the rate-by-volume local application method to determine the carbon dioxide flow rate (R) for a small newspaper printing press in a very large room, mounted to a solid floor, protected by a high-pressure carbon dioxide system. The press is 4 ft. in width (W), 3 ft. in length (L), and 7 ft. in height (H) and is not located near any walls. Determine the total amount of carbon dioxide required (W) if the required duration (D) is 30 sec.

Solution

L is given as 3 ft., W is given as 4 ft., H is given as 7 ft., and D is given as ½ minute. The imaginary volume ($V_{imaginary}$) for an object mounted to a solid floor is computed as follows:

$$V_{imaginary} = (\text{length} + 4') \times (\text{width} + 4') \times (\text{height} + 2')$$
$$= (3' + 4') \times (4' + 4') \times (7 + 2')$$
$$= (7') \times (8') \times (9')$$
$$= 504 \text{ ft.}^3$$

The rate of discharge is computed as follows for a high-pressure local application system:

$$R = (V_{imaginary}) \times (1 \text{ lb/min/ft.}^3) \times (1.4)$$
$$= (504 \text{ ft.}^3) \times (1 \text{ lb/min/ft.}^3) \times (1.4)$$
$$= 504 \text{ lb/min} \times (1.4)$$
$$= 705.6 \text{ lb/min}$$

The total weight of liquid carbon dioxide required is computed as follows:

$$W = R \times D$$
$$= (705.6 \text{ lb/min}) \times (1/2 \text{ min})$$
$$= 352.8 \text{ pounds of carbon dioxide required}$$

RATE-BY-AREA LOCAL APPLICATION: DESIGN PROCEDURE

rate-by-area method
a method of applying carbon dioxide to a two-dimensional surface area based on the capability of listed nozzles to discharge a given amount of carbon dioxide over a fixed area of coverage

diptank
a vat used for dipping, coating, or stripping an object in a flammable liquid

drainboard
an object that collects flammable liquid residue that drips from the dipped item onto an inclined surface, allowing the flammable liquid residue to drain back to the diptank

The **rate-by-area method** is best for two-dimensional surface fires on flat horizontal surfaces, usually involving flammable liquids. This method is based on the capability of listed nozzles to discharge a prespecified, listed amount of carbon dioxide over a fixed area of coverage, as determined by the nozzle manufacturer and the testing agency.

The minimum allowable duration of discharge for rate-by volume local application is 30 sec, and for high-pressure systems, NFPA 12 requires that the carbon dioxide weight requirement be increased by 40%.

Rate-by-area local application carbon dioxide systems frequently are used on diptanks and drainboards, shown in Figure 9-8. The **diptank** is the tank at the bottom of the figure, and the drainboard is to the left of the diptank. Diptanks are used for dipping, coating, or stripping an object in a flammable liquid. A **drainboard** collects flammable liquid residue that drips from the dipped item onto an inclined surface, allowing the flammable liquid residue to drain back to the diptank. A robotic operation ordinarily dips an object in the diptank, then moves it over the drainboard slowly until the dripping stops, then transports the object to the next step in the fabrication or assembly process.

Nozzle coverage for the diptank and drainboard are determined by a square area of protection, meaning that if the width of a diptank is 5 ft., the nozzle is designed to cover an area of 5 ft. × 5 ft., or 25 sq.ft. The area covered by one nozzle varies in accordance with listed nozzle capabilities but often is within the 16- to 25-sq ft. per nozzle range. Area of coverage and flow rate determination differ for each nozzle manufacturer. **Figure 9-12** shows a graph that can be easily used to make quick determinations for estimating purposes.

The procedure for using the graph is as follows:

1. Determine the maximum width of coverage. Note that two scales exist for coated and liquid surfaces at the "y" axis of the graph at the left of the figure. The liquid surface graph is for the diptank, and the coated surface graph is for the drainboard. The nozzle is listed to cover a square area of protection.

2. For liquid surfaces (diptanks), select a width of coverage from the liquid surface graph, calibrated on the right side of the "y" axis, and extend a line horizontally until it intersects with one of the two nozzle graphs, either the 5½-in. nozzle, or the 4-in. nozzle, at the top center of the figure. For coated surfaces (drainboards), extend a line from the coated surface graph, horizontally until it intersects with one of the two nozzle graphs, either the 5½-in. nozzle, or the 4-in. nozzle, at the center of the graph. Note that by continuing this horizonal line to the vertical axis on the right, the square footage of nozzle coverage can

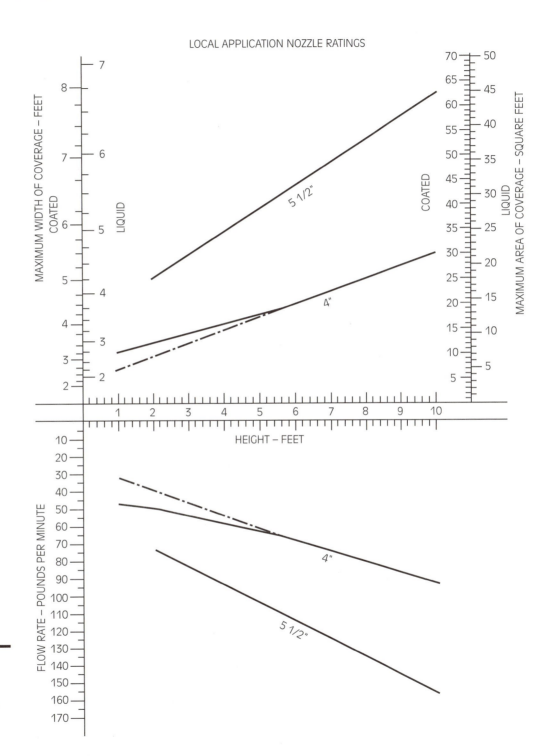

LOCAL APPLICATION NOZZLE RATINGS

Figure 9-12
*Nozzle flow rate
determination
(1 ft. = 0.3048 m).
(Courtesy of
Chemetron, Inc.)*

be read. This square footage is always a square area, such as 5′ × 5′, or 25 sq ft.

3. From the point of intersection on the nozzle graph, extend a line vertically to one of the nozzle graphs at the bottom of the page. A line extended from a 4-in. nozzle at the top of the page must be extended to a 4-in. nozzle at the bottom of the page, and likewise for the 5-in. nozzle.

4. From the point of intersection of the lower nozzle graph, extend a horizontal line to the flow rate graph on the bottom left of the page.

5. From the graph, read the flow rate per nozzle, in pounds per minute per nozzle.

The minimum flow rate of carbon dioxide required to protect the diptank (FR_{liquid}) is determined by multiplying the number of nozzles protecting the liquid surface (N_{liquid}) by the flow rate per nozzle, determined by Figure 9-12 for liquid surfaces (F_{liquid}).

$$FR_{liquid} = (N_{liquid}) \times (F_{liquid})$$

The minimum flow rate of carbon dioxide required for protection of the drainboard (FR_{coated}) is determined by multiplying the number of nozzles protecting the coated surface (N_{coated}) by the flow rate determined by Figure 9-12 for coated surfaces (F_{coated}).

$$FR_{coated} = (N_{coated}) \times (F_{coated})$$

The total carbon dioxide flow rate (FR) is determined by adding the flow rates for liquid (FR_{liquid}) and coated (FR_{coated}) surfaces. For high-pressure systems, this total is multiplied by a factor of 1.4 to account for the 40% extra required for high-pressure local application carbon dioxide systems in accordance with NFPA 12.

$$FR = FR_{liquid} + FR_{coated}$$

To determine the minimum weight of liquid carbon dioxide required, multiply the total flow rate (FR) by the duration (D). Local application high-pressure carbon dioxide systems require a 40% increase.

High-pressure systems:

$$W = (FR) \times (D) \times (1.4)$$

Low-pressure systems:

$$W = (FR) \times (D)$$

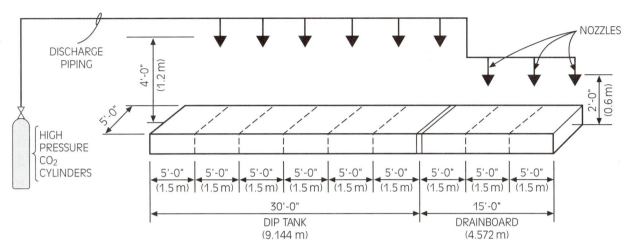

Figure 9-13   *Illustration of Example 9-3.*

EXAMPLE 9-3

Rate-By Area Local Application Carbon Dioxide Calculation

Use the rate-by-area method to estimate the minimum amount of carbon dioxide required for a high-pressure system protecting a diptank 30 ft. long and 5 ft. wide, and a drainboard 15 ft. long and 5 ft. wide. Treat the diptank as a liquid surface and the drainboard as a coated surface, and use a minimum discharge time of 30 seconds.

Solution

Figure 9-13 illustrates the problem. Because the diptank is 5 ft. wide and the drainboard is 5 ft. wide, the width of coverage for each hazard was selected to be 5 ft., for a total area of coverage of 25 ft.^{2} per nozzle. Six nozzles are therefore required to protect the diptank, and three are required for the drainboard. For the graph in Figure 9-13 to provide accurate results, the area that each nozzle covers must be perfectly square in this example, and the area that each nozzle covers is 5 ft. $\times$ 5 ft., or 25 sq.ft., as shown on Figure 9-13.

Figure 9-14 shows the graphic solution of the problem.

Width of coverage, diptank = 5 ft. (Length of nozzle coverage also must be
5 ft. and area of coverage must be 25 sq ft.)

Width of coverage, drainboard = 5 ft. (Length of nozzle coverage also must be 5 ft.)

From Figure 9-14

Nozzle height, diptank (liquid surface) = 4 ft. (5½-in. nozzle)

Nozzle height, drain board (coated surface) = 2 ft. (5½-in. nozzle)

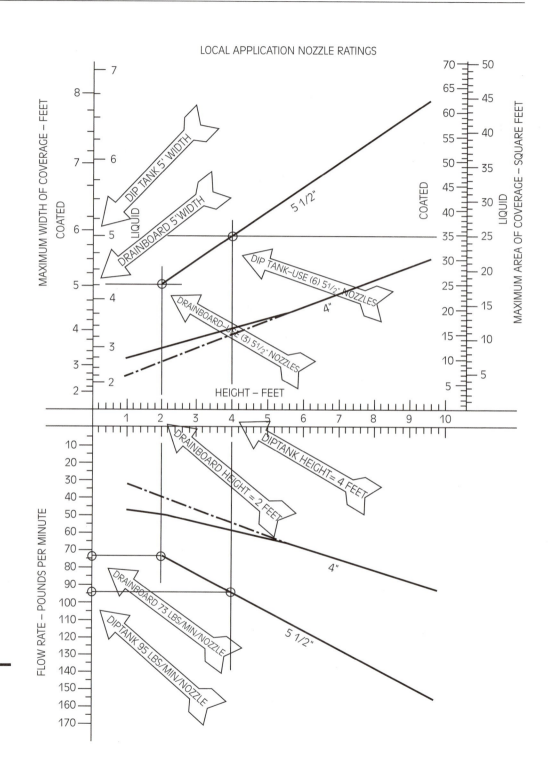

Figure 9-14

Graphic solution to Example 9-3. (Courtesy of Chemetron, Inc.)

From bottom graph in Figure 9-14
Discharge rate per nozzle, diptank = 95 lbs per minute (5½-in. nozzle)
Discharge rate per nozzle, drainboard = 73 lbs per minute (5½-in. nozzle)
Number of nozzles, diptank = (30 ft.) ÷ (5 ft. tank width) = 6 nozzles each covering a square area of 5′ × 5′.
Number of nozzles, drainboard = (15 ft.) ÷ (5 ft. tank width) = 3 nozzles, each covering a square area of 5′ × 5′.

System Discharge Rate:
Diptank:
$$FR_{liquid} = (N_{liquid}) \times (F_{liquid})$$
$$= (6 \text{ nozzles}) \times (95 \text{ lbs/min}) = 570 \text{ lbs/min}$$

Drainboard:
$$FR_{coated} = (N_{coated}) \times (F_{coated})$$
$$= (3 \text{ nozzles}) \times (73 \text{ lbs/min}) = 219 \text{ lbs/min}$$

Total:
$$FR = FR_{liquid} + FR_{coated}$$
$$= (570 \text{ lbs/min}) + (219 \text{ lbs/min})$$
$$= 789 \text{ lbs/min}$$

Quantity of CO_2 required:
$$W = (FR) \times (D)$$
$$= (789 \text{ lbs per min}) \times (½ \text{ min})$$
$$= 395 \text{ lbs carbon dioxide}$$

NFPA 12 requires a 40% increase in this value for high-pressure local application carbon dioxide systems:
$$W = (395 \text{ lbs}) \times (1.4) = 553 \text{ lbs carbon dioxide}$$

TOTAL FLOODING CARBON DIOXIDE SYSTEM DESIGN PROCEDURE

Total flooding systems involve analysis not only of the expected fire but also of the integrity of the enclosure.

Evaluate Enclosure Integrity

A thorough survey of the room being protected is required to ensure that it will hold carbon dioxide for the required duration. A list to use for this survey includes the following items:

1. Clip down and seal all acoustical ceiling tiles.
2. Install automatic electromagnetic closers on all doors. These devices hold doors open when wired to the carbon dioxide system control unit and are released to the closed position on system activation.

3. Once closed, egress doors must permit egress from the room but must be locked to prohibit reentry to the room.

4. Close and lock windows, or install automatic window closers.

5. Accurately measure all other openings that do not have doors or hatches.

6. Provide automatic closure to all openings where possible.

7. If it is not possible to close or seal openings, add gas later in this design procedure to compensate for gas lost through these openings.

8. Conduct a thorough inspection of all floor openings and tightly seal them.

9. Seal any wall joints that might leak gas.

10. Inspect wall rigidity and integrity. The pressure applied by the gas, especially at floor level, could damage the wall. If it is possible economically to reinforce the wall, do so. If not, consider a system other than carbon dioxide.

11. Conduct a thorough inspection of the HVAC system. This system consists of two separate features, the supply air system and the return air system. Before gas is discharged, the HVAC system must be shut off by connecting the fire alarm panel to the HVAC control panel. The supply air duct must be closed off automatically with a duct damper that closes upon detecting a fire. The return air system is either ducted return air or nonducted plenum air. Ducted return air ducts must be closed off automatically with electrically actuated dampers upon detection of a fire. A plenum return air system places the space above a suspended ceiling under negative pressure with a fan that draws stale air from the room, through a ceiling grille, and into the plenum space above the ceiling. This air is exhausted from the plenum space to the outside. An automatic damper can be installed on the ceiling grille or plenum exhaust duct to prevent gas from entering the plenum space.

12. Consider a fan test of the room. A specialized fan with a calibrated pressure-reading device can be placed in the room to measure the air-holding integrity of the room and additional avenues for leakage can be identified. It is possible in some cases to perform a full-flow test of the carbon dioxide system.

Evaluate Personnel Hazards

Carbon dioxide systems should not be specified for normally-occupied areas. When considering a room for carbon dioxide protection, a careful study of maintenance or inspection personnel who may require access to the room and a thorough analysis of their ability to escape upon system actuation must be performed. For an existing room, the minimum evaluation of personnel must include the following:

1. Make a record, over the course of at least one week, of all personnel who enter the room. Record the time of day, length of stay, and purpose of the

visit. Also consider other visits that might happen infrequently, such as a computer repair person or an electrician.

2. If a room is occupied, generally or occasionally, CO_2 should not be used.

3. Accurately record the time necessary for a person to exit the room under normal circumstances, and again under an alarm condition.

4. Perform a test to determine the actual egress time, based upon a worst-case scenario. An example of a space where egress time could be lengthy could involve a person in a vault or pit. A test of such scenarios must be evaluated under alarmed test conditions. This test should consider that a delay may occur between initiation of the alarm and commencement of egress. If it is noticed that the alarm is ignored or not heard, or that a significant delay is noticed between initiation of alarm and commencement of egress, consider a louder or more noticeable alarm, a voice alarm, or an intense period of personnel training. Consider the possibility that an untrained transient person may be in the room and may not know how to interpret a system preactuation warning alarm. An alarm supplemented by a recorded voice message should be seriously considered in such cases.

5. As part of the egress time test mentioned previously, perform an egress door recognition test. In some rooms, the exit door is not immediately obvious, and time may be needed by persons unfamiliar with the room to search for an exit. This is a potentially serious problem that must be solved before a carbon dioxide system is installed. A voice alarm, extra signs, or additional exits may be useful in these cases. Tests of relative egress time using the voice alarm and tests not involving voice instructions must be performed and compared.

6. Determine the effectiveness of warning signs through testing and postegress interviews.

7. Review all information contained within the discussion of this chapter under the heading "Personnel Hazards."

8. Remember that a carbon dioxide system is intended to be a life- and property-saving system and should not under any circumstance increase the hazard to personnel.

Evaluate Fire Scenario of the Expected Fire

Probabilities for ignition and initiation of an expected fire must be evaluated thoroughly. In so doing, ignition-mitigation procedures and other modifications to working or manufacturing procedures that would reduce the probability of fire can be ascertained. The procedure could consist of

- control of smoking materials
- grounding of electrical short circuits
- control of electrostatic discharges

- reduction or isolation of potential combustibles
- control of cutting or welding operations
- lightning current control
- radiant heat transfer evaluation
- control of chemical heat-producing reactions
- identification and control of potential sources for spontaneous ignition
- identification and control of potential sources of frictional heating
- identification and protection of surfaces whose surface temperature may ignite adjacent combustibles

Another important evaluation that must be made is whether the expected fire would be a surface fire or a deep-seated fire. This evaluation affects the percentage of carbon dioxide that must be injected into the enclosure.

Accurately Measure Room Volume

The raw room volume is computed by multiplying its dimensions:

$$V = (L) \times (W) \times (H)$$

where V = room volume, in ft.^{3} (m^3)
        L = length of room, in ft. (m)
        W = width of room, in ft. (m)
        H = height of room, in ft. (m)

This volume may be reduced by the volume of objects impervious to the absorption of gas, such as concrete pedestals. Electronic cabinets and other mechanical or electronic devices must not be deducted from the room volume because they could fill with gas.

EXAMPLE 9-4

Room Volumetric Calculation for a Total Flooding Carbon Dioxide System

Determine the volume of a room that is 15 ft. long, 15 ft. wide, and 12 ft. high, which contains a solid concrete pedestal measuring 4 ft. long by 4 ft. wide by 4 ft. high and electrical switchgear cabinets spaced throughout the room. The room contains a 10-ft. suspended ceiling, but dampers cannot be provided for isolation of the plenum space from the room volume below the ceiling. The supply ducts have been dampered.

Solution

The volume calculated must include the plenum space because dampers or other sealing measures cannot be provided:

$$V_{room} = (L) \times (W) \times (H)$$
$$= (15') \times (15') \times (12')$$
$$= 2700 \text{ cubic feet (ft.}^3)$$
$$V_{pedestal} = (L) \times (W) \times (H)$$
$$= (4') \times (4') \times (4')$$
$$= 64 \text{ ft.}^3$$
$$V = (V_{room}) - (V_{pedestal})$$
$$= (2700 \text{ ft.}^3) - (64 \text{ ft.}^3)$$
$$= 2636 \text{ ft.}^3$$

Determine Type of Combustible

An accurate evaluation of the hazard includes identification of all potential combustibles. Combinations of combustibles, such as tall stacks of paper adjacent to bulk storage of kerosene, should be avoided by separating such items and relocating them into enclosures separated from each other by rated fire walls. In performing this analysis, it is important to note whether an expected fire would be a surface fire or a deep-seated fire.

Determine Minimum Design Concentration

Using **Table 9-2**, select the minimum design carbon dioxide concentration, in percent. Note that two columns of percentages are shown, one for theoretical minimum concentration and the other for minimum design concentration. The theoretical minimum concentration is the concentration found to be effective under controlled laboratory testing conditions. Most rooms do not match the enclosure integrity or ambient conditions in a laboratory; therefore for design, the minimum design concentration (the column on the far right) is always used. Note that minimum design concentrations range from a minimum concentration of 34% to as high as 75%. Most authorities having jurisdiction will not permit design percentages less than 34%.

The theoretical minimum concentrations are reference numbers obtained from testing and not intended to be used as design concentrations. An AHJ will universally expect total flooding systems to meet the minimum design concentration for the specific commodity listed on Table 9-2.

Material	Theoretical Minimum CO_2 Concentration (%)	Minimum Design CO_2 Concentration (%)
Acetylene	55	66
Acetone	27*	34
Aviation gas grades 115/145	30	36
Benzol, benzene	31	37
Butadiene	34	41
Butane	28	34
Butane-I	31	37
Carbon disulfide	60	72
Carbon monoxide	53	64
Coal or natural gas	31*	37
Cyclopropane	31	37
Diethyl ether	33	40
Dimethyl ether	33	40
Dowtherm	38*	46
Ethane	33	40
Ethyl alcohol	36	43
Ethyl ether	38*	46
Ethylene	41	49
Ethylene dichloride	21	34
Ethylene oxide	44	53
Gasoline	28	34
Hexane	29	35
Higher paraffin hydrocarbons $C_n H_{2m} + 2m - 5$	28	34
Hydrogen	62	75
Hydrogen sulfide	30	36
Isobutane	30*	36
Isobutylene	26	34
Isobutyl formate	26	34
JP-4	30	36
Kerosene	28	34
Methane	25	34
Methyl acetate	29	35
Methyl alcohol	33	40
Methyl butene-I	30	36
Methyl ethyl ketone	33	40
Methyl formate	32	39
Pentane	29	35
Propane	30	36
Propylene	30	36
Quench, lube oils	28	34

Table 9-2 *Minimum carbon dioxide concentration for extinguishment.*

Source: NFPA 12 (2005), Table 5.3.2.2. Reprinted with permission from NFPA 12, *Carbon Dioxide Extinguishing Systems*, Copyright © 2005, National Fire Protection Association, Quincy, MA 02169. This reprinted material is not the complete and official position of the National Fire Protection Association on the referenced subject which is represented only by the standard in its entirety.

Note: The theoretical minimum extinguishing concentrations in air for the materials in the table were obtained from a compilation of Bureau of Mines, Bulletins 503 and 627, *Limits of Flammability of Gases and Vapors*.

*Calculated from accepted residual oxygen values.

EXAMPLE 9-5

Determination of Minimum Design Concentration for Total Flooding Carbon Dioxide Systems

Determine the minimum total flooding design concentration for ethane.

Solution

Using Table 9-2, the minimum design concentration is 40%.

Determine Volume Factor

volume factor

a value used to determine the amount of carbon dioxide required to be injected into a room at the minimum design concentration of 34%

The **volume factor** is used to determine the amount of carbon dioxide required to be injected into a room at the minimum design concentration of 34%. **Table 9-3** is used to select the volume factor for surface fires, and **Table 9-4** is used to select the volume factor for deep-seated fires.

Note that for deep-seated fires, the design concentration percentages shown on Table 9-2 are not used. Table 9-4 describes a variety of typical deep-seated hazards, and provides design concentrations and volume factors that must be used for those specific hazards.

EXAMPLE 9-6

Determine the Total Flooding Carbon Dioxide Volume Factor for a Surface Fire

Using the room described in Example 9-4, and assuming the hazard described in Example 9-5, determine the volume factor.

Solution

In Example 9-4, we determined that the net room volume is 2636 cubic feet. Example 9-5 assumed the presence of ethane, which indicates a surface fire. Table 9-3 indicates that for a surface fire of this volume, the volume factor is 18 ft.^{3}/lb CO_2 or 0.056 lb CO_2/ft.^{3}, with a calculated quantity in pounds no less than 100 lb CO_2. Note that 0.056 lb CO_2/ft.^{3} is the inverse of 18 ft.^{3}/lb CO_2, as can be seen by observing the reversing of the units.

To determine the inverse of 18 ft.^{3}/lb CO_2:

$$\frac{1}{18 \text{ ft.}^3/\text{lb}} = 0.055555 \text{ lb/ft.}^3$$

Both values are listed on Table 9-2 for ease of calculation. Remember that these values are flooding factors that assume a surface fire and a total flooding carbon dioxide design concentration of 34%.

(A)	(B) Volume Factor		(C) Calculated Quantity (lb) (Not Less Than)
Volume of Space (ft^3)	ft^3/lb CO$_2$	lb CO$_2$/ft^3	
Up to 140	14	0.072	—
141–500	15	0.067	10
501–1600	16	0.063	35
1601–4500	18	0.056	100
4501–50,000	20	0.050	250
Over 50,000	22	0.046	2500

Table 9-3 *Flooding factors for surface fires.*

Source: NFPA 12 (2005), Table 5.3.3(a) & (b). Reprinted with permission from NFPA 12, *Carbon Dioxide Extinguishing Systems*, Copyright © 2005, National Fire Protection Association, Quincy, MA 02169. This reprinted material is not the complete and official position of the National Fire Protection Association on the referenced subject which is represented only by the standard in its entirety.

Flooding Factors (SI Units)

(A)	(B) Volume Factor		(C) Calculated Quantity (kg) (Not Less Than)
Volume of Space (m^3)	m^3/kg CO$_2$	kg CO$_2$/m^3	
Up to 3.96	0.86	1.15	—
3.97–14.15	0.93	1.07	4.5
14.16–45.28	0.99	1.01	15.1
45.29–127.35	1.11	0.90	45.4
127.36–1415.0	1.25	0.80	113.5
Over 1415.0	1.38	0.77	1135.0

Table 9-4 *Flooding factors for specific hazards.*

Design Concentration (%)	Volume Factors				Specific Hazards
	ft^3/lb CO$_2$	m^3/kg CO$_2$	lb CO$_2$/ft^3	kg CO$_2$/m^3	
50	10	0.62	0.100	1.60	Dry electrical hazards in general [spaces less than 2000 ft^3 (56.6 m^3)]
50	12	0.75	0.083 (200 lb minimum)	1.33 (91 kg minimum)	Dry electrical hazards in general [spaces greater than 2000 ft^3 (56.6 m^3)]
65	8	0.50	0.125	2.00	Record (bulk paper) storage, ducts, covered trenches
75	6	0.38	0.166	2.66	Fur storage vaults, dust collectors

Source: NFPA 12 (2005), Table 5.4.2.1. Reprinted with permission from NFPA 12, *Carbon Dioxide Extinguishing Systems*, Copyright © 2005, National Fire Protection Association, Quincy, MA 02169. This reprinted material is not the complete and official position of the National Fire Protection Association on the referenced subject which is represented only by the standard in its entirety.

Determine the Basic Quantity of Carbon Dioxide

The basic quantity of carbon dioxide, assuming a concentration of 34%, is determined by multiplying the net room volume times the applicable volume factor:

$$Q_{basic} = (V) \times (\text{volume factor, in lb/ft.}^3)$$
$$Q_{basic} = (V) \times (\text{volume factor, in kg/m}^3)(\text{Metric})$$

EXAMPLE 9-7

Determining the Basic Quantity of Total Flooding Carbon Dioxide for a Surface Fire

Using the room and hazard described in Examples 9-4, 9-5, and 9-6, determine the basic quantity of carbon dioxide required at 34% design concentration.

Solution

Our room volume is 2636 ft.^{3} from Example 9-4; our hazard is ethane and the design concentration is 40% from Example 9-5; and our volume factor is 0.056 lb CO_2/ft.^{3} from Example 9-6. We determine the basic quantity of carbon dioxide assuming a design concentration of 34%, but because the minimum design concentration for ethane is 40%, will have to adjust for this difference in Example 9-8. The basic quantity of carbon dioxide, assuming a design concentration of 34%, is:

$$Q_{basic} = (V) \times (\text{flooding factor in lb/ft.}^3)$$
$$= (2636 \text{ ft.}^3) \times (0.056 \text{ lb } CO_2/\text{ft.}^3)$$
$$= 147.6 \text{ lb } CO_2$$

Note that this figure exceeds the minimum required calculated quantity of 100 lb CO_2 per Table 9-3. Note also that we also could have calculated the basic quantity in this manner:

$$Q_{basic} = (V)/(\text{volume factor in ft.}^3/\text{lb})$$
$$= (2636 \text{ ft.}^3)/(18 \text{ ft.}^3/\text{lb})$$
$$= 146.4 \text{ lb } CO_2$$

The difference in the two values is attributable to rounding of significant digits. The inverse of 18 ft.^{3}/lb, as calculated previously, is 0.055555 lb/ft.^{3}. Using the rounded value of 0.056 lb/ft.^{3} from Table 9-3 results in a slightly higher value.

---

material conversion factor

a dimensionless number that increases the basic quantity of carbon dioxide for hazards where the minimum design concentration exceeds 34%

Determine the Material Conversion Factor

The **material conversion factor**, shown on **Figure 9-15**, is a dimensionless number that increases the basic quantity of carbon dioxide for hazards where the minimum design concentration obtained from Table 9-2 for surface fires exceeds 34%. The material conversion factor is multiplied by the basic quantity to obtain the minimum quantity required for the specified design concentration.

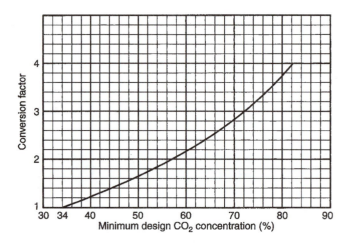

Figure 9-15   *Material conversion factors.*

NFPA 12 does not require application of the material conversion factor for deep-seated fires. Extinguishment of deep-seated fires are more difficult to predict because of the variance in the ability of carbon dioxide to penetrate the combustible material, and because of the variance in the thermal insulating effects of differing combustible materials.

EXAMPLE 9-8

Determining the Total Flooding Carbon Dioxide Material Conversion Factor for a Surface Fire

Using the basic quantity calculated in Example 9-7 and the concentration determined in Example 9-5, use the material conversion factor to adjust the basic quantity required.

Solution

In Example 9-7 we calculated a basic quantity of carbon dioxide of 147.6 lb CO_2, based on the minimum design concentration of 34%. In Example 9-5 we determined that for ethane the minimum design concentration is 40%.

Using Figure 9-15, we can see that for a design concentration of 40%, the material conversion factor is 1.2.

> The basic quantity is adjusted by multiplying the material conversion factor by the basic quantity.
>
> $$(Q_{basic} \text{ at } 40\%) = (Q_{basic} \text{ at } 34\%) \times (\text{Material Conversion Factor})$$
> $$= (147.6 \text{ lb CO}_2) \times (1.2)$$
> $$= 177.12 \text{ lb CO}_2$$

Adjust Basic Quantity for Temperature

Carbon dioxide volume is affected by very high and very low ambient temperature. For enclosures where the ambient temperature exceeds 200°F, and for enclosures where the ambient temperature is below 0°F, the basic quantity of carbon dioxide must be increased as follows:

1% increase for each additional 5°F above 200°F

1% increase for each 1°F below 0°F

Note that NFPA 12 requires that high-pressure cylinders be stored between 0°F (-18°C) and 130°F (54°C), so the temperature adjustment recognizes that the hazard area may be of an ambient temperature different from the storage temperature.

EXAMPLE 9-9

Adjusting Carbon Dioxide Total Flooding Quantity to Account for Temperature

Adjust the basic quantity of carbon dioxide calculated in Example 9-8, assuming an ambient room temperature of -10°F.

Solution

Because the ambient temperature is 10°F below zero, NFPA 12 requires a 1% increase for each 1°F below 0°F; therefore, an additional 10% must be added to the basic quantity of 177.12 lb CO_2.

$$Q_{basic} = (177.12 \text{ lb CO}_2) \times (1.10) = 194.83 \text{ lb CO}_2$$

Adjust Basic Quantity for Unclosable Openings

Doors, hatches, and other openings that cannot be closed by manual or automatic means must be accommodated by adding extra gas. NFPA 12 makes clear that if the amount of carbon dioxide lost through an unclosable opening exceeds the basic quantity calculated previously, a total flooding system is inappropriate and a local application carbon dioxide system, or a system other than carbon dioxide, must be used.

To determine the additional amount of gas needed, we must determine the area of the opening and the distance from the center of the opening below the ceiling of the enclosure. For a given design concentration, the rate of loss through an opening increases not only with increased opening size but also with increased distance

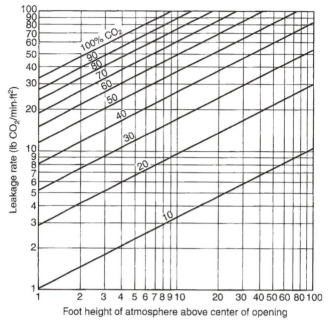

Figure 9-16
Calculated CO$_2$ loss rate based on an assumed 70°F (21°C) temperature within the enclosure and 70°F (21°C) ambient outside.

For SI units, 1 ft = 0.305 m; 1 lb/min·ft² = 4.89 kg/min·m².

below the ceiling, attributable to the greater pressure of the heavier-than-air gas at elevations closer to the floor. **Figure 9-16** can be used to quickly determine the amount of additional gas required for the specified design concentration. Note that for a given concentration, the leakage rate increases as the distance from the ceiling to the center of an opening increases. It is important to that the designer understand that Figure 9-16 is based upon the distance from the *center* of the opening to the ceiling.

Note that Figure 9-16 assumes that the ambient temperature on both sides of the enclosure is 70°F.

Carbon Dioxide Total Flooding Discharge Duration

NFPA 12 requires a minimum discharge time of 1 minute for surface fires, and 7 minutes for deep-seated fires.

EXAMPLE 9-10

Determination of Loss Through Openings for a Total Flooding Carbon Dioxide System

The enclosure for the hazard described in Examples 9-1 through 9-9 contains a 5-sq-ft. opening whose center is 5 ft. below the ceiling of the enclosure. Determine the additional amount of agent required and the adjusted quantity of agent required.

Solution

In Example 9-5 we determined that the design concentration for ethane is 40%. Figure 9-16 shows values along the "x" axis, or the bottom of the graph, that represent the distance of the center of the opening below the ceiling.

1. In this example the center of the opening is 5 ft. below the ceiling, so the vertical line intersecting the "5" on the "x" axis is chosen.

2. Our enclosure requires a 40% design concentration, represented by the fourth angular graph up from the "x" axis.

3. The vertical line from the value "5" on the "x" axis intersects 40% graph at a point.

4. Draw a horizontal line from the point determined on the 40% line, and extend to the "y" axis, which shows a value representing a leakage rate of about 19 lb CO_2 per minutes per square foot of opening, read from the values along the "y" axis of the graph.

5. The leakage rate is a rate per square foot of opening. The leakage rate determined from Figure 9-16 is therefore multiplied by the total square footage of the opening, assuming a 1-minute discharge:

$$(5 \text{ ft.}^2) \times (19 \text{ lb } CO_2/\text{min/ft.}^2) \times (1 \text{ min}) = 95 \text{ lb } CO_2$$

6. The amount of CO_2 that must be added to the basic quantity of CO_2 is 95 lb and assumes an ambient temperature of 70°F, and should be adjusted by 10% for the actual temperature of -10°F, specified in Example 9-9, similar to the way we adjusted the basic quantity:

$$(95 \text{ lb } CO_2) \times (1.10) = 104.5 \text{ lb } CO_2$$

7. For a total flooding system to be considered, the amount of carbon dioxide added to compensate for nonclosable openings must be less than the basic quantity; 104.5 lb CO_2 is less than the basic quantity of 194.83 lb CO_2, and a total flooding carbon dioxide therefore is appropriate for of the enclosure in these examples.

8. The additional quantity required to compensate for unclosable openings may now be added to the basic quantity:

$$(CO_2 \text{ requirement}) = (\text{basic quantity}) + (\text{leakage quantity})$$
$$= (194.83 \text{ lb } CO_2) + (104.5 \text{ lb } CO_2)$$
$$= 299.33 \text{ lb } CO_2$$

Consider Other Scenarios for Loss of Gas

If a nondampered heating, ventilating, and air conditioning (HVAC) supply or return duct is in the room, the airflow in the duct should be shut down by the fire alarm control unit and, and if dampers cannot be installed at the duct openings, the interior volume of the duct will fill with carbon dioxide, and this volume must be added to the total basic quantity obtained above and adjusted for temperature. If the supply air cannot be shut down, the volume of fresh air introduced to the room must be accounted for by the addition of gas during the period of carbon dioxide discharge.

EXAMPLE 9-11

Compensating for HVAC Air Flow During Carbon Dioxide Discharge

A supply duct diffuser distributes 100 cfm and cannot be shut down. Determine the additional carbon dioxide required for the enclosure described in Examples 9-4 through 9-10, assuming a discharge time of 1 minute.

Solution

100 cfm represents 100 ft.^{3} of fresh air discharging during the 1-minute system discharge period; 100 cubic feet is multiplied by the same flooding factor for ethane that we used in Examples 9-4 through 9-10 to determine the basic quantity:

$$Q_{air} = (100 \text{ ft.}^3) \times (0.056 \text{ lb CO}_2/\text{ft.}^3)$$
$$= 5.6 \text{ lb CO}_2$$

This amount then is modified by the material conversion factor of 1.2 for a 40% concentration, as we did for the basic quantity:

$$Q_{air} = (5.6 \text{ lb CO}_2) \times (1.20)$$
$$= 6.72 \text{ lb CO}_2$$

Adjusting for an ambient room temperature of $-10°F$, the compensatory factor becomes:

$$Q_{air} = (6.72 \text{ lb CO}_2) \times (1.10)$$
$$= 7.39 \text{ lb CO}_2$$

This amount required to compensate for fresh air introduced to the enclosure now can be added to the carbon dioxide requirement calculated in Example 9-10:

$$Q_{total} = (\text{CO}_2 \text{ requirement}) + Q_{air}$$
$$= (299.33 \text{ lb CO}_2) + (7.39 \text{ lb CO}_2)$$
$$= 306.72 \text{ lb CO}_2$$

Although this compensatory quantity accounts for gas lost during the discharge period, it does not compensate for loss during the holding period. For this situation, an extended discharge system would be specified. An extended discharge system is specified for rooms whose normal leakage is excessive, and for deep-seated fires where extended holding times are mandated.

Extended Rates of Total Flooding Carbon Dioxide Application

extended discharge system

a separate system of small pipes and nozzles that provides a rate of discharge after the primary discharge system ceases operation, to compensate for the amount of gas projected to be lost during the required holding period

Several avenues of leakage have been identified in this chapter, and calculations have been performed to provide compensatory amounts of extra carbon dioxide to account for these leaks. Rooms having avenues not previously mentioned, such as cracks in walls or ceilings, spaces between doors and their frames, or other room scenarios that are expected to leak significant quantities of gas during the holding period, may require an **extended discharge system**. This is a separate system of small pipes and nozzles that provides a rate of discharge after the primary discharge system ceases discharge of all gas in its cylinders, to compensate for the amount of gas that is projected to be lost during the required holding period.

Table 9-5 *Strength and allowable pressures for average enclosures.*

Type Construction	Windage (mph)	Pressure (lb/ft^2)	Water		
			in.	psi	kPa
Light building	100	25*	5	0.175	1.2
Normal building	140	50†	10	0.35	2.4
Vault building	200	100	20	0.70	4.8

*Venting sash remains closed.
†Venting sash designed to open freely.

Source: NFPA 12 (2005), Table A.5.6.2. Reprinted with permission from NFPA 12, *Carbon Dioxide Extinguishing Systems*, Copyright © 2005, National Fire Protection Association, Quincy, MA 02169. This reprinted material is not the complete and official position of the National Fire Protection Association on the referenced subject which is represented only by the standard in its entirety.

The extra gas is stored in separate carbon dioxide cylinders and is controlled by a separate selector valve. The extended discharge system is programmed to commence discharge after the basic supply is exhausted, and continue discharge throughout the required holding period.

Calculate Pressure Relief Venting Area

Rooms that are very tight could develop a pressure on interior walls and equipment that may be damaging. To prevent this, a relief vent may be required. To calculate the area of the vent required, NFPA 12 specifies that the following formula be used:

$$X = \frac{Q_{\text{total}}}{1.3P^{1/2}}$$

where X = free venting area, in^2
Q_{total} = calculated carbon dioxide flow rate, lb/min
P = allowable strength of the enclosure, lb/ft^2

P is chosen from **Table 9-5** from the column entitled "Pressure."

EXAMPLE 9-12

Calculation of Pressure Relief for Total Flooding Carbon Dioxide Systems

The enclosure described in Examples 9-4 through 9-11 is determined to be of light construction. Determine the minimum required venting area in square inches.

Solution

The basic quantity was calculated as 306 lb/CO_2 in Example 9-11. Table 9-5 shows that P is equal to 25 lb/ft.^{2} for a light building.

$$X = \frac{Q_{total}}{1.3P^{1/2}}$$

$$= \frac{306}{(1.3) \times (25^{1/2})}$$

$$X = 47\ in^2$$

This value is very small—about the size of a crack around the perimeter of a door or the space between a door and its frame. Therefore, special relief venting probably is not required for this enclosure, because sufficient natural venting probably exists.

Conversely, if the required pressure venting area is very large, gas lost through the relief vent could significantly lower the concentration of total flooding carbon dioxide in the space. In such cases, a gaseous suppression system may be inappropriate.

Select Carbon Dioxide Containers

Because the calculated quantity is significantly less than our rule-of-thumb amount of 4000 lbs minimum suggested for a low-pressure container, we use high-pressure cylinders.

EXAMPLE 9-13

Determination of the Number of High-Pressure Carbon Dioxide Cylinders

Determine the number of high-pressure carbon dioxide cylinders required for the storage of agent for the system described in Examples 9-1 through 9-12, assuming a 50-lb cylinder and a requirement of an equal amount of gas in reserve.

Solution

In Example 9-11 we determined that 307 lb of CO_2 is required.

The number of 50-lb cylinders required for the main carbon dioxide supply is:

$$N = (307\ lb)/(50\ lb/cylinder)$$

$$= 6.14\ cylinders$$

Because a fractional cylinder does not exist, we round up to the nearest whole number:

$$N = (7)\ 50\text{-lb cylinders for the main carbon dioxide supply.}$$

The total number of cylinders required is 7 for the main supply and 7 for the reserve supply, for a total of 14 cylinders. Note that total flooding carbon dioxide systems are not increased by 40% for this total flooding application, as we did for local application systems. The 40% increase is specified by NFPA 12 for local application high pressure systems only.

Determine Number of Nozzles

A convenient rule of thumb that can be used to determine the minimum number of nozzles required, assuming nozzles are ceiling mounted, is to figure one nozzle per 400 ft.^{2} (37.16 m^2) of ceiling area, with nozzles spaced no farther than 20 ft. (6.096 m) apart and no farther than 10 ft. (3.048 m) from any wall or obstruction. A nozzle commensurate with these attributes must be selected.

EXAMPLE 9-14

Determining the Number and Placement of Nozzles on a Total Flooding Carbon Dioxide System

Determine the minimum number of nozzles required for the protection of the enclosure described in Examples 9-4 through 9-13, and sketch the system layout and cylinder location.

Solution

From Example 9-4, the ceiling area is 15 ft. long by 15 ft. wide, or 225 sq. ft. For this application:

$$N = \frac{225 \text{ ft.}^2}{400 \text{ ft.}^2/\text{nozzle}}$$

$$= 1 \text{ nozzle}$$

A sketch is shown on **Figure 9-17**.

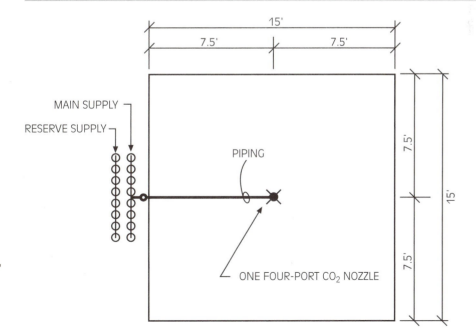

Figure 9-17 *Plan view sketch illustrating Example 9-14.*

1. VOLUME OF SPACE TO BE PROTECTED . . . . _____ CU. FT.

2. TYPE OF COMBUSTIBLE . . . . _____

3. DESIGN CONCENTRATION . . . . _____ % CO_2
    (USE FIGURE 9–2)

4. FLOODING FACTOR . . . . LB. CO_2/FT^3
    (USE FIGURE 9–3 FOR SURFACE FIRE, USE FIGURE 9–4 FOR DEEP-SEATED FIRE)

5. BASIC QUANTITY OF CARBON DIOXIDE . . . . _____ POUNDS
    (MULTIPLY #1 BY #4)

6. MATERIAL CONVERSION FACTOR. . . . . _____ VALUE
    (USE FIGURE 9–15)

7. REVISED BASIC QUANTITY . . . . _____ POUNDS
    (MULTIPLY #6 BY #5)

8. SPECIAL CONDITIONS: (SPECIFY) _____
    _____
    _____

9. OPENINGS: (USE FIGURE 9–16)
    AIR INLET _____ SQ. FT.   BELOW CEILING _____ FT.
    AIR INLET _____ SQ. FT.   BELOW CEILING _____ FT.

10. ADJUST BASIC QUANTITY FOR OPENINGS . . . . _____ POUNDS

11. ABNORMAL TEMPERATURE CONDITIONS:_____
    (1% FOR EACH 5°F INCREMENT OVER 200°F OR 1% FOR EACH DEGREE BELOW 0°F)

12. ADJUST BASIC QUANTITY FOR TEMPERATURE CONDITIONS . . . . _____ POUNDS

13. ADDITIONAL CARBON DIOXIDE REQUIRED: (SHOW CALCULATIONS). . . . . _____ POUNDS
    _____
    _____

14. TOTAL CARBON DIOXIDE REQUIRED _____ POUNDS

15. RATE OF APPLICATION (SPECIFY NON UNIFORM RATES) _____ LBS./MIN.

16. EXTENDED RATE OF APPLICATION: (STATE SPECIAL CONDITIONS AND TIME)

$$X = \frac{Q}{1.3\sqrt{P}}$$

17. PRESSURE RELIEF VENTING . . . . _____ SQ. IN.

18. STORAGE SUPPLY AND LOCATION (SPECIFY)
    A. HIGH PRESSURE (IF LOWER THAN 4000 LBS.)_____
    B. LOW PRESSURE (IF HIGHER THAN 4000 LBS.)_____

19. PROVIDE SKETCH OF PIPING SYSTEM

20. SELECT DETECTION SYSTEM
    _____
    _____

Figure 9-18 *Carbon dioxide system calculation form.*

Select Detection System

A smoke, heat, or flame detection system, described in Chapters 11 through 15, can be selected and designed to serve as a releasing mechanism for the carbon dioxide system.

Use Carbon Dioxide System Calculation Form

The form shown as **Figure 9-18** may be used to calculate the carbon dioxide system review questions in this text, using the methodology covered in this chapter.

SUMMARY

Designers must carefully consider personnel hazards when specifying a carbon dioxide system for an enclosure that is occupied permanently or sporadically by personnel. Carbon dioxide is stored in its liquid form by maintaining the pressure and temperature of the carbon dioxide within the liquid range of the phase diagram. High-pressure cylinders store liquid carbon dioxide at room temperature and at high-pressures, whereas low-pressure containers store refrigerated liquid carbon dioxide at a low-pressure.

Carbon dioxide systems can be designed for total flooding, local application, hand hose lines, and standpipe systems with mobile supply. Local application carbon dioxide system design may be performed using the rate-by-volume or the rate-by-area methods. Total flooding design is performed by evaluating the commodity protected and compensating for factors such as unclosable openings, temperature, fresh-air infusion, and natural leakage. The pressure of the gas on the walls is a consideration that may be overcome by designing relief venting for the enclosure.

REVIEW QUESTIONS

1. Explain the phase diagram for carbon dioxide. Why is carbon dioxide stored in liquid form?

2. What criteria are used to select high-pressure cylinders versus low-pressure containers?

3. List and explain applications for which carbon dioxide is effective and applications where carbon dioxide is not effective.

4. A valuable painting is in a display case 5 ft. wide, 6 ft. long, and 5 ft. tall, elevated 1 ft. above a solid floor. The case is centered in a 10 ft. long by 10 ft. wide enclosure with 8 ft.

high walls on all sides. Use the rate-by-volume method of local application to calculate flow rate and the total amount of carbon dioxide required, assuming a discharge time of 30 sec.

5. The case in problem 4 has been moved to a very large room with no walls near the case, and with the case mounted directly to the floor. Use the rate-by-volume method of local application to calculate flow rate and total amount of carbon dioxide required, assuming a discharge time of 45 sec.

6. Use the rate-by-area method to determine the minimum amount of carbon dioxide

required to protect a diptank 45 ft. long and 5 ft. wide, and a drainboard 10 ft. long and 5 ft. wide, using a discharge time of 30 sec.

7. A room 20 ft. long by 25 ft. wide by 12 ft. high contains an acetylene flammable liquid pump that sits on a 10 ft. long by 5 ft. wide by 3 ft. tall concrete pedestal. Determine the minimum quantity of carbon dioxide required if the ambient room temperature is 205°F and a 4-square-ft. unclosable opening is located 2 ft. below the ceiling. Determine whether high pressure or low pressure is appropriate, and sketch the system. Assume a discharge time of 1 minute.

DISCUSSION QUESTIONS

1. Why should designers be extremely careful when specifying carbon dioxide suppression systems, and how does the code of ethics become critical to a successful design?

2. What purpose does the use of an "imaginary volume" serve in the application of a rate-by-volume carbon dioxide system design?

3. Discuss personnel hazards that must be addressed when designing a carbon dioxide system for an enclosure. Under what conditions would a carbon dioxide system be inappropriate for an occupied space?

ACTIVITIES

1. Visit a carbon dioxide system distributor or testing firm. Tour the warehouse and become familiar with the components used on a carbon dioxide system. Accompany a service vehicle to a job site and observe a carbon dioxide system inspection.

2. Meet with a specifying engineer, a consulting engineer, or a responsible fire protection professional with a firm that uses carbon dioxide systems. What criteria are used to evaluate whether a hazard is protected by a carbon dioxide system or by another type of fire protection system?

3. Witness a carbon dioxide discharge test. List the procedures that are followed to ensure a tight enclosure and to ensure personnel safety during the test.

Chapter
10

DRY CHEMICAL AND WET CHEMICAL EXTINGUISHING SYSTEM DESIGN

Learning Objectives

Upon completion of this chapter, you should be able to:

■ Compare and contrast the types of dry chemical agents used for automatic fire protection systems and the extinguishment mechanisms involved with each.

■ Discuss the advantages and disadvantages of dry chemical agents for fire protection.

■ Identify the dilemma that could face authorities having jurisdiction when responsible for ensuring that dry chemical systems are capable of functioning as intended.

■ Understand and describe the sequence of operation for a dry chemical system.

■ List and discuss the types of dry chemical systems recognized by NFPA 17.

■ Estimate the quantity of dry chemical needed for a total flooding system.

■ Describe the differences between dry chemical and wet chemical systems.

■ Discuss wet chemical protection of cooking equipment.

■ List the operating sequence and extinguishment mechanism of wet chemical extinguishing systems.

NFPA

NFPA 17, Standard for
Dry Chemical
Extinguishing Systems

NFPA

NFPA 17A, Standard for
Wet Chemical
Extinguishing Systems

The National Fire Protection Association publishes NFPA 17, *Standard for Dry Chemical Extinguishing Systems*, and NFPA 17A, *Standard for Wet Chemical Extinguishing Systems*, the applicable national standards for wet and dry chemical design.

It is possible to gain a good understanding of the basics of wet and dry chemical design by reading this chapter. Nevertheless, obtaining and using these NFPA standards will enhance and deepen the reader's understanding of these systems.

DRY CHEMICAL SYSTEMS

Readers of this book are unlikely to design and calculate a dry chemical system, because the vast majority of dry chemical systems are either pre-engineered package units or are designed by the manufacturer of the dry chemical system. Still, readers need to know the basics of dry chemical design to be able to competently specify, approve, supervise, or test a dry chemical system installation.

Because dry chemical systems are commonly specified for the protection of a widening variety of applications, design professionals must be competent with respect to these systems. New dry chemical systems have not been manufactured for cooking hazard areas since 1994 and no longer are listed in accordance with UL 300 for this application. (Wet chemical systems, discussed later in this chapter, meet the listing requirements of UL 300 for fire testing of fire extinguishing systems for the protection of restaurant cooking areas.)

NFPA standards do not cover all design details relative to dry chemical system design because the dry chemical agents and dry chemical systems differ greatly among manufacturers in accordance with their listing, and most design details are the proprietary property of the manufacturer.

Dry Chemical Agent Analysis

dry chemical
a power consisting of small particles suspended in a gaseous medium, which permits distribution of the powder to a hazard

A **dry chemical** is a powder that consists of very small particles that are suspended in a gaseous medium, which permits distribution of the powder to a hazard. Dry chemical agents fall into three general categories: sodium bicarbonate-based, potassium-based agents, and multipurpose dry chemicals. These agents should not be mixed with each other because of the potential for destructive chemical reactions.

Sodium Bicarbonate-based Dry Chemicals Sodium bicarbonate dry chemicals can successfully extinguish flammable liquid fires (Class B) and electrical fires (Class C), and may be effective with certain solid-surface fires. Sodium bicarbonate is related to the commercial baking soda that historically has been used to extinguish household cooking fires.

Some commercially available sodium bicarbonate agents have been treated with silicone to enhance their compatibility with protein-based foams that may have been distributed simultaneously with the distribution of the sodium bicarbonate. The widespread use of AFFF and other non–protein-based foams makes unnecessary the use of foam-compatible sodium bicarbonate-based dry chemicals.

Potassium-based Dry Chemicals Potassium-based dry chemical agents generally are more effective than sodium bicarbonate-based dry chemicals for protecting against Class B and Class C fires. Types of potassium-based agents include potassium bicarbonate ($KHCO_3$), known by its registered trade name of *Purple K*; potassium chloride (KCl), known by its registered trade name of *Super K*; and urea-based potassium bicarbonate ($KC_2N_2H_3O_3$), known by its registered trade name of *Monnex*.

Multipurpose Dry Chemicals Multipurpose dry chemicals consist of monoammonium phosphate ($NH_4H_2PO_4$) and are suitable for most Class A fires involving ordinary cellulosic combustibles, and also for Class B and C fires. These agents form a molten residue on the surface of a burning object to prevent oxygen from contributing to the combustion reaction. Multipurpose dry chemicals are not effective for protecting delicate electrical equipment where temperatures exceed 250°F (121°C) or where relative humidity exceeds 50%, because deposits after discharge can be difficult to remove, especially in areas where accessibility is a problem.

Advantages of Dry Chemical Agents

Dry chemicals provide rapid knockdown of a flame and break the combustion chain, reducing the damage done by the flame and also reducing the probability of the fire spreading to other areas. Another advantage to the use of dry chemicals is its surface-coating capability of Class A materials. Surface coating cuts off oxygen and is highly effective in preventing reignition.

Surface coating by dry chemical is especially effective for three-dimensional fires involving burst pipes, flammable liquid-spray fires, elevated objects, and objects with vertical surfaces. In cases where other agents would run down and provide only partial protection of such surfaces, dry chemical is more likely to adhere to the surface and provide coating and insulation for the object.

Disadvantages of Dry Chemical

Because dry chemical is a powder and one of its main attributes is the coating of three-dimensional objects, the residue left after a fire can be messy. The powder can be washed away with a hose, but environmental restrictions may prohibit mixtures of water and powder from entering bodies of water or from seeping into the groundwater system.

Powder may be vacuumed using a vacuum cleaner with a HEPA filter, provided that the vacuum bag is handled properly in accordance with the applicable environmental regulations. The clean-up may involve a shutdown of the affected area, and in some cases may be time consuming and difficult.

Readers who have used sodium bicarbonate as a household fire protection agent or refrigerator air freshener know that when the material is exposed to a moist atmosphere over time, the powder solidifies into a solid block. Therefore, a phenomenon unique to dry chemical among all fire protection agents is its propensity to solidify into caked clumps when exposed to moist atmospheres. Those responsible for inspecting dry chemical systems must ensure that the dry chemical remains a powder so it can mix successfully with gas for distribution to the hazard.

To ensure that dry chemical remains in particulate form, NFPA has created a definition that is unique among all NFPA documents. How do we determine if dry chemical is caked? According to the 1990 edition of NFPA 17, dry chemical is caked if it does not crumble when dropped from a height of 4 inches (10.16 cm). The "4-inch drop" criterion was removed from the 1994 edition of NFPA 17 and replaced by a description of caking as being evidence of the presence of lumps of dry chemical within a storage container. This less restrictive definition could render a higher percentage of stored dry chemical as being caked. Caked dry chemical must be replaced.

Personnel and testing concerns outlined in the next two sections also may be considered as disadvantages.

Personnel Hazards with Dry Chemical

Dry chemical is considered to be nontoxic if breathed in its pure form. Most dry chemical particles are too large to pass deep into the lungs. The particles are more likely to attach themselves to the bronchial tubes, become mixed with mucus, and be expelled by cilia. Cilia are small hairs lining the bronchial tubes that clear foreign objects from the bronchial tube surfaces and minimize the possibility that those objects will enter the lungs. This process could be uncomfortable and irritating physically, and some particles may not be expelled by the cilia.

Some dry chemical agents decompose in flame, forming breathable chemical byproducts of decomposition that can be harmful to personnel in confined areas. Breathing decomposition products could result in postfire illnesses ranging from mild to serious, depending on the quantity of decomposition byproduct to which a person had been exposed. When specifying a dry chemical agent, a designer must obtain the Material Safety Data Sheet (MSDS) for the dry chemical used and consider exposure mitigation procedures for protection of personnel. Exposure mitigation must include post-incident care for those who breathe powder and for those who breathe decomposition byproducts.

Procedures for protection of personnel include many that apply to other agents covered previously. These procedures include predischarge alarms, signs, personnel training, egress analysis, and the provision of self-contained breathing apparatus (SCBA).

Another method of personnel protection is the most important, yet the most difficult, as discussed next. Testing of dry chemical systems is the best way to ensure their proper function, and the best way to ensure that they perform in a manner that will not be harmful to personnel.

Testing of Dry Chemical Systems

As with any fire protection system, regular testing is the best way to know for certain that the system functions as intended. Consider the dilemma of an authority having jurisdiction (AHJ) responsible for testing a large dry chemical system that protects a marine pumping facility in a public harbor. Discharging the system could result in a large powder cloud that could drift over populated areas, coating houses and cars, causing breathing discomfort, and possibly panic. Would you order the system to be discharged for testing knowing that you may be featured as the culprit on the local 6 o'clock news? Most AHJs decline to perform a full discharge test, not only for the large systems described previously but sometimes for smaller systems as well.

The question must be raised as to whether we can be completely assured of proper performance without full functional testing. Although inspection procedures performed without testing are of some value, the only way to completely ensure system reliability and effectiveness is to conduct a full discharge test. If this is not possible, functional testing of all components, combined with periodic maintenance, should provide satisfactory performance in most cases.

Dry Chemical Extinguishing Mechanism

We have discussed the coating and smothering properties of dry chemical. These properties block oxygen from contributing to and continuing the combustion reaction. The powder coating also absorbs heat from the fire in an endothermic reaction that reduces flame temperature.

The rapid knockdown of flame evidenced by dry chemical discharge reduces the flame radiation that could heat and ignite adjacent objects. This occurs when particles coat a fuel surface and interfere with vaporization of the fuel.

Dry chemicals extinguish primarily by using the fourth point on the fire tetrahedron (see Chapter 3), interference with the combustion chain reaction via free radical scavenging. This reaction happens when the sodium (Na) and potassium (K) salts attach themselves to free H and OH radicals. When those radicals become attached, they cannot contribute to continuance of the combustion process. Because testing shows that the extinguishing mechanism for dry chemical, as reported by the National Association of Fire Equipment Distributors (NAFED), is short-term, dry chemical is not recommended for use where reignition is expected.

DRY CHEMICAL SYSTEM COMPONENTS

A dry chemical system, an example of which is shown in **Figure 10-1**, consists of a container holding the dry chemical, one or more cylinders containing gas used to pressurize and mix with the dry chemical, a piping system, nozzles, and a detection system.

Dry Chemical Containers

A dry chemical container, as shown in Figure 10-1, is placed as close as possible to the hazard and is designed to hold up to about 3000 pounds of dry chemical under conditions intended to avoid caking. Dry chemical containers must be installed in an area protected from mechanical damage and, if possible, should be kept in a dry, accessible enclosure protected from humidity and heat.

Figure 10-2 shows the parts found on a typical dry chemical skid. A **dry chemical skid** is a pre-assembled assembly that includes pre-piped dry chemical

dry chemical skid
a pre-assembled assembly that includes pre-piped dry chemical storage and pressurization facilities

Figure 10-1 *A dry chemical skid assembly; this unit is equipped with a hose reel, but units designed for piped application systems can be ordered. (Courtesy of Ansul, Inc.)*

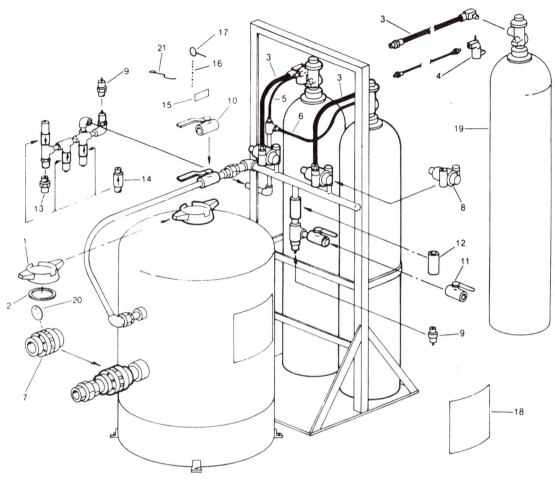

1	FILL CAP	12	VALVE ASSEMBLY
2	GASKET	13	RELIEF VALVE, $\frac{1}{4}$", 265 PSI
3	HOSE AND ADAPTOR ASSEMBLY, 2 REQUIRED	14	CHECK VALVE, $\frac{1}{4}$", 3 REQUIRED
4	QUICK-OPENING ACTUATOR ASSEMBLY, 2 REQUIRED	15	BRACKET NAMEPLATE
5	HOSE ASSEMBLY, 18" LG.	16	RING PIN CHAIN
6	HOSE ASSEMBLY, 36" LG.	17	RING PIN
7	SAFETY HEAD, 3"	18	NAMEPLATE
8	PRESSURE REGULATOR, 2 REQUIRED	19	NITROGEN CYLINDER, 400-B QUICK OPENING,
9	POPPET VALVE, 2 REQUIRED		2 REQUIRED
10	BALL VALVE, $\frac{3}{4}$"	20	BURSTING DISK, 3"
11	BALL VALVE, $\frac{1}{2}$"	21	VISUAL SEAL

Figure 10-2   *Dry chemical system components. (Courtesy of Ansul, Inc.)*

storage and pressurization facilities. The skid is delivered to the site as a unit and requires only that it be connected to a piping delivery system.

Calculations are performed to ensure that sufficient dry chemical is stored in the container to extinguish an expected fire in accordance with the requirements of the hazard, with a reserve supply for a second application. NFPA 17 does not require a reserve supply for a single hazard, but a reserve supply should be considered seriously, especially in remote locations where replenishment could be lengthy. Systems protecting multiple hazards do require a reserve supply.

Expellant Gas Cylinders for Dry Chemical Systems

fluidization
mixing of an expellant gas with dry chemical powder for ease of distribution to the hazard

An expellant gas, usually nitrogen but sometimes carbon dioxide, is required to be mixed with the dry chemical powder for ease of distribution to the hazard. The process of mixing an expellant gas with dry chemical powder is called **fluidization**. The gas cylinders are mounted adjacent to the dry chemical container, and piping is run from the gas cylinders to the dry chemical container. No gas is discharged until a detection signal is received.

Dry Chemical System Piping and Nozzles

Because dry chemical suspended in a gas is likely to settle and lose momentum as it travels to the hazard, piping must be designed so that dry chemical is remixed at each tee, as shown in **Figure 10-3**. The pipe must be capable of withstanding the pressures encountered, and piping material must be compatible with the agent and resistant to chemical or ambient corrosion. Cast iron pipe and fittings may be subject to corrosion and are not permitted.

Nozzles come in numerous varieties, as shown in **Figure 10-4**, and are selected for the best application of dry chemical on a specific hazard. Nozzles

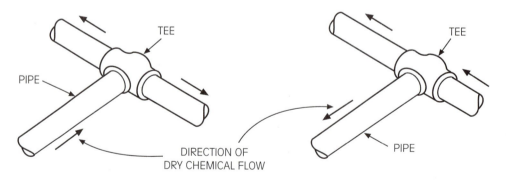

Figure 10-3 *Dry chemical piping methods. (Courtesy of Ansul, Inc.)*

TEE

PIPE

DIRECTION OF DRY CHEMICAL FLOW

CORRECT
DRY CHEMICAL HITS TEE AND RE-MIXES

TEE

PIPE

INCORRECT
VERY LITTLE RE-MIXING CAN OCCUR HERE

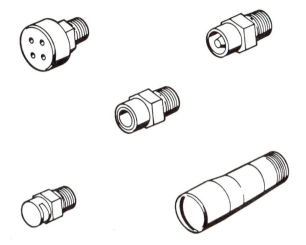

Figure 10-4 *Dry chemical nozzles. (Courtesy of Ansul, Inc.)*

must be capable of withstanding the system pressures anticipated and are made of brass, stainless steel, or other corrosion-resistant material. Nozzles must be marked permanently so they can be identified in the field and re-searched readily, and provided with blowoff caps in dusty atmospheres and in areas where insects could build nests or other sources of nozzle clogging could occur.

Detection Systems for Dry Chemical Systems

A detection system and fire alarm control unit, designed in accordance with national standards, using the methodology shown in Chapters 11 through 15 in this book, is required to initiate the discharge of expellant gas and commence fluidization. The detectors must be chosen carefully to identify the type of combustion expected in the hazard area, and must be capable of screening out sources of false detection events.

DRY CHEMICAL SYSTEM SEQUENCE OF OPERATION

Figure 10-5 shows the sequence of operation for a dry chemical system. The following numbered list corresponds to the sequence-of-operation numbers cir-cled in Figure 10-5.

1. *Detection.* A flame, heat, or smoke detector senses a fire condition. (Sub-sequent chapters in this book provide detail on selection of detectors.)
2. *Transmission.* If fire conditions reach the threshold level of the detector, the detector sends an electronic signal to the control panel through wires connecting the detector to the control unit.

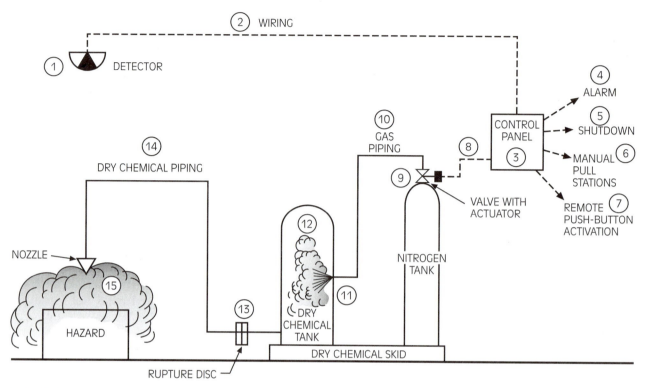

Figure 10-5   *Sequence of operation for a dry chemical system; numbers in circles correspond to numbered list under the heading "Dry Chemical System Sequence of Operation." (Courtesy of Ansul, Inc.)*

3. *Interpretation.* The fire alarm control unit interprets the detection signal. If the signal is interpreted as a fire condition, the following events occur simultaneously.

4. *Alarm.* An alarm is sounded (local and/or remote).

5. *Shutdown.* Equipment is shut down, and supply and return air are shut down.

6. *Manual pull station activation (alternative).* In place of the foregoing automatic detector actuation, the system could be activated manually by local manual pull stations.

7. *Manual push-button activation (alternative).* In place of the previously listed automatic actuation, the system could be activated manually by remote push-button activation. With this option the detection system notifies an operator, who pushes a manual push-button to initiate discharge.

8. *Nitrogen actuator notification.* A signal is sent to an actuator on the nitrogen tank.

9. *Nitrogen actuation.* The nitrogen cylinder valve is opened.

10. *Nitrogen release.* Pressurized nitrogen is piped to the dry chemical tank.

11. *Fluidization.* Nitrogen enters the dry chemical tank and mixes with the dry chemical powder (fluidization).

12. *Fluidization pressure.* Pressure builds up as volume increases in the dry chemical tank.

13. *Rupture disc bursts.* Pressure increases to a point where it bursts the rupture disc on the dry chemical storage tank discharge outlet.

14. *Dry chemical flows.* Fluidized dry chemical/nitrogen mixture flows through the dry chemical system piping under pressure.

15. *Dry chemical application.* One or more nozzles apply dry chemical to the hazard.

APPLICATIONS FOR DRY CHEMICAL SYSTEMS

Engineered dry chemical piping systems can be designed for petrochemical applications such as offshore liquefied natural gas (LNG) loading platforms, as well as marine loading docks and machine spaces. Manufacturing applications include paint spray booths and dip tanks. Utility applications include transformer protection, electric generator protection, and turbine protection. Mining applications include conveyor drives and lubrication areas.

Automobile gas stations sometimes are protected by a dry chemical suppression system, as are industrial fuel-dispensing facilities. In addition, industrial paint spray booths and large vehicles, such as mining, logging, landfill, and transit vehicles, are successfully protected by dry chemical suppression systems. **Figure 10-6** is a schematic diagram of a vehicle dry chemical system, and **Figures 10-7** and **10-8** are, respectively, photos of dry chemical containers and a vehicle dry chemical test.

Dry Chemical Systems for LNG Hazards

NFPA

NFPA 59A, Production, Storage, and Handling of Liquefied Natural Gas

Liquefied natural gas (LNG) is defined by NFPA 59A, *Production, Storage, and Handling of Liquefied Natural Gas*, as "a fluid in the cryogenic liquid state that is composed predominantly of methane and that can contain minor quantities of ethane, propane, or nitrogen." LNG is a significant gaseous energy resource used to provide power.

Because LNG liquid can vaporize, protection of both liquid and vapor LNG hazards should be considered. Dry chemical can protect many LNG applications.

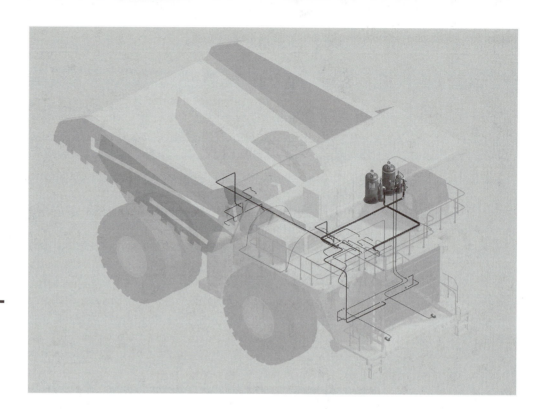

Figure 10-6
Schematic view of a vehicle dry chemical system. (Courtesy of Ansul, Inc.)

Figure 10-7 *Dry chemical containers installed on a vehicle. (Courtesy of Ansul, Inc.)*

Figure 10-8 *Test of a vehicle-mounted dry chemical system. (Courtesy of Ansul, Inc.)*

Ansul, Inc. has published a booklet entitled "Fire Protection Solutions for Lique-fied Natural Gas," Form No. F-75158-2, (2007), which provides valuable advice for designers, including the following:

- *Vaporized LNG:* fires in an enclosed volume; total flooding carbon dioxide is recommended as the best solution.
- *Vaporized LNG:* pressure and pool fires; dry chemical is recommended as the best solution.
- *Liquid LNG:* spill fires; dry chemical or a combination of dry chemical, followed by the application of high expansion foam, is recommended as the best solution.
- *Liquid LNG:* pressurized and pool fires; dry chemical or a combination of dry chemical, followed by the application of high expansion foam, is recommended as the best solution.

NFPA 59A provides requirements for emergency shutdown of LNG supplies, fire and leak detection, gas detection, water suppression systems, and portable or wheeled fire suppression equipment.

TYPES OF DRY CHEMICAL SYSTEMS

Four types of dry chemical systems are recognized by NFPA 17.
1. Total flooding
2. Local application
3. Hand hose line
4. Pre-engineered systems

This chapter concentrates primarily on total flooding dry chemical systems.

TOTAL FLOODING DRY CHEMICAL SYSTEMS

A total flooding system, shown in **Figure 10-9**, is designed to fill an enclosed volume, such as a paint spray booth, with a minimum concentration of dry chemical/expellant gas mixture. Deep-seated fires must be protected with a multi-purpose dry chemical capable of reaching the surface of the smoldering combustible. The enclosure must be sealed, and the enclosure ventilation system must be shut down in a manner similar to that used in gaseous agent systems.

Unclosable Openings in Total Flooding Dry Chemical Applications

Criteria for the evaluation of compensation for unclosable openings are as follows:

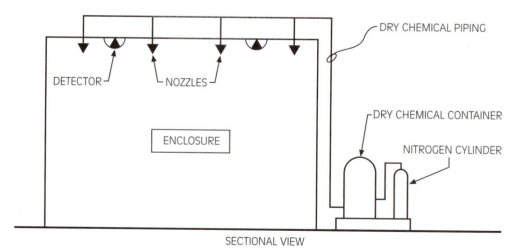

Figure 10-9 *Total flooding dry chemical system.*

- When the area of unclosable openings is within 1% to 5% of the total enclosure interior surface area, an additional 0.5 lb (2.44 kg/m^2) of dry chemical per square foot of opening is required.

- When the area of unclosable openings is between 5% and 15% of the enclosure interior surface area, an additional 1.0 lb (4.88 kg/m^2) of dry chemical per square foot of opening is required.

- The total area of unclosable openings is not permitted to exceed 15% of the total interior surface area for total flooding systems, except in cases where pre-engineered systems are used.

The interior surface area of a room is defined as the area of the floor, plus the area of the ceiling, plus the area of all four walls. The percentage of the area of unclosable openings is the ratio of the area of unclosable openings to the total interior surface area of the room.

Total Flooding Quantity Calculation

NFPA 17 does not specify a method to determine the amount of dry chemical required for a total flooding application. Manufacturers' criteria, in accordance with their listing, apply to the design of dry chemical systems. Each manufacturer has criteria that apply only to their dry chemical, nozzles, and equipment. Ansul Fire Protection uses the following design criteria for its equipment:

$$Q_b = (V \text{ hazard}) \times (0.0385 \text{ lb/ft.}^3)$$

$$Q_{b(\text{metric})} = (V \text{ hazard}) \times (0.01746 \text{ kg/m}^3)$$

where Q_b = basic minimum dry chemical required, in pounds (kg)
        V = volume of enclosure protected, in cubic feet (m^3)

Additional agent is added to the total above to account for unclosable openings.

EXAMPLE 10-1

Determining Dry Chemical Total Flooding Quantity

Using NFPA 17 criteria, determine whether a total flooding system or a local application system is appropriate to protect an enclosure 10 ft. wide, 20 ft. long, and 8 ft. high, with 90 sq ft. of unclosable openings. If total flooding is appropriate, use Ansul criteria to determine the minimum amount of dry chemical required.

Solution

The volume of the hazard is:

$$V = (L) \times (W) \times (H)$$

$$= (20') \times (10') \times (8')$$

$$= 1600 \text{ ft.}^3$$

The interior surface area of the enclosure is:

ceiling: $(20') \times (10') = 200$ ft.^{2}

floor: $(20') \times (10') = 200$ ft.^{2}

wall: $(20') \times (8') = 160$ ft.^{2}

wall: $(20') \times (8') = 160$ ft.^{2}

wall: $(10') \times (8') = 80$ ft.^{2}

wall: $(10') \times (8') = 80$ ft.^{2}

Therefore, total interior surface area is 880 ft.^{2}.

The total area of unclosable openings is 90 ft.^{2}, as stated in the problem. To determine the percentage of unclosable openings relative to the total interior surface area:

$$\frac{90 \text{ ft.}^2}{880 \text{ ft.}^2} = 0.102 \text{ or } 10.2\%$$

Because the percentage of unclosable openings as related to the total interior surface area is less than 15%, a total flooding system is appropriate for this application.

Using Ansul criteria, the minimum basic quantity (Q_b) of dry chemical required is:

$$Q_b = (V) \times (0.0385 \text{ lb/ft.}^3)$$
$$= (1600 \text{ ft.}^3) \times (0.0385 \text{ lb/ft.}^3)$$
$$= 61.6 \text{ lb dry chemical}$$

The total of unclosable openings falls between 5% and 15% of the total interior surface area, so 1 pound of dry chemical per square foot of unclosable opening must be added to the basic quantity calculated previously. The problem stated that the area of unclosable openings is 90 ft.^{2}. The additional quantity required for the openings (Q_o) is:

$$Q_o = (1 \text{ lb/ft.}^2) \times (90 \text{ ft.}^2) = 90 \text{ lb}$$

The total minimum dry chemical required (Q_t) for this total flooding application is equal to the basic quantity (Q_b) plus the quantity required to compensate for the openings (Q_o):

$$Q_t = Q_b + Q_o$$
$$= (61.6 \text{ lb}) + (90 \text{ lb})$$
$$= 151.6 \text{ lb}$$

This quantity is required to be distributed to the hazard within 30 seconds. It should be noted that a pre-engineered dry chemical suppression likely would be specified for this hazard, using these calculations as the basis for specification. Note that the amount of dry chemical flowing through openings exceeds the basic flooding quantity. This may be a concern for designers, but the calculations demonstrate that the preceding procedure is permissible.

Nozzle Spacing and Location for Total Flooding Systems

The minimum nozzles required for a total flooding system can be estimated by specifying one nozzle for each 500 ft.^{3} (14.16 m^3) of enclosure volume:

$$N = (V)/(500 \text{ ft.}^3)$$
$$N = (V)/(14.16 \text{ m}^3)$$

where N = minimum estimated number of nozzles (whole number)
V = volume of hazard, ft.^{3} (m^3)

In most cases, nozzles may be spaced 7'-6" (2.286 m) apart and 5'-0" (1.524 m) from a wall or obstruction, assuming a 20'-0" (6.096 m) maximum ceiling height.

EXAMPLE 10-2

Determining the Number and Spacing of Nozzles for a Total Flooding Dry Chemical System

Determine the minimum number of nozzles for the enclosure described in Example 10-1, and sketch the nozzle layout.

Solution

Room volume is 1600 ft.^{3}, so the number of nozzles is:

$$N = (V)/(500 \text{ ft.}^3)$$
$$= (1600 \text{ ft.}^3)/(500 \text{ ft.}^3)$$
$$= 4 \text{ nozzles}$$

Figure 10-10 is a sketch of the system.

After these estimates are performed, more detailed design, involving flow calculations, nozzle and cylinder selection, and pipe size determination, can be performed in accordance with the manufacturer's recommendations.

LOCAL APPLICATION DRY CHEMICAL SYSTEMS

NFPA 17 requires that a hazard protected by a local application system be isolated from other hazards so it can be considered and protected individually without affecting or being affected by other hazards. Exterior local application systems must consider wind dispersion of dry chemical and possible

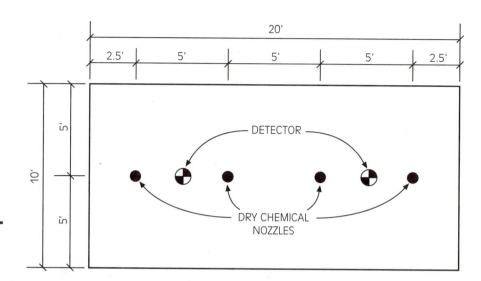

Figure 10-10
*Illustration for
Example 10-2.*

environmental effects of accidental or purposeful discharge. Protection of liquid surfaces must consider nozzle placement that would minimize splashing or otherwise disturbing the liquid surface. Discharge times and rates are determined by the authority having jurisdiction (AHJ) and the manufacturer. NFPA 17 does not provide design criteria for local application dry chemical extinguishing systems.

HAND HOSE LINE DRY CHEMICAL SYSTEMS

Manually actuated hand hose line systems are permitted as a supplement but not as a replacement for automatic fixed extinguishing systems. If the possibility exists that hand hose line systems could be discharging dry chemical on a hazard protected by a dry chemical system, a separate dry chemical supply must be provided for the hand hose line system. The agent supply capable of supporting discharge of dry chemical from a hand hose line must consider a supply capable of supporting a minimum duration of 30 seconds for each hand hose line operated from a fixed source.

pre-engineered systems
packaged units in which supply quantity, nozzle selection, pipe size, and detector selection are predetermined for a range of volumes, areas, or applications, and are listed as a unit

PRE-ENGINEERED DRY-CHEMICAL SYSTEMS

Pre-engineered systems are used primarily for protecting industrial processes and paint booths, vehicle-fueling service stations, and mobile equipment. They may be designed for local application, total flooding, or hand hose line systems.

These systems are packaged units in which supply quantity, nozzle selection, pipe size, and detector selection are predetermined for a range of volumes, areas, or applications, and are listed as a unit.

WET CHEMICAL SYSTEM DESIGN

wet chemical
a solution of water and chemical to form an extinguishing agent

A **wet chemical** is a solution of water and organic or inorganic salts, or a combination thereof, to form an extinguishing agent. The water, in a manner similar to the gas for dry chemical, provides a medium for ease of transport of the chemical, through piping and nozzles, to the hazard and provides additional cooling. The chemical used is primarily potassium carbonate or potassium acetate.

Dry chemical extinguishing systems, designed in accordance with NFPA 17, *Standard for Dry Chemical Extinguishing Systems*, once were commonly used for the protection of restaurant hoods and cooking equipment, but new dry chemical systems have not been manufactured for this application since 1994, when their listing for cooking applications was removed from UL 300, *Fire Testing of Fire Extinguishing Systems for the Protection of Restaurant Cooking Areas*.

Wet chemical extinguishing systems, designed in accordance with NFPA 17A, *Standard for Wet Chemical Extinguishing Systems*, have predominately replaced dry chemical systems as a UL 300-listed extinguishing system for protection of cooking applications in restaurants, hotels, schools, nursing homes, hospitals, and other facilities that feature cooking equipment, particularly deep-fat fryers and other cooking operations involving significant amounts of heated grease.

The major advantage in using wet chemical extinguishing systems is the extinguishment of UL 300 test fires without reflash. Dry chemical systems fail these tests. A secondary advantage to wet chemical is the ease of clean-up and reduced restaurant downtime after discharge. Wet chemical is less prone to migration and dispersion from its intended target than is dry chemical, and poses less hazard to kitchen personnel.

Cooking Grease and Wet Chemical

The primary hazard facing system designers in cooking equipment protection is cooking grease. Cooking grease is not a concern when used at room temperature because insufficient vapor is released for ignition but can become a significant problem when heated. Heated grease offers the following hazardous scenarios:

- Cooking grease, when heated to its autoignition temperature, will ignite spontaneously at approximately 685°F (363°C) without the need for a pilot flame to come in contact with the surface area of the grease. This hazard can be reduced by using preprogrammed temperature controls that regulate temperatures below the autoignition temperature of the grease.

- Cooking grease, when it contains flammable material such as french fries, potato chips, fried chicken, or doughnuts, will have an altered autoignition temperature with solid fuel added to the flammable liquid that differs with each type and quantity of food.
- Cooking grease, when it spills or boils over into a piloted flame such as a gas or electric stove-top burner, will ignite and perhaps spread in a potentially violent fire.
- Exhaust hoods and exhaust ducts, when they become coated with cooking grease, can create a dangerous fire condition that requires direct application of wet chemical.

Wet Chemical Extinguishing Mechanism

Wet chemical extinguishing agents extinguish cooking fires by the following suppression methodologies:

- Wet chemicals react chemically with hot grease (saponification) to form a soap-like layer of foam on the surface that excludes oxygen from the combustion process.
- Wet chemicals cool the flame via the heat of vaporization.
- Wet chemicals provide some extinguishing advantage by smothering the flame with extinguishant.
- Extinguishment is enhanced when wet chemical is applied to a grease fire at a velocity and with a droplet size that will not cause the grease to splash and spread from its container.
- Distribution of wet chemical to a hazard in a fine spray or mist delivers the extinguishant to the hazard without splash and without thermal shock to a hot cooking appliance.

Reignition

Reignition is a concern with cooking-grease fires. To prevent reignition, the power to the cooking appliance must be turned off automatically by the fire alarm control unit upon detection of a fire, and the wet chemical suppressant must remain on the surface of the grease until the temperature of the grease falls below its autoignition temperature.

Pre-Engineered Wet Chemical Systems

Most wet chemical extinguishing systems are listed as a pre-engineered unit, with detectors, heat-sensitive fusible links, control panel, wet chemical tanks,

Figure 10-11

Components of a wet chemical extinguishing system. (Courtesy of Ansul, Inc.)

wet chemical extinguishing agent, piping, and nozzles included in a pre-engineered package, as shown in **Figure 10-11**. The installed pre-engineered wet chemical system permits both automatic and manual operation.

Nozzles are installed in grease ducts, exhaust hoods, plenums, and filters, and are directed at range tops, griddles, broilers, ovens, and deep fat fryers, as shown in **Figure 10-12**. Pre-engineered systems are offered in a variety of sizes that correspond to the delivery of wet chemical to the required surface areas of a given hazard. **Figure 10-13** shows a cooking operation being protected by a wet chemical system.

Sequence of Wet Chemical System Operation

The operation of a wet chemical extinguishing system is as follows:

- The system is initiated by heat detection, by operation of a heat-sensitive fusible link, or by manual operation. Heat detectors generally are not used because they can become loaded with grease residue and their response times can be increased. Fusible links, usually employing eutectic solder, which melts at a precise temperature, constitute the predominant release mechanism for wet chemical extinguishing systems. A series of fusible

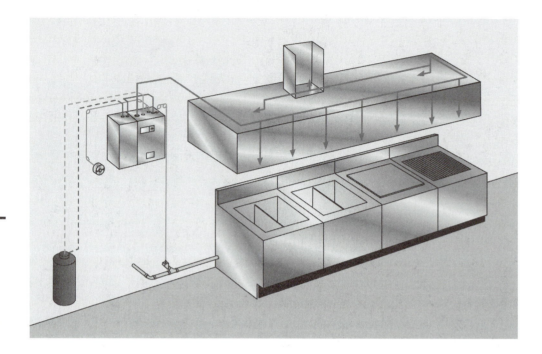

Figure 10-12   *Wet chemical extinguishing system layout for restaurant applications. (Courtesy of Ansul, Inc.)*

Figure 10-13   *A range and range hood are protected by a wet chemical system; wet chemical nozzles can be seen at the top of the photo, and a wet chemical extinguisher is shown at left. (Courtesy of Ansul, Inc.)*

links are connected by a cable, arranged to trip an actuator when a eutectic link separates, which also can send a signal to the fire alarm control unit to initiate a local or remote alarm.

- The cooking equipment is shut down by mechanical means, usually operated by a cable-actuated apparatus.
- When connected to a fire alarm control unit, the alarm is initiated to local and/or remote alarms.
- Other functions initiate, including door closure, exhaust-fan initiation or cessation, or duct damper operation.
- Wet chemical is discharged by automatic opening of the wet chemical cylinder release mechanism, with pressure created by an expellant gas.
- Wet chemical extinguishing agent is distributed through system piping to the nozzles, which are directed at specific points of hazard.
- Wet chemical is applied to surface areas of the hazard.
- Hazardous surfaces are coated.
- Saponification of grease fires occurs.
- The fire is extinguished. Extinguishment occurs above the autoignition temperature by excluding oxygen from the combustion process. Grease may remain above the autoignition temperature for minutes thereafter during this phase.
- Grease fire temperature is lowered to below its autoignition temperature.
- The hazard is cleaned up and the agent replenished.
- Detectors and fusible links are replaced as required.
- The system is reset and placed back into service.
- Normal cooking operations resume.

SUMMARY

Dry chemical fire protection systems use sodium bicarbonate-based, potassium-based, and multipurpose dry chemicals. These agents provide rapid flame knockdown and coating of the combustible and break the combustion chain reaction. Dry chemical can become solidified when exposed to moisture.

Clean-up of expended dry chemical may be a problem, and exposure of personnel to dry chemical agents or their byproducts of decomposition are of considerable concern. Dry chemical systems sometimes are difficult to test because of the residue they leave or because of public health concerns for large exterior dry chemical systems. Dry chemical is fluidized with nitrogen to enable the powder to flow through pipes to a hazard. Dry chemical systems are designed for total flooding or local application in accordance with the manufacturer's listing.

Wet chemicals saponify grease fires and have taken over as the predominant restaurant fire suppression system. Wet chemical systems have been found to be ideal for saponification of grease fires, and extinguish by smothering, flame cooling, and chemical reactions, without propensity for reflash. Most wet chemical suppression systems are selected from listed pre-engineered packages.

REVIEW QUESTIONS

1. List and compare the available types of dry chemical used for fire protection.

2. Describe the extinguishing mechanism used by dry chemicals in fire protection.

3. Explain the sequence of operation of a dry chemical system. Why does nitrogen have to be mixed with the dry chemical powder? Would you think that fluidized dry chemical flowing to a hazard through a straight, mile-long pipe would be effective? Why or why not?

4. A room measures 50 ft. long, 35 ft. wide, and 20 ft. high. An unclosable opening measuring 10 ft. long and 10 ft. high is situated on one wall of the room. Is this room amenable for the application of a total flooding dry chemical system, or does NFPA 17 require a local application system?

5. If the room in Review Question 4 is acceptable for a total flooding system, determine the basic quantity of dry chemical required using Ansul criteria, determine the amount of dry chemical that must be added to compensate for the unclosable opening, and determine the total minimum quantity of dry chemical needed.

6. If the system in Review Questions 4 and 5 must be designed for total discharge of agent in 30 sec, what is the dry chemical flow rate, in lb/min?

7. For the system in Review Questions 4, 5, and 6, determine the number of nozzles required, and sketch the system.

8. Describe the mechanism for extinguishment for a wet chemical system.

DISCUSSION QUESTIONS

1. List and discuss the advantages and disadvantages of dry chemical as a fire protection agent.

2. What personnel hazards are associated with dry chemical systems? Would total flooding be more of a personnel hazard than local application systems? Why? Give an example of a room or situation in which a dry chemical discharge could create personnel hazards.

3. What is the problem associated with testing of dry chemical systems? If you were the AHJ, would you test a large exterior dry chemical system adjacent to a public area? Why or why not? If not, what would you do to ensure reliability?

ACTIVITIES

1. Inspect a dry or wet chemical system for a range hood or other application. Where are the nozzles and dry chemical cylinders located? What kind of detection is used? Has a full discharge test ever been conducted? If not, what was done to ensure proper function of the system? Has the system ever discharged? If so, was the system effective, and was clean-up a problem? Obtain records of the testing and inspection of the system and evaluate them for conformance with NFPA 17.

2. Meet with a consulting or specifying engineer and determine the criteria used for specification of a dry or wet chemical system for a hazard, as opposed to systems employing other agents.

Unit

4

SPECIAL HAZARD DETECTION, ALARM, AND RELEASING FIRE ALARM SYSTEMS

Chapter

11

FIRE DETECTION AND ALARM SYSTEMS

Learning Objectives

Upon completion of this chapter, you should be able to:

- List, compare, and contrast the types of fire alarm systems.
- Properly select a decibel rating for a household warning smoke alarm.
- Choose the best smoke alarm for a household fire alarm system, and defend your choice.
- Identify fire alarm requirements for a system that complies with the Americans with Disabilities Act.
- List and discuss ways to maximize the reliability of a fire alarm system.
- Discuss performance objectives that can be met by a fire alarm system.
- Evaluate the functions of a fire alarm system, and compare them to achievable performance objectives.
- List and evaluate concerns relative to smoke alarm notification with respect to the waking of sleeping children.
- Discuss the intent and uses of a mass notification system.

An individual in good health, possessing all five senses, can be a good fire detector but may not always be a good fire alarm system. A person may not identify a fire alarm signal correctly or may otherwise react inappropriately and fail to notify others, or ignore the detection of smoke until it is too late to initiate a successful evacuation of all occupants from the building. For this reason, building and fire codes require automatic fire detection and alarm systems to be designed and installed in accordance with NFPA 72, the *National Fire Alarm Code*. The latest edition of NFPA 72 usually is preferable.

The latest edition of NFPA 72 includes requirements and information that takes into consideration new products, technology advances, research findings, trends, and current thinking in fire protection and fire safety. One must be aware, though, that not all jurisdictions adopt the latest editions of national codes and standards, so working with the appropriate editions is imperative. NFPA 72 does not apply unless the authority having jurisdiction mandates a building code, such as NFPA 5000, NFPA 101, or the International Building Code, all of which invoke the use and applicability of NFPA 72.

However comforting it may be when a fire alarm system is installed, it must be remembered that fire detection and alarm systems do not extinguish or control fires. The primary purpose of a fire alarm system is to provide local and remote notification and to initiate evacuation or relocation of occupants in a building. Analysis of the level of fire safety in a building usually is not complete unless an automatic fire suppression system has been considered.

Sometimes a fire spreads too rapidly for the safe egress of occupants, even when they receive proper notification. Consider the case of persons connected to intravenous devices or life support systems on the fifth floor of a nursing home in an aging wooden structure. Or occupants may ignore alarm signals or stay in the fire area attempting to extinguish the fire or to recover possessions. The concept of ownership often goes hand-in-hand with the concept of protecting that which is owned. The wise fire protection engineer or technician considers a fire detection and alarm system as a companion to a fully functioning automatic fire suppression system.

NFPA 72 (2007) recognizes the following types of fire alarm systems:

1. Household Fire Alarm Systems
2. Protected Premises (Local) Fire Alarm Systems
   - Building Fire Alarm Systems
   - Dedicated Function Fire Alarm Systems
   - Releasing Fire Alarm Systems
3. Public Fire Alarm Reporting Systems
   - Municipal Fire Alarm Systems
   - Auxiliary Fire Alarm Systems
4. Combination Fire Alarm Systems

5. Supervising Station Fire Alarm Systems
- Central Station Service Fire Alarm Systems
- Proprietary Supervising Station Fire Alarm Systems
- Remote Supervising Station Fire Alarm Systems

HOUSEHOLD FIRE ALARM SYSTEMS

Most household fire warning systems consist of self-contained detection and alarm units that are positioned to provide accurate detection and the notification necessary to awaken sleeping occupants and initiate evacuation. In the United States, most fire deaths and injuries occur in the home. This does not necessarily mean that a home is intrinsically more hazardous than a chemical processing plant. Consider a worker who is prohibited from smoking in the workplace but falls asleep in bed at home while smoking, or an employee who is required to follow strict fire safety protocol at work but leaves a frying pan unattended on the stove at home. Perhaps without workplace restrictions and protocol, we let down our guard at home. Employers would not knowingly permit an employee to place a kerosene space heater next to drapes, a mattress, or flammable substances, but this occurs with frightening regularity in homes during the winter months.

What is particularly frightening about deaths from home fires is that the weakest among us are the primary victims of residential fires. Children younger than 5 years of age and adults older than 65 years of age are the predominant victims of residential fires. People in good health with the strength and agility to escape have a considerably better chance of surviving a fire. Fire sometimes places people in an unreasonable position, forcing them to make life-or-death decisions under duress and with little time. Is there time to save the baby, the pets, the in-laws, the wedding photo album, the computer? If not, which should be saved and which should be abandoned, and in what order?

The Consumer Product Safety Commission, in its report "1999 Residential Fire Loss Estimates," reports that the primary sources of ignition in residential fires are

- cooking equipment fires (ranges and ovens): 29%
- heating and cooling equipment fires: 14%
- electrical distribution system components (wiring and lighting): 12%
- by item first ignited: upholstered furniture ignition was involved in the most deaths
- by heat source: smoking materials were the largest contributor to deaths from residential fires

A relatively recent phenomenon is the fire caused by halogen lamps near drapes, particularly in college dormitories. This phenomenon might tend to increase the incidence of fire with respect to hot objects including properly operating electrical equipment. The issue of fires in dormitories has become a recent hot topic. How do you protect a 75-year-old woodframe fraternity house after an all-night keg party? Will smoke detectors awaken sleeping occupants? Will one existing interior open stairway be adequate to evacuate an overly occupied facility with impaired occupants? With recent fatalities in college housing creating national news, colleges are increasingly protecting these structures with automatic sprinkler systems in addition to detection and alarm systems.

The public has some common misunderstandings about fire, some of which result in injuries and deaths that could have been prevented. These include the following:

- "It appears to be a small fire, mostly smoke. I've got time to go back into the building and fetch the computer, my golf clubs, and the wedding album." Fires can be deceptive if they are confined in a room with limited air, or if they are spreading vertically in walls or blind spaces.

- "I can outrun the fire." Depending on the location and type of combustibles involved, the fire may have a head start, especially if the fire is on the first floor and you are on the third floor. An alternative exit plan is needed in some instances.

- "It won't happen here." Denial is dangerous and lowers our awareness of hazardous situations.

- "Flashover—what's that?" A flashover is a sudden and violent combustion of all combustibles in an enclosed space. A room in flashover is likely to spread flame quickly to adjacent spaces and doom the entire structure.

- "I don't need to replace the battery on my smoke detector. If there's a fire, I'll smell the smoke." Smoke contains carbon monoxide and other potentially toxic gases that can prevent a person from awakening.

- "As long as I stay away from the flame, I'm okay." Most deaths from fire are the result of inhalation of toxic smoke and gases, such as carbon monoxide. This can occur at a significant distance from the point of fire origin.

- "Time spent trying to extinguish the fire is more valuable than time spent calling the fire department." The opposite is true in most cases.

NFPA 72 Requirements for Household Fire Alarm Systems

As a minimum, NFPA 72 (2007) requires the following:

- at least one smoke alarm located on each level of a residence
- one smoke alarm located in each bedroom
- one smoke alarm located in the corridor outside each sleeping area
- in a split-level home, one smoke alarm is sufficient for both ceiling elevations, provided that smoke is free to distribute between levels.

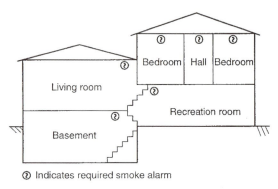

ⓣ Indicates required smoke alarm

SPLIT LEVEL ARRANGEMENT.

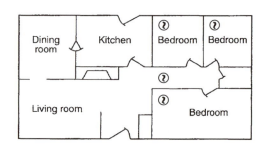

A SMOKE ALARM SHOULD BE LOCATED BETWEEN
THE SLEEPING AREA AND THE REST OF THE
DWELLING UNIT AS WELL AS IN EACH BEDROOM.

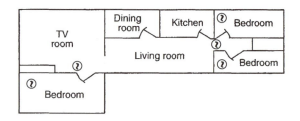

IN DWELLING UNITS WITH MORE THAN ONE SLEEPING AREA,
A SMOKE ALARM SHOULD BE PROVIDED TO PROTECT
EACH SLEEPING AREA IN ADDITION TO SMOKE
ALARMS REQUIRED IN BEDROOMS.

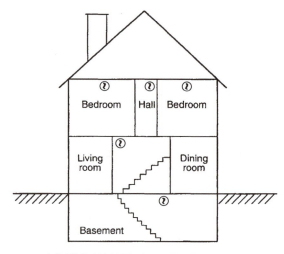

A SMOKE ALARM SHOULD BE LOCATED ON
EACH LEVEL IN ADDITION TO EACH BEDROOM.

Figure 11-1 *Residential fire alarm systems—detector arrangement.*

Source: NFPA 72 (2007), Figure A-11-5.1. Reprinted with permission from NFPA 72, *National Fire Alarm Code*®, Copyright © 2007, National Fire Protection Association, Quincy, MA 02169. This reprinted material is not the complete and official position of the National Fire Protection Association on the referenced subject which is represented only by the standard in its entirety.

Figure 11-1 illustrates this concept.

Your home may not meet the above requirements or the arrangement shown in Figure 11-1. The requirement that smoke alarms be installed in bedrooms was added for the first time in the 1993 edition of NFPA 72 and is not retroactive; thus, homes built before the requirement was instituted are not required to be modified.

The performance objective of a residential fire alarm system is notification for life safety and egress.

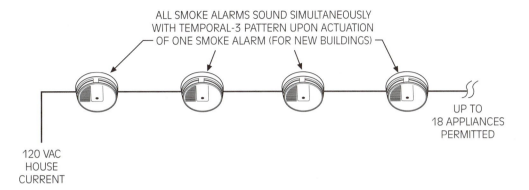

ALL SMOKE ALARMS SOUND SIMULTANEOUSLY
WITH TEMPORAL-3 PATTERN UPON ACTUATION
OF ONE SMOKE ALARM (FOR NEW BUILDINGS)

UP TO
18 APPLIANCES
PERMITTED

120 VAC
HOUSE
CURRENT

Figure 11-2 *Parallel wiring of smoke alarms; individually activated devices with self-contained alarm.*

Household Detection and Alarm System Arrangement

Figure 11-2 shows a parallel wiring smoke detection schematic for most homes, with up to 18 individually activated detectors with self-contained alarms wired to house current, 120 volts alternating current (VAC), sounding the temporal-3 pattern (discussed later in the chapter). NFPA 72 permits a maximum of 18 appliances, including up to 12 smoke alarms, in series. All smoke alarms are required to sound simultaneously upon actuation of one smoke alarm when installed in new construction. Electric smoke alarms also are available with a self-contained battery standby power source, which supplies power to the appliance when house power is lost.

More sophisticated fire detection and alarm configurations can be installed with a fire alarm control unit (FACU), as shown in **Figure 11-3**. The fire alarm control unit arrangement is more reliable because it has battery backup for the entire system and constantly monitors the integrity of each circuit, can issue a trouble alarm if a circuit fault is detected, and can permit notification appliances to be placed in locations other than where the detectors are placed. Additional benefits of the fire alarm control unit are the availability of a trouble signal, a secondary power source, and remote notification.

Another arrangement permitted by NFPA 72 (2007) is the use of battery-powered smoke alarms without a connection to a permanent power supply. These appliances are not recommended because batteries have a lifespan of approximately one year and tend not to be replaced. Several national fire organizations are promoting a "change your clock, change your smoke detector battery" awareness program to coincide with spring and fall changes from standard time to daylight savings time.

In the 1970s, when awareness of and installation of residential detectors increased dramatically, deaths from residential fires decreased significantly as battery-powered detector give-away programs became widespread. Once batteries began to expire in large numbers, however, residential fire deaths increased slightly. Even when batteries are working properly, detector operation depends

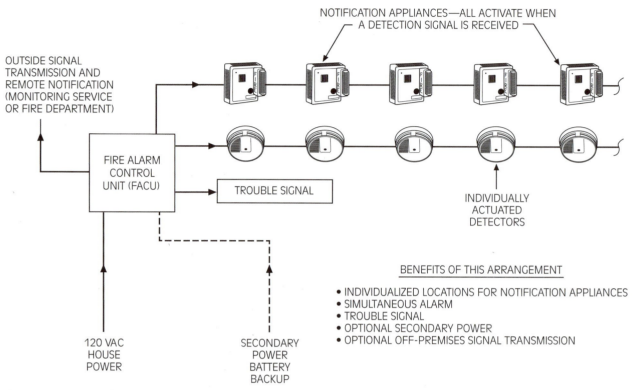

Figure 11-3   *Optional arrangement for residential fire alarm systems with a control unit; smoke alarms can eliminate the need for a separate system of notification appliances.*

on the location and installation of the detector. A lucky family was saved by a battery-powered smoke alarm, received in a give-away program that sounded even though it was still in its original box, placed in a drawer by the bed. In sum, fire protection professionals serve the public better when permanently wired smoke alarms are required.

Household Detection Device Selection

Smoke detectors are effective in most residential fires because these fires traditionally involve a period of smoldering before a flame is evident, as exemplified by a cigarette burning slowly between two seat cushions. Because smoke detectors are specifically designed to detect smoldering and fires in their incipient phase, a smoke detector is most likely to provide the most timely notification.

Smoke detectors are effective early warning devices when they are placed properly. They are likely to produce unwanted alarms if they are placed too close to sources of ambient smoke, such as fireplaces, garages, or stoves. Unwanted alarms also may be activated by steam, such as found within or directly outside a bathroom, shower room, or sauna.

Household Audible Notification Requirements

decibel (dB)
a unit of sound intensity at a given distance

Audible notification appliances are rated in **decibels (dB)**, a unit of sound intensity at a given distance, such as 84 dB at 10 feet. A decibel is the ratio of a sound to a reference sound of 2×10^{-4} dyne per square centimeter, equal to one-tenth bel or one decibel. If P_1, and P_2 are two sound levels, P_1 is considered N decibels greater than P_2 as shown in **Figure 11-4**, if

$$N = 10 \log_{10}(P_1/P_2)$$

The human ear does not hear all sounds that are emitted, only those within frequencies perceivable by humans, usually in the 1 kHz to 4 kHz range. For this reason, sound meters often are equipped with a filter designed to replicate the

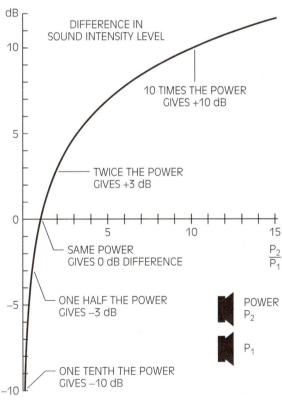

Figure 11-4
Logarithmic decibel power output chart.

performance levels of the human ear. A filter uses weighted measurements, given in dBA, which are widely used in the fire alarm industry.

Audible notification appliances are required to project sound to the pillow sufficient to awaken a sleeping occupant. Most people will be awakened by a signal of 15 dB above the ambient sound level. For individuals taking medication or under the influence of alcohol, higher decibel levels may be necessary for arousal from a deep sleep. Research in this vital area is ongoing.

Recently, children have become the focus of research relative to their ability to awaken to smoke alarms. A number of alarm tests have failed to awaken children, and children in some tests are awakened by the alarm but return to bed. These findings have prompted research into the development of smoke alarms that increase the probability of successful notification of sleeping individuals. Some research findings demonstrate that vocal notification from smoke alarms, perhaps using a recording of a mother's voice, is more effective than a standard signal in awakening children. Among the many factors influencing this phenomenon are the deep-sleep patterns of children, the inability of children to decipher an alarm signal, the lack of knowledge to decide on and initiate a course of egress, and perhaps impaired judgement in the presence of carbon monoxide. Because of children's deep sleep patterns, however, they may not awaken and react appropriately to smoke alarms no matter how loud or whose voice is played. Parents and guardians of children should be aware of this finding and should know that children cannot be expected to awaken and self-evacuate. Therefore, parents or guardians are responsible for ensuring the evacuation of children and teens.

Ambient sound levels in a residence may consist of noise made by a heat pump, fan, television, or other electronic device, the rush of air from air-distribution vents, a humidifier, or the roar of a self-contained window air-conditioner. The most accurate way of specifying a notification appliance decibel rating is to measure the ambient sound level in the room in which the appliance is to be located, measured during the worst-case scenario. The typical ambient sound level in most residences is 55 dB, but for especially noisy residences, a higher appliance decibel level is in order.

Doors reduce the sound level perceived from the side opposite a notification appliance. The sound reduction through most hollow-core residential doors is 15 dB, assuming that the audible appliance is close to the door. For thick, solid-core doors, a higher decibel loss should be considered.

EXAMPLE 11-1

Household Decibel Determination Including Loss Through Door

Determine the minimum decibel rating of a residential notification appliance located in a corridor directly outside a bedroom if the bedroom door is hollow-core, and if the ambient sound level in the bedroom is 55 dB, assuming that 20 dB above ambient is required to awaken a sleeping occupant in the bedroom, and the minimum decibel rating capable of awakening is 75 dB, per NFPA 72.

Solution

In this case, 20 dB above ambient is required to awaken a sleeping occupant in the bedroom. The ambient noise level in the bedroom is 55 dB, and the hollow-core door reduces the sound level of the appliance by 15 dB. The problem is solved by adding these minimum sound levels: 20 dB above ambient is required to awaken a sleeping occupant plus 15 dB required to overcome sound loss through the hollow core door plus 55 dB required to overcome the ambient sound level, for a total of 90 dB. This value represents the minimum decibel requirement for the notification appliance to awaken a sleeping occupant with the appliance placed directly outside the bedroom door.

Calculation including loss through door:

55 dB ambient noise level
20 dB to awaken sleeping individual
<u>15 dB</u> to account for loss through door
= 90 dB total minimum appliance rating

EXAMPLE 11-2

Household Decibel Determination—No Loss Through Door

Determine the minimum decibel level for a notification appliance located in a bedroom with an ambient sound level of 55 dB, assuming that the minimum decibel rating capable of awakening a person is 75 dB, per NFPA 72.

Solution

Because no intervening door will reduce the sound from the notification appliance, the decibel level of the appliance is permitted to be less than calculated in Example 11-1. In this case, 20 dB above ambient is required to awaken a sleeping occupant plus 55 dB required to overcome the ambient sound level, for a total of 75 dB, the minimum decibel requirement for a notification appliance to awaken a sleeping occupant with the appliance placed within the bedroom.

Calculation not requiring loss through door:

55 dB ambient noise level
20 dB to awaken sleeping individual
<u>0 dB</u> to account for loss through door
= 75 dB total minimum appliance rating

As can be understood by comparing Examples 11-1 and 11-2, NFPA 72 requires smoke alarms in bedrooms to increase the probability of awakening, and to remove variables associated with door design and construction, including door decibel loss values.

In new construction, NFPA 72 (2007) requires all notification appliances to sound when any one detector is actuated. A fire on the lower level, therefore, will initiate smoke alarms within the bedrooms, or a fire in any one of the bedrooms will initiate smoke alarms in all other bedrooms.

Sufficient power is required to be supplied to all notification appliances on the system for 4 minutes. Although this requirement is the same for AC-powered and battery-operated detectors, a detector with an expended battery is incapable of meeting this performance objective. Permanently wired AC-powered detectors are the most reliable means of meeting this requirement.

STANDARDIZED NOTIFICATION SIGNAL

NFPA 72 (2007) requires that audible notification appliances produce a standard signal, as shown in **Figure 11-5**. The three-pulse temporal pattern has been determined to be more recognizable as an alarm signal than is a continuous sound that could be perceived as part of the ambient noise level and could be easier to ignore. This NFPA 72 signal standardization became effective in July 1997. Figure 11-5 also shows the temporal pattern combined with a voice alarm.

In her work with the National Research Council of Canada, Guylene Proulx[1] has found that standard fire alarm signals often are ignored for the following reasons:

- failure to recognize the signal as a fire alarm
- loss of confidence in the system because of nuisance alarms
- failure to hear the signal

Narrow-Band Signaling

The 2002 edition of NFPA 72 introduced a new audible signaling requirement that is permitted to be used on fire alarm systems. Narrow-band signaling permits dB requirements based on octave band instead of dBA measurements. The fire alarm signal emitted is required to exceed the ambient noise by a specified amount in at least one octave band. Large, noisy facilities may be available for the application of narrow-band signaling.

Notification of the Hearing Impaired

Under the **Americans with Disabilities Act (ADA)**—civil rights legislation that ensures that persons with disabilities are accommodated—people with a hearing impairment have the right to demand a specialized notification signal. Any person requesting such notification must be accommodated with alternative notification in accordance with ADA, as well as NFPA 72 (2007). A hearing-impaired individual could be alerted effectively with visible strobe alarms. Tactile appliances in pillows are capable of arousing individuals with hearing and/or visual impairments. A **tactile appliance** provides notification by generating vibrating signals that can be sensed by an individual possessing the appliance.

Americans with Disabilities Act (ADA) legislation that ensures that persons with disabilities are accommodated

tactile appliance a device that provides notification by generating vibrating signals that can be sensed by an individual possessing the appliance

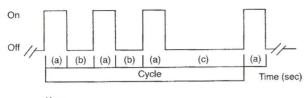

Key:
Phase (a) signal is on for 0.5 sec ±10%
Phase (b) signal is off for 0.5 sec ±10%
Phase (c) signal is off for 1.5 sec ±10% [(c) = (a) + 2(b)]
Total cycle lasts for 4 sec ±10%

TEMPORAL PATTERN PARAMETERS.

Key:
Phase (a) signal is on for 0.5 sec ±10%
Phase (b) signal is off for 0.5 sec ±10%
Phase (c) signal is off for 1.5 sec ±10% [(c) = (a) + 2(b)]
Phase (c) signal can incorporate voice notification.
Total cycle lasts for 4 sec ±10%

TEMPORAL PATTERN PARAMETERS WITH
1.5-SECOND VOICE ALLOWANCE.

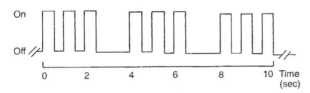

TEMPORAL PATTERN IMPOSED ON SIGNALING APPLIANCES
THAT EMIT A CONTINUOUS SIGNAL WHILE ENERGIZED.

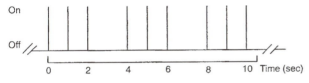

TEMPORAL PATTERN IMPOSED ON A SINGLE-
STROKE BELL OR CHIME.

Figure 11-5   *Temporal pattern notification signal.*

Olfactory notification

a form of notification that uses odors to arouse sleeping individuals

Research by the NFPA[2] has shown that tactile alerting, such as oscillating fans aimed at a person's head while sleeping, were capable of awakening 82% of deaf subjects in the study. Vibratory appliances such as bed-shakers, when placed under the mattress, awakened 92% of participants with hearing, 82% of the hard-of-hearing, and 93% of the deaf, when the appliance is placed under the mattress. This was an improvement over the performance of a 110 candela strabe. **Olfactory notification**, which uses odors to arouse sleeping individuals, has received scant research.

PROTECTED PREMISES (LOCAL) FIRE ALARM SYSTEMS

protected premises
buildings or enclosures that are supervised by a fire alarm and detection system

Protected premises are buildings or enclosures that are supervised by a fire detection and alarm system. A protected premises fire alarm system provides initiation, notification, and control within or directly outside the premises.

Objectives of Protected Premises Fire Alarm Systems

The objectives of protected premises fire alarm systems, illustrated in **Figure 11-6**, are:

- life safety protection: initiation of occupant evacuation or relocation
- property protection: early warning to enhance the probability of manual suppression by the fire service, or to release automatic fire suppression systems

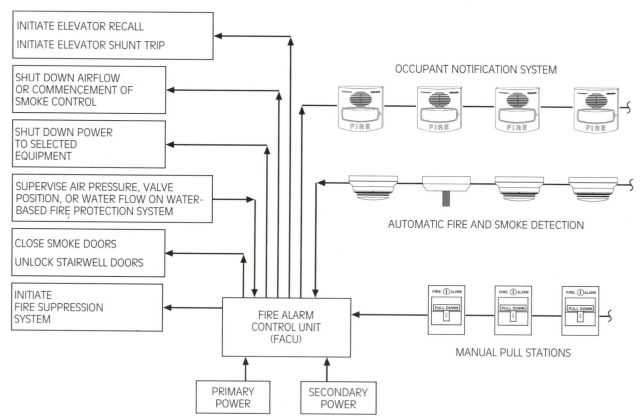

Figure 11-6   *Protected premises (local) fire alarm system. (Courtesy of Bruce Fraser.)*

- control of mechanical systems and equipment: shutting down electric power and HVAC systems, starting pressurization in stairwells, door unlocking, and commencement of the delivery of agent from a fire suppression system
- monitoring dangerous processes: explosive process observation, monitoring, toxic gas shutdown, and warning notification
- supervision of a fire alarm system: circuit malfunction or ground fault trouble indication
- investigation and prediction of an expected fire: in engineered systems, fire alarm system performance objectives are designed to be congruent with the attributes of an expected fire. For example, a flammable liquid fire is less smoky than a sofa fire and would burn at a different temperature, yielding an engineered detection and suppression design for this specific expected fire
- monitoring of air pressure, valve position, and waterflow for a water-based fire protection system
- elevator control, recall, and elevator main line power disconnect (shunt trip)

Reliability of a Protected Premises System

Failure to detect a fire and notify occupants in a timely manner can be catastrophic. The explosions perpetrated by terrorists in tall, occupied buildings clearly demonstrate that loss of a fire alarm system can result in a chaotic evacuation scenario. As exemplified by the 1993 bombing of the World Trade Center in New York, exiting a building that has lost emergency lighting, smoke control systems, or alarm notification is frightening and potentially dangerous.

NFPA 72 (2007) defines methods to limit damage to fire alarm systems by requiring redundancy or other means so the loss of one evacuation zone of the system does not compromise the proper function of any other notification zones. A **notification zone** is a floor or portion of a floor of a building, or a discrete area or wing of a building, where fire alarm notification can be provided.

Protection of the power source to a fire alarm system can be accomplished by installing the main power and backup power wiring in separate locations so the destruction of one source is less likely to affect performance of the other. Power wiring can be separated by a fire-rated wall capable of preventing fire spread through the wall for 2 hours. Experience with explosions in buildings may prompt some designers to consider a physical separation capable of protecting one of the circuits from a blast of the expected magnitude.

notification zone
a floor or portion of a floor of a building, or a discrete area or wing of a building, where fire alarm notification can be provided

Fire alarm circuits that provide notification for the purpose of partial evacuation or relocation of occupants are required by NFPA 72 to be survivable. This topic is discussed later in the chapter.

EMERGENCY VOICE ALARM COMMUNICATION SERVICE

In previous chapters, voice alarm service was recommended when verbal instructions given after receiving a detection signal could expedite evacuation or relocation in advance of discharge of the gaseous or particulate agent. This feature also is commonly used to provide direction to occupants in high-rise buildings by floor, in rooms where special relocation procedures must be followed, or where the location of the exit is not immediately obvious.

Two basic types of voice alarm service can be considered:

1. *Voice alarm (one-way).* As shown in **Figure 11-7**, this is a system that provides a recorded message giving precise instructions subsequent to receiving a detection signal.

2. *Firefighters' phone system (two-way).* A local authority having jurisdiction using a building code or NFPA 101, *Code for Safety to Life in Buildings and Structures*, may require a firefighters' phone system, as shown in

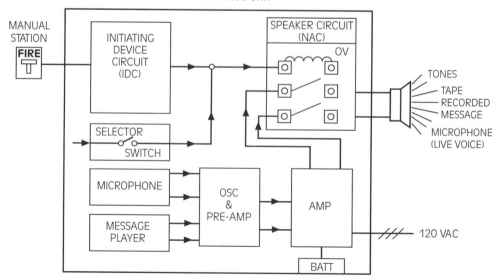

Figure 11-7 *Fire alarm/voice system. (Courtesy of Simplex Grinnell.)*

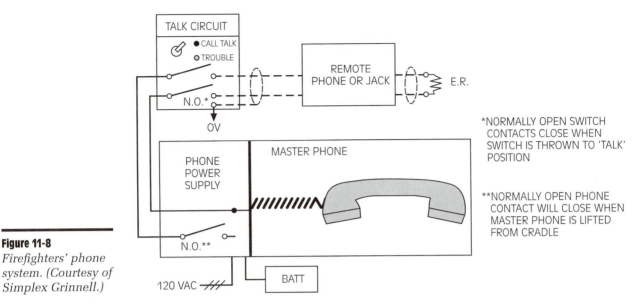

Figure 11-8

Firefighters' phone system. (Courtesy of Simplex Grinnell.)

Figure 11-8. This system permits two-way communication between firefighters in different evacuation zones. Also available are wireless phones and radios for firefighters.

MASS NOTIFICATION SYSTEMS

The President's National Strategy for Homeland Security established the case for informing citizens of emergencies in times of war and peace by way of a fully integrated national emergency response system. This understanding created the impetus for use of integrated fire alarm systems that include provisions for mass notification systems. Annex E of NFPA 72 (2007) includes new information on this initiative, which permits mass notification systems to be separate from fire alarm systems, or fully integrated.

The Annex, which is not a part of the body of the Standard, includes life-safety advice on integrating fire alarm systems as part of a national emergency notification network, using interfaces capable of performing this function. Integrated fire alarm systems are having a growing impetus with respect to NFPA 72.

The intent of a mass notification function in NFPA 72 is to provide prompt, accurate, easily understandable, and useful messages to occupants and first responders, instructing them relative to the status of natural disasters, utility failures or other human-caused dangers, biological and hazardous chemical leaks, accidents, or terrorist attacks, and advises a course of action. The systems are intended to provide warning early in the incident, and also to provide instructions that are accurate and easy to follow. The systems are capable of providing information through speakers installed within a building, as well as mass-communication speakers outside a building. Wireless or fiberoptic fire alarm networks can be used to notify command centers and initiate notification. Most research has concluded that voice notification is a more effective means of relaying information than is the traditional horns and bells notification, especially with respect to information of a critical nature. The combination of voice and other methods of audible and visible notification can increase the overall effectiveness of a notification methodology.

NFPA 72 requires that mass notification voice communications be intelligible and provides requirements for voice intelligibility. NFPA 72 gives advice on securing a mass-notification fire alarm command center and its power supply from the event about which it is warning.

WIRELESS FIRE ALARM SYSTEMS

Wireless (low-power radio) fire alarm systems are permitted by NFPA 72. These systems are permitted to be operated by a battery power supply provided that a number of steps are taken to prevent battery failure. The systems transmit an alarm signal to a receiving station capable of initiating action commensurate with the alarm. Applications include protection of asbestos environments and historic buildings.

AUXILIARY FIRE ALARM SYSTEMS

auxiliary fire alarm system
a system of manual pull boxes located throughout a jurisdiction with each box wired to a circuit connected directly to the fire department

The **auxiliary fire alarm system**, shown in **Figure 11-9**, consists of coded manual pull boxes located throughout a jurisdiction protected by a fire department, with each manual pull box wired to a circuit connected directly to the fire department. This feature also can be installed with a parallel circuit. Each box is given a unique code, and the monitoring station interprets the code to determine the exact location of the signal.

Although they were once prominent in urban areas, municipal fire alarm systems have become obsolete in some cities because of government downsizing and the prevalence of building detection systems that are monitored by private firms. Some municipal systems use radio frequency to transmit codes to the municipal alarm receiving center.

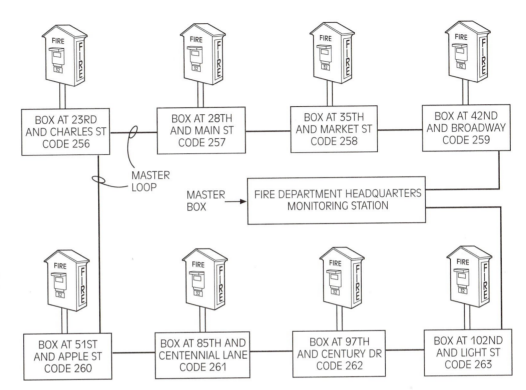

Figure 11-9 *Auxiliary fire alarm systems, in which each box sends a distinctive coded signal to the fire department monitoring station.*

REMOTE SUPERVISING STATION FIRE ALARM SYSTEMS

remote supervising station fire alarm system
a system that transmits alarm, supervisory, and trouble signals from one or more protected premises to a distant location where appropriate action is initiated

A **remote supervising station fire alarm system**, shown in **Figure 11-10**, transmits alarm, supervisory, and trouble signals from one or more protected premises to a distant location where appropriate action is taken.

Owners of protected properties pay a monthly fee to a monitoring company. This feature is becoming increasingly popular and is mandated by some building codes.

CENTRAL STATION SERVICE FIRE ALARM SYSTEMS

central station service fire alarm system
a fire alarm system that is supervised and constantly monitored by a privately owned company

A **central station service fire alarm system**, shown in **Figure 11-11**, is a fire alarm system for a protected property that is signaled to, supervised, and monitored constantly by a listed central station that takes action pursuant to the signal. Fire alarm signals are transmitted to the central station, where the signal is checked and the fire department is called by. A central station can monitor the signals of numerous fire alarm system owners.

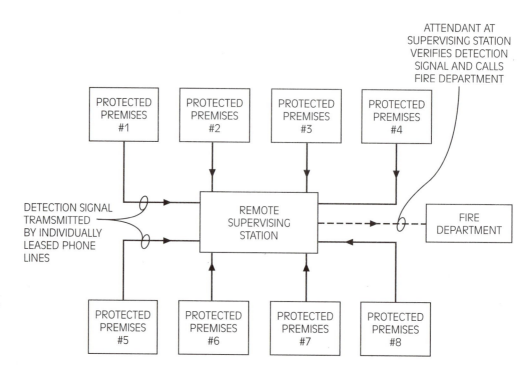

Figure 11-10 *Remote supervising station fire alarm system.*

Central station service often is provided in highly protected risk (HPR) applications, with the majority of offsite monitoring performed by remote monitoring stations. Central station service has been expanded to supervise a wide variety of protected risks, including the monitoring, recording, and supervision of residential fire alarm and intrusion alarm systems. The "service" is what differentiates remote supervising station fire alarm systems and central station service fire alarm systems. The following elements comprise central station service:

- installation of fire alarm transmitters
- alarm, guard, supervisory, and trouble signal monitoring
- retransmission
- associated record keeping and reporting
- testing and maintenance
- runner service

proprietary supervising station fire alarm system
a system that supervises properties owned by the same company

PROPRIETARY SUPERVISING STATION FIRE ALARM SYSTEM

A **proprietary supervising station fire alarm system**, shown in **Figure 11-12**, constantly monitors contiguous and noncontiguous properties under one ownership from a proprietary supervising station located at the protected property

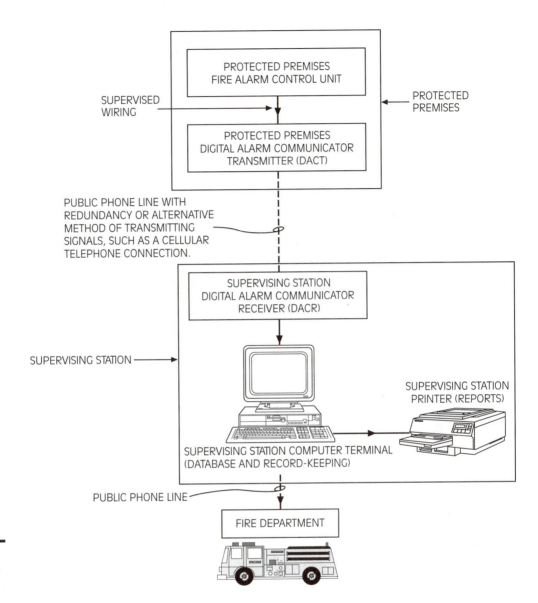

PROTECTED PREMISES
FIRE ALARM CONTROL UNIT

SUPERVISED WIRING

PROTECTED PREMISES

PROTECTED PREMISES
DIGITAL ALARM COMMUNICATOR
TRANSMITTER (DACT)

PUBLIC PHONE LINE WITH
REDUNDANCY OR ALTERNATIVE
METHOD OF TRANSMITTING
SIGNALS, SUCH AS A CELLULAR
TELEPHONE CONNECTION.

SUPERVISING STATION
DIGITAL ALARM COMMUNICATOR
RECEIVER (DACR)

SUPERVISING STATION

SUPERVISING STATION
PRINTER (REPORTS)

SUPERVISING STATION COMPUTER TERMINAL
(DATABASE AND RECORD-KEEPING)

PUBLIC PHONE LINE

FIRE DEPARTMENT

Figure 11-11

*Supervising station
fire alarm system.*

at which trained personnel are in constant attendance. These systems are owned, operated, and maintained by owners of the protected properties. The supervising station receives a detection signal, verifies it, and calls either an onsite fire brigade or an offsite fire department. A typical example of a proprietary supervising station fire alarm system is a college campus, with numerous buildings, and is often equipped with its own fire brigade.

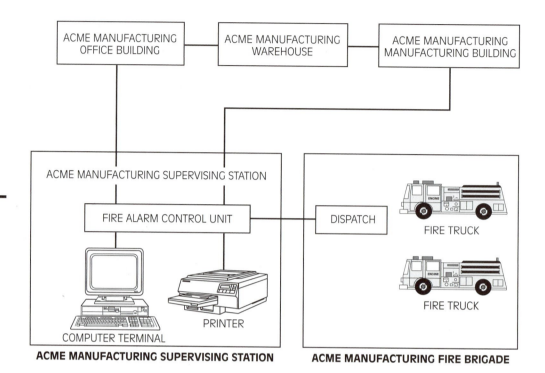

Figure 11-12
Proprietary supervising station fire alarm system; some companies have fire brigade services on their premises, and others rely on an offsite fire department.

PERFORMANCE OBJECTIVES FOR FIRE ALARM SYSTEMS

Several performance objectives for an engineered fire alarm system must be considered to provide assurance that the system is effective for the specific occupancy for which it is designed. Some examples of performance objectives that might become a part of an engineered fire alarm system analysis are shown in **Table 11-1**.

RELIABILITY OF FIRE ALARM SYSTEMS

Designing fire alarm systems to be reliable requires that each component of the system perform in accordance with its performance objectives.

Reliability of Wiring

NFPA

NFPA 70, National Electric Code (NEC)

All wiring must installed in compliance with NFPA 70, the *National Electric Code* (NEC). The NEC is an amazing document that can be appreciated more fully by travelling abroad and seeing the lack of conformity in electrical systems. In the United States an electric shaver can be plugged into any electrical outlet and

Table 11-1 *Performance objectives for fire alarm systems.*

Goals of Fire Alarm System	Examples of Performance Objectives
1. Detection	• Detect smoke particles larger than _____ microns • Detect fire at _____ °F (°C) • Detect fire within _____ seconds of ignition • Detect fire before the heat release of the fire exceeds _____ kw
2. Notification	• Provide direct notification of fire department • Provide staged evacuation or relocation of occupants • Serve occupants with sight or sound disabilities • Provide internal communication with firefighters

3. Fire safety functions (controls)	• Elevator control
	  — Return elevator to 1st floor (recall elevators)
	  — Stop elevators immediately upon detection
	  — Delay power shutdown unitil recall is completed
	  — Stop elevator at first available floor
	• Unlock stairwell doors
	• HVAC control
	  — Shut down air and close doors
	  — Exhaust smoke with fans and relief hatches
	  — Pressurize stairwells
	• Release magnetically held smoke doors

4. Suppression	• Delay automatic suppression for _____ seconds until occupants leave • Notify responsible authorities for manual suppression • Shut down suppression system after _____ seconds • Re-initiate suppression upon receipt of re-verified signal • Ensure enclosure integrity by closing doors, windows, and other closable openings

the power will be a reliable 120 VAC. In many other countries the shaver cannot be plugged in because the prongs on the plug do not match the available outlet, or the shaver might burn out if more than 120 VAC are supplied, or could perform poorly if less than 120 VAC is supplied.

The NEC is universally adopted in the United States and is becoming the electrical standard in many other countries as well. Fire protection professionals should note that this code is published by the NFPA, not by an electrical association or organization, because some early electrical systems used bare wires, ungrounded cables, or nonstandard power levels, which resulted in numerous serious fires. The NFPA took action to standardize electrical wiring to ensure proper electrical function and also to ensure life safety.

Of specific relevance is Article 760 of the National Electric Code (NEC), which applies directly to requirements of fire alarm wiring installation. NFPA 72 provides circuit descriptions and their performance characteristics, as well as survivability requirements for those circuits.

Reliability of Power Supplies

In addition to requiring wiring in accordance with the NEC, the *National Fire Alarm Code* requires no fewer than two power supplies, a primary and a secondary power, the latter of which sometimes is referred to as "backup" power, to ensure operability of the systems when commercial power is lost.

Primary Power Supply   The primary power supply is required to be constant and reliable. In the early days of electrification of the United States, some systems were AC, (alternating current), and some competing systems were DC (direct current), with proponents of each pointing out the evils and dangers of the other. The NFPA put this competition to rest by standardizing 120 VAC for general power distribution.

Although 120 VAC is supplied to houses, it is not always absolutely constant. In severe storms a lightning strike could cause a power surge that has the potential to destroy electrical components. Most of us have surge suppressors for our computers and television sets. Surge suppressors are no less important for a life safety system. Underwriters' Laboratories standard UL 864 requires built-in surge suppression for a fire alarm system.

A fire alarm control unit (FACU) notifies occupants and provides signals to the fire service. It must be placed in an enclosure that protects internal components from moisture, temperature, humidity, and physical damage. The room in which an FACU is placed is required to be accessible by the fire service and maintenance personnel, and must be protected by smoke detection if no sprinkler system is installed in the room.

Metering and undervoltage detection (brownout) devices monitor the power level coming into a fire alarm control unit, initiate a trouble signal, and transfer power to the secondary power supply when loss or undervoltage of the primary electric power is detected.

Secondary Power Supply   A secondary power supply usually is a rechargeable storage battery but also could be an approved engine-driven generator. The secondary power is required to meet the following performance objectives:

- The secondary power supply is required to initiate service within 30 seconds after detecting the loss of primary power.
- The secondary power supply must be sufficient to operate the system in a supervisory mode for a minimum of 24 hours upon loss of primary power.

Supervision of some items (such as amplifiers) does not occur during the standby period. When the panel is in standby, it must be capable of detecting sources of trouble in the system and displaying a trouble light.

- After 24 hours of standby using secondary power, the control unit must be capable of initiating an evacuation alarm and operating all audible and visible notification appliances for 5 minutes. Emergency voice alarm systems must be capable of operating on secondary power for 15 minutes, and the firefighters' phone system must be capable of operating on secondary power for no shorter than 2 hours of intermittent service.

- The secondary power supply for supervising station facilities shall be capable of supporting operations for a minimum of 24 hours.

Survivability of Fire Alarm Systems

Fire alarm systems that have been designed for use in the partial evacuation or relocation of occupants are required to be designed so an attack by fire within the signaling zone shall not impair the functioning of notification appliances within another zone. NFPA 72 requires that fire alarm circuits be protected by any of the following methods:

- 2-hour rated cable or cable system
- 2-hour rated enclosure
- performance alternative approved by the AHJ

FUNCTIONS OF FIRE ALARM SYSTEM SIGNALS

In performing its function, a fire alarm system is designed to emit the following signals:

- *Local alarm signals and functions* associated with performance objectives shown in Table 11-1. The alarm signal must be distinguishable from ambient building sounds and, with the exception of mass notification inputs, must take priority over all other signals. Mass notification requirements, added in the NFPA 72, 2007 edition, significantly expands the traditional role of a fire alarm system.

- *Remote transmission of alarm signals.* These are the signals sent off-premises. They usually consist of alarm, supervisory, and trouble signals and generally are not coded. These signals must be distinctive from other signals transmitted from a building. They can be noncoded alarm signals or coded alarm signals. Public or private coded signals usually refer to *interior* signals. Public mode is meant for the general public, and private mode is meant for persons concerned directly with implementing action.

- *Trouble signals,* visible and audible. These signals identify sources of trouble that may compromise the reliability of a fire alarm system. Such abnormal conditions include loss of primary power, grounded wiring, loose terminal connections, low power, or improperly mounted devices. A silencing switch that disables an audible signal can be provided, but visual trouble indication must remain until the problem is corrected.

- *Supervisory signals.* These signals indicate the status of equipment supervised by a fire alarm system, such as sprinkler valve position, air pressure, low temperature, or low water level. It is also common to monitor other life-safety systems or devices such as carbon monoxide (CO) alarms, hazardous or toxic gas detectors, and ventilation systems. Also, some fire and smoke detectors are permitted to initiate a supervisory signal, such as duct smoke detectors, smoke detectors used solely for releasing smoke doors, and smoke detectors used exclusively to initiate elevator recall. Trouble signals provide information relative to normal status and relay audible or visible trouble signals when a supervisory condition is abnormal.

- *Annunciation signals.* These signals are provided to give visible notification to the fire service or maintenance person relative to the location of an alarm or trouble signal. Visible indication can be in the form of a text message on a computer terminal or on the fire alarm panel. Annunciation also is commonly performed by either a system of lights on the panel labeled by floor and zone, or by a map of a building that visually highlights the zone actuated and indicates whether the alarm signal is from a detector or the sprinkler system, or both.

- *Presignal service.* This form is not recommended but can be provided by a fire alarm system with the express approval of the authority having jurisdiction. A presignal is a signal that is sent to a control room, hotel front desk, or security headquarters, and is sent without a general public evacuation alarm being sounded. The presignal must be sent to a constantly attended location, and human action is required to actuate a public evacuation alarm manually. To understand why this procedure is not recommended, consider this commonly encountered scenario:

  1. A detector responds to a fire.
  2. A signal is sent to the front desk of a hotel or to the desk of a security guard.
  3. An individual, fearful of sounding what might be a false alarm in a restaurant or all-night casino, where evacuation could cost the owner a considerable amount of money, clears the signal and goes to the area of detection to investigate.
  4. The individual becomes trapped or seriously delayed by the fire or leaves the building without sounding a general evacuation alarm. This action places all persons in the vicinity of the fire at considerable risk.
  5. With the desk clerk or security guard trapped or delayed, the fire department is not called and an alarm has not been sounded.

6. Once the fire department is called by a person other than the one responsible for responding to the presignal, the fire department arrives and finds an unattended presignal monitoring station with a cleared or deactivated presignal monitoring panel. No location for a detection signal can be obtained.

7. The first task of the fire service in cases like this is to find and rescue the trapped or missing presignal investigator, then search from floor to floor for the fire, which by now may be out of control.

As incredible as this scenario seems, it actually describes numerous fires involving the death of occupants or fire service personnel.

A positive alarm sequence, which must be approved by the AHJ, features transmission of an alarm signal to a location where the signal must be acknowledged within 15 seconds to initiate an investigation phase. Failing to respond to the signal in 15 seconds initiates building notification automatically. Trained personnel who respond within the 15-second interval have 180 seconds to investigate and reset the system. If the system is not reset within the 180-second interval, the building notification system will activate automatically. If a second detector or other initiating device is actuated, notification will commence.

SUMMARY

The types of fire alarm systems available include household, protected premises (local), public fire alarm reporting, combination, and supervising station fire alarm systems. Household warning systems are designed to awaken a sleeping person. These systems use smoke detectors because most household fires involve a period of smoldering before the fire reaches the flaming stage. The Americans with Disabilities Act requires that fire alarm systems accommodate hearing and visually impaired individuals through the use of visible or tactile devices.

Designing for reliability involves proper wiring, supervision of primary power, and the provision for a secondary (backup) power supply. Designing a fire alarm system to meet code requirements and specific performance objectives ensures that the system will responds appropriately when needed.

REVIEW QUESTIONS

1. Why is the *National Electric Code* published by the NFPA rather than an organization of electrical engineers or contractors? Research the history of this noteworthy historical development.

2. A residence requires smoke alarms within a bedroom and in the corridor directly outside the bedroom. The ambient noise level in the bedroom is 60 dB, and a solid-core door reduces the noise level by 25 dB. Determine the

minimum decibel ratings for the bedroom smoke alarm and the corridor smoke alarm.

3. A residence was built in 1980 in accordance with NFPA standards of that year. An owner wishes to upgrade the residence to be in accordance with NFPA 72, the National Electric Code, 1996 edition. What changes are required?

4. A fire protection engineer is designing a fire alarm system for an existing high-rise residential building. How will the engineer determine where visual or tactile fire alarm notification appliances are required?

5. List the functions and performance objectives for a fire alarm system, and explain the purpose of each.

6. The space shuttle has numerous backup systems to ensure redundancy. How can redundancy and reliability be designed into a fire alarm control system?

7. How does a proprietary supervising station service fire alarm system differ from a central station service fire alarm system?

DISCUSSION QUESTIONS

1. Compare and contrast the types of fire alarm systems, and give examples of each system.

2. List and discuss the heat sources identified by the NFPA as being the primary causes of ignition for residential fires.

3. What advantages would there be to using a fire alarm control unit for a residential fire alarm system as opposed to a system without a control unit?

ACTIVITIES

1. Survey and draw the detection system in your home, using accurate measurements, and compare the detection coverage with NFPA 72, 2007 edition. What differences do you find?

2. Visit a large school complex or manufacturing facility. How many of the types of fire alarm systems covered in this book can you find? Speak to the facility's fire protection engineer and determine the criteria used to determine the type of system installed. What performance objectives is the fire alarm system meeting?

NOTES

1. Construction Technology Update, Number 42, December 2000.

2. "Reducing Fire Deaths in Older Adults: Optimizing the Smoke Alarm Signal," NFPA

Fire Protection Research Foundation, May 2006, NFPA, One Battery March Park, Quincy, MA.

Chapter 12

FIRE ALARM SYSTEM INITIATING DEVICES

Learning Objectives

Upon completion of this chapter, you should be able to:

- List and compare the available types of manually operated fire alarm devices.
- Discuss the methodology used to increase the reliability of a fire alarm system.
- Evaluate the differences between the available types of heat detectors.
- Compare the capabilities of the various types of smoke detectors.
- Define stratification and explain how it can be predicted.
- Propose a methodology for detecting a fire where stratification is predicted.
- Discuss the "pressure sandwich," and explain how this principle is accomplished by a fire detection system.
- Compare and contrast flame detectors and their applications.
- Give examples of where gas detectors and pressure detectors are appropriate for use.
- List and discuss the fire protection system supervisory functions that can be provided by a fire alarm system.

A fire alarm system consists of initiating devices, notification appliances, control function interfaces, and a fire alarm control unit (FACU). An **initiating device** originates an input signal that is sent to the fire alarm control unit. The control unit interprets the signal and transmits the signal to the appropriate output on the fire alarm system including notification appliances, such as horns, bells, and strobes, for occupant notification or relays for control functions. Initiating devices include:

- manually operated devices
- automatic fire detectors
- fire protection system flow and supervisory switches

A signal from an initiating device actuates notification appliances and performs additional operations, such as initiating discharge of a fire suppression system, closing doors, and shutting down air-handling equipment.

MANUALLY OPERATED DEVICES

Manually operated fire alarm devices require the action of an individual to initiate a fire alarm signal. The signal is transmitted to the fire alarm control unit, and the appropriate notification appliances are activated.

Manual pull stations are used by the building occupants to electronically notify others of a fire. **Figure 12-1** shows several types of manual fire alarm stations, including

- non-coded manual stations, Figure 12-1(a)
- coded manual stations, Figure 12-1(b)
- presignal or general alarm manual stations
- single-action or double-action manual stations
- breakglass or nonbreakglass manual stations

Coded Manual Stations

Coded manual stations have an electrically driven or a spring drawn escapement that is wound when the station is actuated, sending a set of unique and distinctive time-pulsed signals to the fire alarm control unit three or more times, indicating the exact location of the manual station. Knowing the exact location of a manually transmitted alarm can expedite fire service response to the location of the fire. As an example, code 5-3-2 could designate the fifth floor, third zone (east wing), second device (manual station next to stairwell 2).

Coded manual stations are likely to be found only in some existing fire alarm systems. They have been replaced largely by addressable fire alarm systems,

initiating device
a mechanism that originates a signal that is sent to a fire alarm control unit

manually operated fire alarm devices
mechanisms that require the action of an individual to initiate a fire alarm signal

manual pull stations
devices for use by building occupants to electronically notify others of a fire

coded manual stations
stations that send a set of unique and distinctive time-pulsed signals to the fire alarm control unit three or more times, indicating the exact location of the manual station

NON-CODED
DOUBLE ACTION
STATION (BREAKGLASS)

NON-CODED
SINGLE ACTION STATION

**NON-CODED
MANUAL STATION
FOR HAZARDOUS
LOCATIONS**

NON-CODED
SINGLE ACTION
MANUAL STATION

NON-CODED
DUAL ACTION
MANUAL STATION

NON-CODED
SINGLE ACTION
STATION WITH
INSTITUTIONAL COVER

NON-CODED
DOUBLE ACTION
STATION (PUSH TYPE)

NON-CODED SINGLE
ACTION MANUAL
STATION WITH WHITE
NEW YORK CITY STRIPE

NON-CODED
EXPLOSION-PROOF AND
WEATHER-PROOF
MANUAL STATION

NON-CODED
LOCAL FIRE ALARM
COVER OPTION

NON-CODED DOUBLE ACTION
PUSH TYPE MANUAL RELEASE STATION
(SHOWN WITH SAMPLE RELEASE LABEL)

Figure 12-1   *Non-coded and coded stations. (Courtesy of Simplex Grinnell.)*

Figure 12-1b
(continued)

2095-9001
LOCAL ENERGY MASTER BOX

which are capable of electronically identifying the location of a manual station or fire detector indicating an alarm condition.

Non-Coded Manual Stations

Some manual initiating devices are non-coded and do not deliver a distinctive signal. The location of the device can be determined by wise zoning or grouping of **non-coded manual stations** on a circuit. For instance, all manual stations on a particular floor may be grouped together into one circuit so the fire alarm control unit can indicate that a manual station has been activated on that floor. If a floor is very large or complex, separate manual station zones may be needed for each wing or distinctive area of that floor, with a maximum travel distance of 200 feet between stations. In these instances, the location of the fire can be ascertained more accurately.

Presignal and General Alarm Manual Stations

Most manual stations are **general alarm stations**. A general alarm station issues a notification alarm immediately upon activation of a general alarm manual station.

Presignal manual alarm stations do not actuate a general alarm but, instead, send an advance signal to the FACU when actuated or operated. Security personnel then are sent to investigate the signal. If the responding person determines that the signal is valid, a key is used to initiate a general alarm from the presignal device.

Single-Action and Double-Action Manual Stations

Single-action manual stations require only one motion by the user, who activates an alarm by pulling the lever. **Double-action manual stations**, usually labeled "Break Glass, Then Pull Handle," require the operator to perform two separate and distinct functions before an alarm is sounded, such as opening or

non-coded manual stations
devices that do not deliver a distinctive signal and are grouped on a circuit so the control unit can indicate activation

general alarm stations
stations that issue a notification alarm immediately upon activation of a general alarm manual station

presignal manual alarm stations
stations that do not actuate a general alarm but, instead, send an advance signal to the FACU when actuated or operated

single-action manual stations
devices that activate an alarm and require only one motion by the user

double-action manual stations
devices that require the operator to perform two separate and distinct functions before an alarm is sounded

breakglass manual stations
manual stations employing a replaceable slender plastic or glass rod that breaks when the handle on the manual station is pulled

addressable manual stations
stations with a discrete preprogrammed electronic identification that the FACU recognizes as to its type, location, and function

breaking a clear plastic cover before the lever can be pulled. Double-action devices are intended to reduce false alarms resulting from accidental initiation of a manual station.

Breakglass and Nonbreakglass Manual Stations

Breakglass manual stations employ a replaceable slender plastic or glass rod that breaks when the handle on the manual station is pulled. Pulling the lever provides the force necessary to break the rod. Some believe that the breakglass rod is a deterrent to nuisance alarms. Stations without breakglasses initiate an alarm when a button is pushed or when a lever is pulled. Some specifying engineers require an enclosure around the breakglass station to protect it from accidental initiation, and also a keyed entry that permits it to be opened and operated readily by qualified personnel.

Addressable Manual Stations

Addressable manual stations have a discrete preprogrammed electronic identification that the FACU recognizes as to its type, location, and function. Initiation of an addressable manual station can be identified on an alpha-numeric display or annunciation panel and can initiate alarm of control functions.

AUTOMATIC FIRE DETECTORS

Although an automatic fire detector does not have the five senses of a human, nor does it have human intellect, it can be more reliable than a human if it is properly selected, positioned, wired, and maintained. Modern fire detection and alarm systems that are designed using the methods of reliability listed below do not generate nuisance alarms.

Reliability Methodology for Detectors

Methods of ensuring the reliability of an automatic fire detection system include the following:

- *Protection.* When detectors are susceptible to mechanical damage, such as in a gymnasium, wire cages or other protective devices are required, which must be integrally listed with the detector.
- *Wiring.* Wiring of an automatic fire detection system must be in strict accordance with the *National Electric Code*, NFPA 70, Article 760.
- *Support.* Wiring, conduit, junction boxes, and detection devices must be affixed directly to the building structure, and no other item may be supported from the fire alarm system conduit, as shown in **Figure 12-2**.

NFPA

NFPA 70, National Electric Code, Article 760

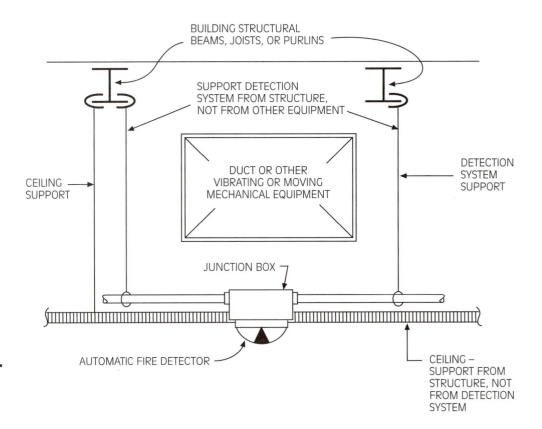

Figure 12-2
Detection system support.

- *Ceiling jet interaction.* An automatic heat or smoke detector must be positioned so it is within the boundaries of the ceiling jet, as shown in **Figure 12-3**. Testing data have demonstrated that placing the detector significantly below the ceiling would place the detector below the ceiling jet boundary, rendering it useless unless the plume is centered directly on the detector. Similarly, placing the detector within ceiling recesses or deep beam pockets would severely handicap or retard the detector's ability to sense a fire. Research currently is being conducted in this area.

- *Testing accessibility.* A ceiling detection system high above an enormous refuse recycling pit, as illustrated in **Figure 12-4**, would likely never be tested, so its reliability may never be known. Extending an air-sampling ceiling detection system to an accessible test detector at the 5-foot elevation level would greatly increase the probability that the detection system would be tested regularly. NFPA 72 (2007) suggests that projected beam detectors may be the best application for this situation, as would air-sampling smoke detectors.

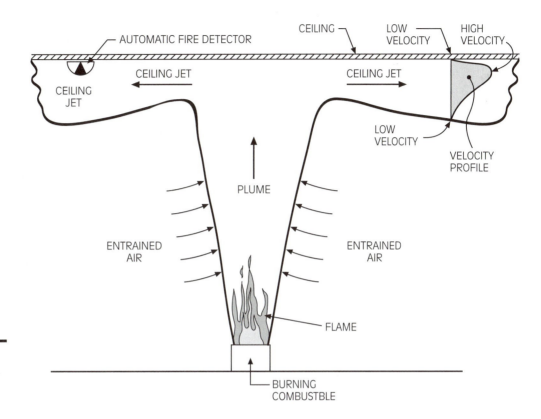

Figure 12-3 *Ceiling jet interaction with an automatic fire detector.*

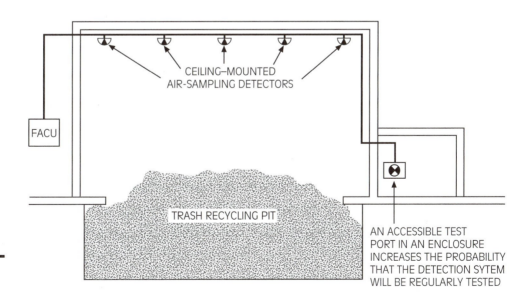

Figure 12-4 *Testing accessibility.*

- *Enclosure evaluation.* The room where detection is desired must be evaluated for unusual humidity, temperature, dust, pressure, light, air-flow conditions, corrosion, or conditions that would negatively affect the probability of system function or testing. Unusual conditions must be addressed or controlled directly, and detectors that will serve adequately in these conditions must be selected carefully. Designers should check the listing documentation for the detector, to ensure that it is suitable for the environment in which it will be installed.

- *Specific points of hazard.* The designer of a detection system must consider detection for known points of hazard, such as a bandsaw used for cutting solid munitions, a kettle of nitroglycerine, or a vessel filled with flammable liquid. Consideration also must be given to placing detectors where the plume and ceiling jet configuration is affected by items such as large tables, platforms, or large mechanical equipment.

- *Anticipated fire.* A designer must be able to accurately predict the nature of a fire that is likely to occur in an enclosure and must select a detector carefully to be appropriate for detecting the output or attributes of an anticipated fire, such as heat release, fire growth rate, smoke production, or gas production. Fire protection engineers use computerized fire models and testing data to replicate an anticipated fire and to model it for engineering analysis.

- *Detection zones.* The detection zones selected should enhance the ability of a fire brigade or the fire service to rapidly identify the fire location and advance quickly to the source of the fire. It also is possible to cross-zone a detection system using two discrete detection zones in an enclosure, with detectors designed such that at least one detector each or no fewer than two separate detection zones are required to actuate before an alarm is actuated.

NFPA 72 frequently uses the phrase "Apply proper engineering judgment." Although it is undefined in the Standard, proper engineering judgment includes, at a minimum, consideration of all of the items listed above. Application of proper engineering judgment will ensure a reliable fire detection and alarm system.

Addressable Detectors

Addressable detectors
devices that have a preprogrammed electronic address assigned to it that the fire alarm control unit will recognize when it actuates or when it issues a trouble or fault signal

Designers increasingly are using addressable fire alarm systems, which identify the initiating device and its location. **Addressable detectors** have become state-of-the-art in detection technology, and their use has become standard for a wide variety of detection applications. An addressable detector has a preprogrammed electronic address assigned to it that the fire alarm control unit will recognize when it actuates or when it issues a trouble or fault signal.

AUTOMATIC FIRE DETECTOR OVERVIEW AND COMPARISON

Automatic fire detectors are composed of

- heat-sensing fire detectors
  - electronic thermistor heat detectors
  - analog heat detectors
- smoke-sensing fire detectors
  - analog smoke detectors
  - particle sampling smoke detectors
- flame-sensing radiant energy fire detectors
- gas-sensing detectors
- pressure detectors
- non-fire detectors (cold detectors, video surveillance systems)

A comparison of some of these detection methods is shown in **Table 12-1**.

Special situations may require nonfire detectors in place of or in addition to one of the types of fire detector listed above. For example, liquified natural gas (LNG) is very cold when it is released, so a detector that senses very low temperatures could warn of an LNG leak and could release suppressing agent in advance of ignition or flame. Similarly, a gas detector could provide notification and initiate the suppressing agent before a flammable gas ignites or could detect a hazardous byproduct of combustion, such as carbon monoxide.

Table 12-1 *Comparison of automatic fire detectors.*

Type of Detector	Relative Detection Speed	Probability of Unwanted Alarms	Relative Detector Cost	Application
Heat	Slow	Low	Low	Flaming fire in confined spaces
Smoke	Fast	Medium	Medium	Smoldering fire in open or confined spaces
Flame	Very Fast	High	High	Flaming fires, deflagration or high value hazard
Particle sampling (aspiration)	Fast	Low-Medium	Medium-High	Open spaces with ambient dust or high value hazard

SPOT-TYPE HEAT-SENSING FIRE DETECTORS

Heat detectors are either spot-type or line-type. Spot-type heat-sensing detectors sense the temperature of gases in a ceiling jet at a specific point on the ceiling, and may be rate-of-rise, fixed-temperature, rate-compensated, electronic thermistor, analog, or combination heat detectors.

Spot-Type Pneumatic Rate-of-Rise Heat Detectors

A pneumatic rate-of-rise spot heat detector initiates an alarm if the temperature rises faster than a predetermined rate. A spot-type pneumatic rate-of-rise detector, shown in **Figure 12-5**, employs an air chamber with a relief orifice to gauge

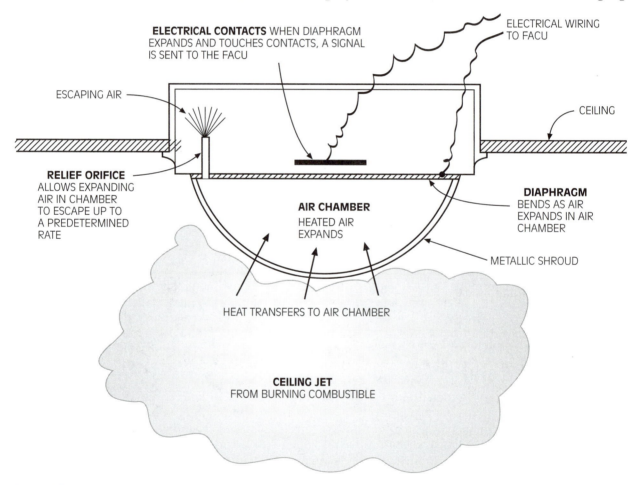

Figure 12-5   *Spot-type pneumatic rate-of-rise heat detector.*

the rate of temperature rise and does not initiate an alarm for slow temperature changes. In the presence of fire, the temperature in the chamber rises rapidly and air in the chamber expands faster than the relief capability of the relief orifice, causing the diaphragm to expand and close a set of electrical contacts. In a non-fire condition, the relief orifice will modulate pressure in the air chamber to account for minor ambient temperature fluctuations.

Spot-Type Thermocouple Rate-of-Rise Heat Detectors

The rate of temperature rise also can be detected by a thermocouple in which the electrical response changes with temperature. A **thermocouple** consists of two dissimilar metals joined at their ends. The voltage developed at these junctions is proportional to the temperature difference between the junctions.

Spot-Type Fixed Temperature Heat Detectors

A **spot-type fixed-temperature heat detector** initiates an alarm when a predetermined temperature is attained by a thermally responsive element. A bimetallic fixed-temperature detector, shown in **Figure 12-6**, uses two metals that expand at different rates to create a bending deformation that closes an electrical contact and completes a circuit. Another type of fixed-temperature spot detector uses a replaceable eutectic alloy, similar to solder, which melts when the eutectic element is heated by the surrounding air to a precise temperature.

Spot-Type Rate-Compensated Heat Detectors

Spot-type rate-compensated heat detectors can compensate for the effects of thermal lag, illustrated in **Figure 12-7**. **Thermal lag** is the temperature differential between a detector and the heated air surrounding it. A spot-type rate-compensated detector, shown in **Figure 12-8**, consists of an outer shell casing that expands along its length at a known rate when heated, with an element that resists expansion of the casing. If the temperature rises slowly, the inner element is heated and resists expansion. If the temperature rises rapidly, the inner element does not have time to be heated and the outer element expands until an alarm sounds. If the temperature rises slowly to a high temperature, both elements are heated and the detector performs like a fixed-temperature device. These detectors are ideal for exterior exposures or interior areas where the ambient temperature changes over time.

Spot-Type Electronic Thermistor Heat Detectors

Spot-type electronic thermistor heat detectors monitor thermistors with temperature sensing capable of detecting when temperature thresholds have been exceeded. The detector monitors the response of a single thermistor and compares

thermocouple
a device consisting of two dissimilar metals joined at their ends

spot-type fixed-temperature heat detectors
devices that initiate an alarm when a predetermined temperature is attained by a thermally responsive element

spot-type rate-compensated heat detectors
devices that compensate for the effects of thermal lag

thermal lag
the temperature differential between a detector and the heated air surrounding it.

Spot-type electronic thermistor heat detectors
devices that monitor thermistors with temperature sensing capable of detecting when temperature thresholds have been exceeded

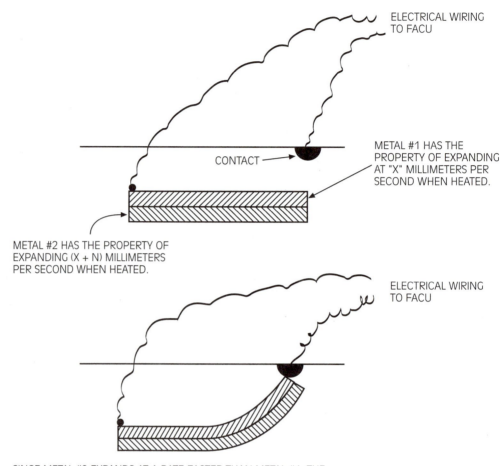

ELECTRICAL WIRING
TO FACU

CONTACT

METAL #1 HAS THE
PROPERTY OF EXPANDING
AT "X" MILLIMETERS PER
SECOND WHEN HEATED.

METAL #2 HAS THE PROPERTY OF
EXPANDING (X + N) MILLIMETERS
PER SECOND WHEN HEATED.

ELECTRICAL WIRING
TO FACU

SINCE METAL #2 EXPANDS AT A RATE FASTER THAN METAL #1, THE
BIMETALLIC STRIP WILL DEFORM TOWARD THE CONTACT WHEN HEATED
UNTIL THE CIRCUIT IS CLOSED, AND A SIGNAL IS SENT TO THE FACU.

Figure 12-6
*Bimetallic fixed
temperature heat
detectors.*

it to the response of a second thermistor, which increases the reliability of a detection signal.

Spot-Type Analog Heat Detectors

Spot-type analog heat detectors can electronically monitor their status and condition and can sense when they become dirty and require cleaning or replacement. Analog heat detectors also can measure and record peak value temperature data for a room, and the sensitivity of the detector can be adjusted to accommodate fluctuating temperature conditions experienced during its recorded history.

Spot-type analog heat detectors
devices that electronically monitor their status and condition and can sense when they become dirty and require cleaning or replacement

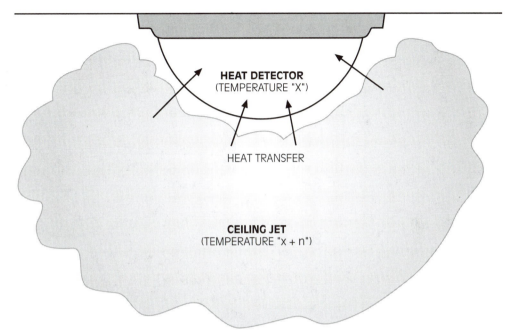

Figure 12-7 *Thermal lag. (Note: Some heat detectors are constructed of materials that permit heat to transfer more rapidly than others; therefore, thermal lag is unique for each type of heat detector.)*

THE TEMPERATURE OF THE CEILING JET EXCEEDS THE TEMPERATURE OF THE DETECTOR. HEAT TRANSFERS FROM THE JET TO THE DETECTOR UNTIL THE TEMPERATURES ARE THE SAME. THERMAL LAG IS A FUNCTION OF THE TEMPERATURE DIFFERENCE AND THE TIME FOR THE DETECTOR TO ASSUME THE TEMPERATURE OF THE CEILING JET.

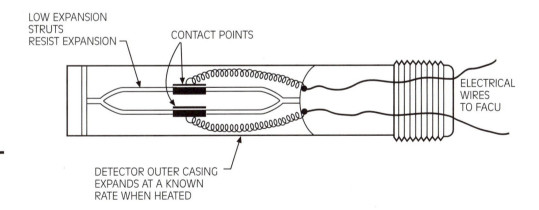

Figure 12-8 *Rate-compensated spot heat detector.*

Spot-Type Combination Heat Detectors

For added reliability, a designer may select a combination rate-of-rise/fixed temperature detector, which possesses both the air chamber apparatus shown in Figure 12-5 and a bimetallic strip within the chamber, shown in Figure 12-6.

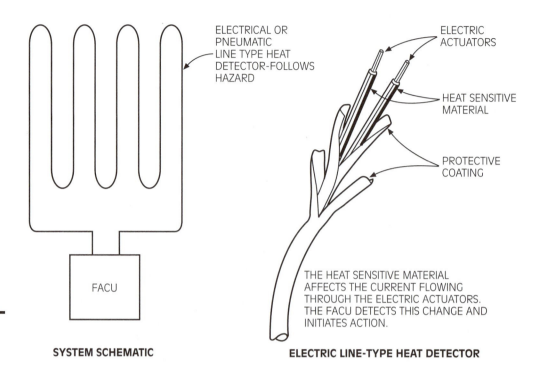

ELECTRICAL OR PNEUMATIC LINE TYPE HEAT DETECTOR-FOLLOWS HAZARD

ELECTRIC ACTUATORS

HEAT SENSITIVE MATERIAL

PROTECTIVE COATING

THE HEAT SENSITIVE MATERIAL AFFECTS THE CURRENT FLOWING THROUGH THE ELECTRIC ACTUATORS. THE FACU DETECTS THIS CHANGE AND INITIATES ACTION.

FACU

Figure 12-9 *Line-type heat-sensing devices.*

SYSTEM SCHEMATIC

ELECTRIC LINE-TYPE HEAT DETECTOR

LINE-TYPE HEAT-SENSING DEVICES

Line-type heat detectors, shown in **Figure 12-9**, are either electric or pneumatic and can be selected to detect either a fixed-temperature or rate-of-temperature rise anywhere along the length of the detector. These detectors are well suited for linear or elongated hazards such as conveyors and electric cable trays, as shown in **Figure 12-10**. An electrical line-type detection system sends current through the detection wire at all times, returning current to the fire alarm control unit for analysis. Two conductors under pressure are separated by a heat-sensitive material. When the heat-sensitive material is heated, it melts and the two conductors initiate an alarm.

Pneumatic line-type devices employ small copper tubes filled with pressurized air. When a tube is heated, the air inside expands and the FACU goes into alarm when the pressure reaches a predetermined level.

SMOKE-SENSING FIRE DETECTORS

Smoke detectors may be selected from the following classifications:

- spot-type ionization smoke detectors
- spot-type photoelectric light scattering smoke detectors

- spot-type photoelectric light obscuration smoke detectors
- air-sampling or air-aspirated smoke detectors
- laser-based air sampling smoke detectors
- linear beam smoke detectors
- air-duct smoke detectors
- analog smoke detectors

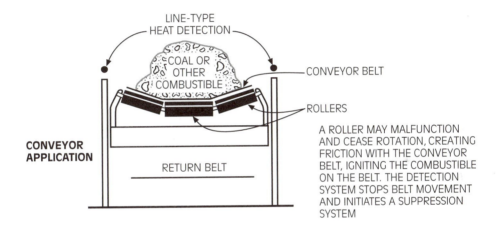

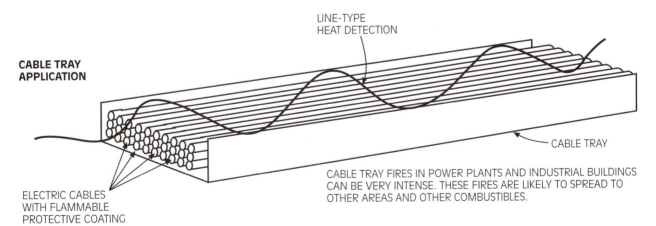

Figure 12-10   *Line-type heat-sensing device: applications. (Note: These detectors are ideal for linear or elongated hazards such as conveyors and cable trays.)*

Spot-Type Ionization Smoke Detectors

spot-type ionization detectors

devices containing a small amount of radioactive material that ionizes the air between a positive and a negative electrode in its sampling chamber to measure conductance

Spot-type ionization detectors, as shown in **Figure 12-11**, contain a small amount of radioactive material that ionizes the air between a positive and a negative electrode. The conductance between the electrodes is measured for the ambient, clean-air state. Introduction of smoke into the detector's sampling chamber reduces the conductance between the electrodes because ionized smoke particles move more slowly and their attraction to the electrodes is less than would be the case for clean ionized particles. When the conductance falls below a predetermined criterion, the detector sends a signal to the fire alarm control unit.

Normal operation time for a smoke detector is totally dependent on the rate of fire growth. The advantage of an ionization smoke detector is that it can detect very small particles from relatively clean-burning or flaming fires, such as a burning Christmas tree. Two varieties of ionization smoke detectors are shown in

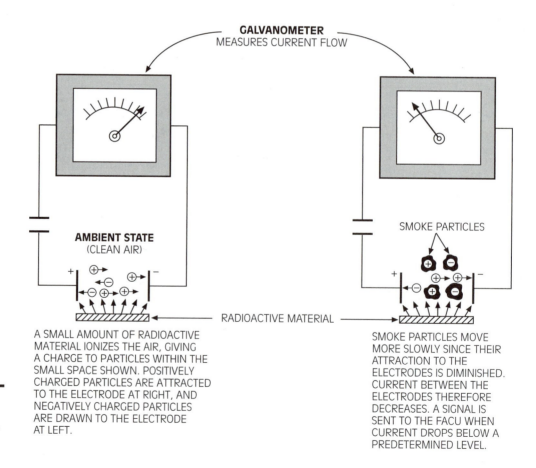

Figure 12-11

Ionization smoke detectors: principle of operation.

GALVANOMETER
MEASURES CURRENT FLOW

AMBIENT STATE
(CLEAN AIR)

A SMALL AMOUNT OF RADIOACTIVE MATERIAL IONIZES THE AIR, GIVING A CHARGE TO PARTICLES WITHIN THE SMALL SPACE SHOWN. POSITIVELY CHARGED PARTICLES ARE ATTRACTED TO THE ELECTRODE AT RIGHT, AND NEGATIVELY CHARGED PARTICLES ARE DRAWN TO THE ELECTRODE AT LEFT.

SMOKE PARTICLES

RADIOACTIVE MATERIAL

SMOKE PARTICLES MOVE MORE SLOWLY SINCE THEIR ATTRACTION TO THE ELECTRODES IS DIMINISHED. CURRENT BETWEEN THE ELECTRODES THEREFORE DECREASES. A SIGNAL IS SENT TO THE FACU WHEN CURRENT DROPS BELOW A PREDETERMINED LEVEL.

IONIZATION SMOKE DETECTORS FOR TWO-WIRE AND FOUR-WIRE BASES

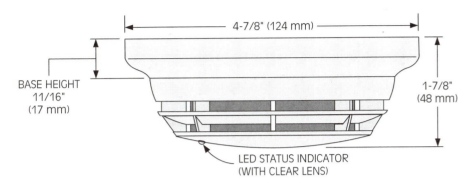

4-7/8" (124 mm)

BASE HEIGHT
11/16"
(17 mm)

1-7/8"
(48 mm)

LED STATUS INDICATOR
(WITH CLEAR LENS)

Figure 12-12 *Spot-type ionization smoke detectors. (Courtesy of Simplex Grinnell.)*

Figure 12-12. Note the smoke sampling ports and the LED (light-emitting diode) "power-on" status indicator on each model. The U.S. Nuclear Regulatory Commission determines the permitted radiation level and regulates detector radiation output.

Spot-Type Photoelectric Light-Scattering Detectors

spot-type photoelectric smoke detectors

detection devices that project light from a source to a light sensor

photoelectric light-scattering smoke detectors

devices that use a light-emitting diode (LED) and a light sensor that does not receive light from the LED under ambient conditions; light reaches the light sensor when it is reflected to the sensor by smoke particles

Spot-type photoelectric smoke detectors, shown in **Figure 12-13**, project light from a light source to a light sensor, sending a signal to the fire alarm control unit when smoke particles of sufficient density create the required change in light transmission.

Spot-type photoelectric light-scattering smoke detectors, as illustrated in **Figure 12-14**, use a light-emitting diode (LED) and a light sensor that does not receive light from the LED under ambient conditions. Light reaches the light sensor when it is reflected to the sensor by smoke particles. When light received by the light sensor increases to a predetermined level, a signal is sent to the FACU. This is the best available technology for detecting smoldering fires and smoke from fires in spaces remote from the detector.

Spot-Type Photoelectric Light Obscuration Smoke Detectors

Spot-type photoelectric light obscuration smoke detectors, as shown in **Figure 12-15**, position the light sensor directly in the beam of a light source. Smoke interferes with light transmission, and light received by the sensor decreases. When light transmission drops to a predetermined level, a signal is sent to the FACU. Some newer smoke detectors use laser-based technology instead of an LED light source.

spot-type photoelectric light obscuration smoke detectors
devices that project light from a source directly to a sensor that measures reductions in the amount of light received in the presence of smoke

PHOTOELECTRIC SMOKE DETECTORS FOR TWO-WIRE AND FOUR-WIRE BASES

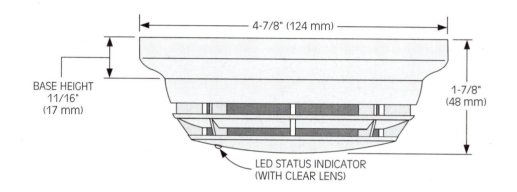

PHOTOELECTRIC SMOKE DETECTOR WITH SMOKE/HEAT DETECTION

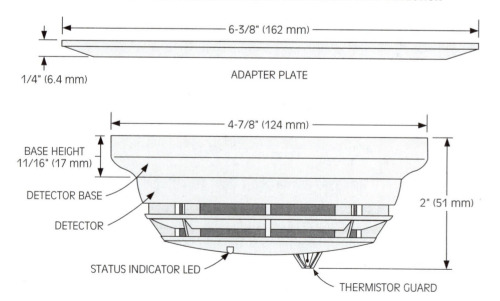

Figure 12-13 *Spot-type photoelectric smoke detectors. (Courtesy of Simplex Grinnell.)*

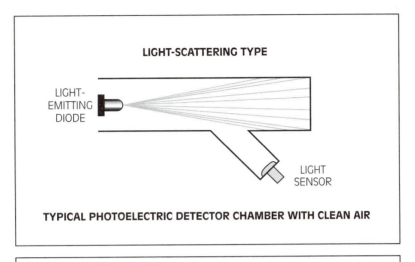

LIGHT-SCATTERING TYPE

LIGHT-EMITTING DIODE

LIGHT SENSOR

TYPICAL PHOTOELECTRIC DETECTOR CHAMBER WITH CLEAN AIR

LIGHT-SCATTERING TYPE

LIGHT-EMITTING DIODE

REFLECTED LIGHT

LIGHT SENSOR

TYPICAL PHOTOELECTRIC DETECTOR CHAMBER WITH SMOKE

Figure 12-14 *Spot-type photoelectric light-scattering smoke detectors; smoke particles (the dark circles on the lower diagram) deflect light beams to the light sensor. (Courtesy of Simplex Grinnell.)*

Air-Sampling or Air-Aspirated Smoke Detectors

air-sampling or air-aspirated smoke detection systems

devices that draw air into a sampling chamber for analysis

Air-sampling or **air-aspirated smoke detection systems**, as shown in **Figure 12-16**, use an aspirating fan to draw air from a room into a sampling chamber for analysis of products of combustion. Specialized analysis programs are available that search smoke characteristics for specific fire signatures. Small-diameter capillary tubes connected to pipe openings are spaced throughout the hazard, and sometimes into high-value electrical cabinets containing sensitive computer or telecommunication equipment.

The location of the combustion event is determined by the pipe delivering the smoke to the detection unit. The ambient or normal condition is measured, and a smoke concentration or particle-level setting is established and adjusted for the sampling system, which can be more sensitive than a spot-type

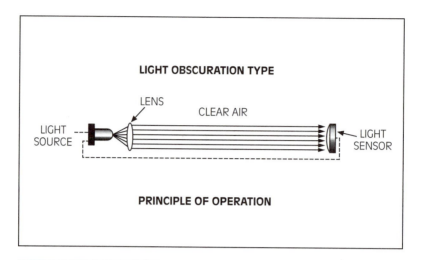

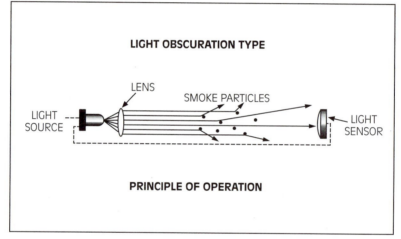

Figure 12-15 *Spot-type photoelectric light obscuration smoke detector. (Courtesy of Simplex Grinnell.)*

smoke detector. The control unit on the system can be programmed with respect to sensitivity, providing rapid detection of incipient fires, or less sensitive, depending on the application and the detection history recorded by the FACU. This type of system may be used for atria, clean rooms, telecommunication equipment rooms, freezers, and other large-volume spaces.

Laser-Based Air Sampling Smoke Detectors

laser-based air-sampling smoke detectors

devices that sample air drawn from a network of pipes or tubes and analyze the sample for comparison with a critical baseline

Laser-based air sampling smoke detectors can be used to achieve very early warning fire detection (VEWFD) for high-risk area such as telecommunications facilities and clean rooms. These detectors sample air drawn from a network of pipes or tubes and analyze the sample for comparison with a critical baseline.

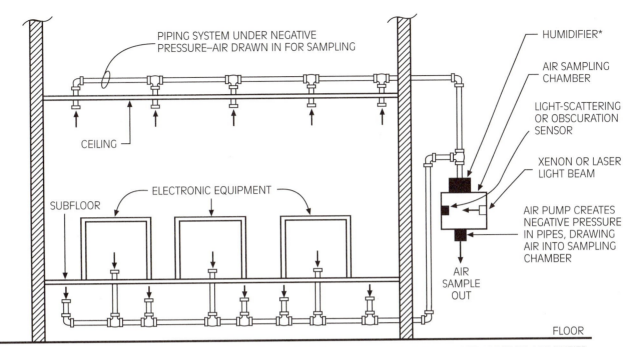

PIPING SYSTEM UNDER NEGATIVE
PRESSURE–AIR DRAWN IN FOR SAMPLING

HUMIDIFIER*

AIR SAMPLING
CHAMBER

LIGHT-SCATTERING
OR OBSCURATION
SENSOR

XENON OR LASER
LIGHT BEAM

AIR PUMP CREATES
NEGATIVE PRESSURE
IN PIPES, DRAWING
AIR INTO SAMPLING
CHAMBER

CEILING

ELECTRONIC EQUIPMENT

SUBFLOOR

AIR
SAMPLE
OUT

FLOOR

*FOR CLOUD CHAMBER TECHNOLOGY, HUMIDIFIED AIR ENTERING THE SAMPLING CHAMBER IS AT 100% HUMIDITY. HUMIDIFIED
AIR CONDENSES INTO WATER PARTICLES. SMOKE ENTERING THE CHAMBER MAKES THE WATER PARTICLES MORE DENSE.
WHEN THE CLOUD INCREASES IN DENSITY TO A PREDETERMINED LEVEL, A SIGNAL IS SENT TO THE FACU.

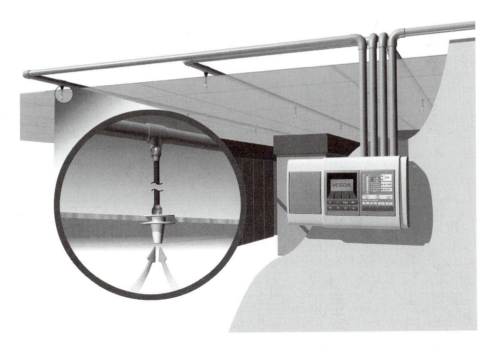

Figure 12-16 *Air-aspirated smoke detection system. (Courtesy of Ansul, Inc.)*

Linear Beam Smoke Detectors

Linear beam smoke detectors, like air-sampling- or air-aspirated smoke detectors, are desirable in spaces where ceiling spot detectors may be difficult to maintain and test and where smoke is subject to **stratification**. As defined by NFPA 92B, *Standard for Smoke Management Systems in Malls, Atria, and Large Areas*, stratification is the predilection or predisposition for smoke in a plume to cease rising when it entrains cool air and cools to the temperature of the ambient air above the plume. Stratification is especially prevalent when the ambient air at the ceiling is significantly warmer than the ambient air at the floor, such as in very tall atria with large glass skylights at the ceiling, as shown in **Figure 12-17**.

NFPA 92B assists fire protection design professionals in making the determination as to whether smoke is likely to stratify by using the following formula:

$$Z_m = (14.7) \times (Q_c^{1/4}) \times (\Delta T/\Delta z)^{3/8}$$

where Z_m = maximum height of smoke rise above fire surface (ft)
   Q_c = convective portion of the heat release rate (Btu). The convective portion of the heat release rate is equal to 70% of the total heat release rate for many fuels. Some fuels (such as alcohols) have a higher convective fraction.
   $\Delta T/\Delta z$ = rate of change of ambient temperature with respect to height (°F/ft).

NFPA 92B lists heat-release data (Q_c) for various commodities, and the temperature differential $\Delta T/\Delta z$ is measured with a thermometer. If Z_m is less than the total ceiling height in a tall building, the smoke is likely to stratify. As can be deduced from the formula, a small fire with a small heat-release rate is more likely than a very large fire to generate stratified smoke in tall spaces. Several computer programs are available that can perform this calculation, such as ASMET (Atria Smoke Management Engineering Tools), developed and published by NIST (National Institute for Standards and Technology).

Stratification can be defeated by exhausting and cooling the warm ambient air at the ceiling, allowing the smoke to rise. A linear beam smoke detection system can initiate actuation of the exhaust fans.

A linear beam smoke detection system, shown in **Figure 12-18**, consists of a sender and a receiver. The sender transmits an infrared light beam to the receiver. In a manner similar to the obscuration photoelectric smoke detector, smoke obscuration beyond a predetermined level initiates a signal from the receiver to the fire alarm control unit (FACU).

The sender and receiver can be mounted either parallel to the floor, as shown in **Figure 12-19**, or at an angle, with the sender mounted to the balcony of one floor and the receiver mounted to a balcony of a higher or lower floor.

stratification
the predilection or predisposition for smoke in a plume to cease rising when it entrains cool air and cools to the temperature of the surrounding air

NFPA
NFPA 92B, Standard for Smoke Management Systems in Malls, Atria, and Large Areas

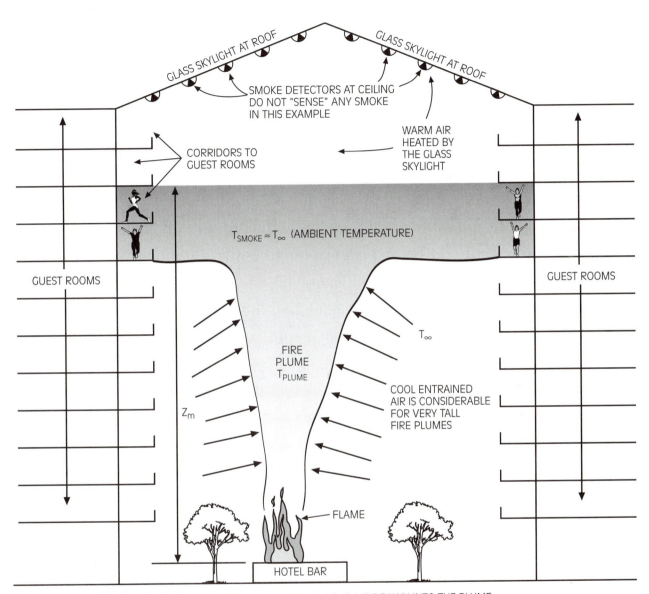

T_{PLUME} (PLUME TEMPERATURE) DECREASES AS COOL AMBIENT AIR (T_∞) IS DRAWN INTO THE PLUME.
THE PLUME CEASES TO RISE WHEN ITS TEMPERATURE APPROXIMATES THE TEMPERATURE OF THE WARM AIR AT THE CEILING.

Figure 12-17   *Smoke stratification in a hotel atrium. (Note: The smoke detectors at the roof are of no value in this case, and guest corridors, the primary means of egress, fill with smoke when stratification occurs. Z_m is the maximum height of smoke rise from the fire surface.)*

SENDER

RECEIVER

Figure 12-18 *Linear beam smoke detector. (Courtesy of Simplex Grinnell.)*

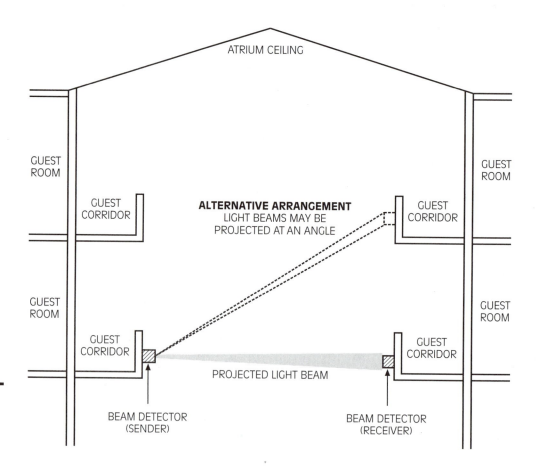

Figure 12-19 *Linear beam smoke detectors: positioning.*

Figure 12-20   *Air-duct smoke detector. (Courtesy of Simplex Grinnell.)*

Air-Duct Smoke Detectors

Heating, ventilating, and air-conditioning (HVAC) systems move fresh supply air into an enclosure and simultaneously exhaust stale return air to the outside. Air-duct smoke detectors, shown in **Figure 12-20**, sample the air in a duct and send a signal to the FACU to shut down or modify air flow to a space. Ionization or photoelectric spot-type smoke detectors are required to be listed specifically as air-duct smoke detectors. HVAC air ducts transport air, and in the process dilute the smoke within the duct. For this reason, NFPA 72 will not permit duct detectors to replace room detection because duct detectors require more smoke to activate.

An effective way to regulate air flow and smoke movement is to create a **pressure sandwich**, as shown in **Figure 12-21**, where smoky air is exhausted from the fire floor while supply air to the fire floor is shut down. All other floors are supplied with fresh air, with return air shut down. This arrangement provides a "sandwich" of pressure, with positive pressure above and below the floor of origin, and negative pressure on the floor of origin. This arrangement is accomplished with zoned ceiling detectors, zoned sprinkler waterflow switches, or zoned duct smoke detectors, where the respective zones match each other. The duct detectors control HVAC pressurization while the automatic sprinkler system simultaneously controls the fire.

Analog Smoke Detectors

Newer fire alarm smoke detection technology involves a system in which the smoke detector (or sensor) transmits smoke-obscuration values to the control

pressure sandwich
a procedure in which a smoke control system exhausts smoky air from the fire floor with supply air to the fire floor shut down, and all other floors are supplied with fresh air with return air shut down

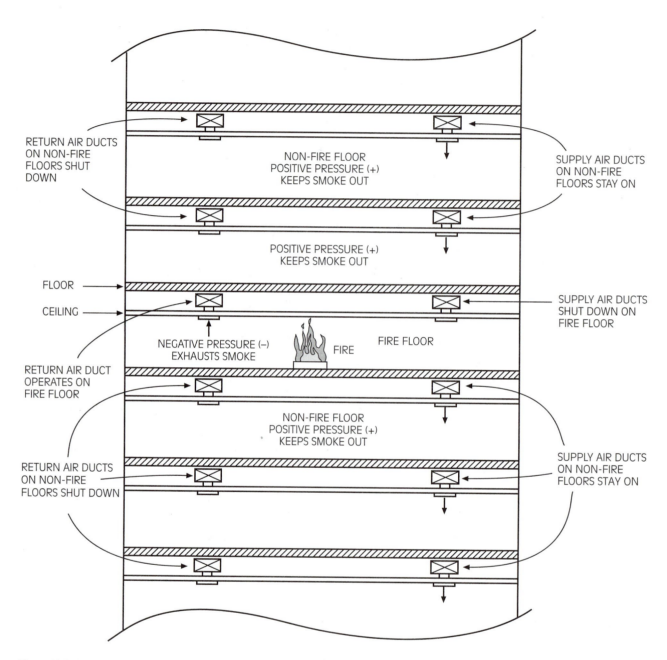

Figure 12-21   *Smoke control in a high-rise building; the "pressure sandwich" is positive pressure on all non-fire floors and negative pressure on the fire floor.*

unit, which analyzes the data and determines whether the system goes into alarm. This arrangement differs from the systems discussed previously in which the detector itself analyzes smoke-obscuration levels and determines whether an alarm signal is sent to the FACU.

One difference is that on the conventional system, each detector requires individual calibration and the detector becomes more sensitive with the accumulation of ambient dust. The analog system achieves calibration at the FACU, which checks the level of calibration constantly. This technology has reduced the number of unwanted nuisance alarms dramatically by compensating for contaminants and by notifying of the need to clean the unit.

FLAME-SENSING RADIANT FIRE DETECTORS

Flame-sensing fire detectors fall within the following categories:

- ultraviolet (UV) flame detectors
- infrared (IR) flame detectors
- combination ultraviolet/infrared (UV/IR) flame detectors
- spark/ember detectors

Ultraviolet (UV) Flame Detectors

ultraviolet (UV) flame detectors
devices designed to detect radiation falling within ultraviolet wavelengths below 4000 angstroms

Ultraviolet (UV) flame detectors use a Geiger-Müeller tube to analyze radiation from a flame and are designed to detect radiation falling within ultraviolet wavelengths below 4000 angstroms, as shown in **Figure 12-22**. Fires emitting radiation with these wavelengths generally have high-intensity flames associated with diffusion flame combustion. Proximity of the detector with respect to the source of the combustion is critical to operation within its intended range of sensitivity.

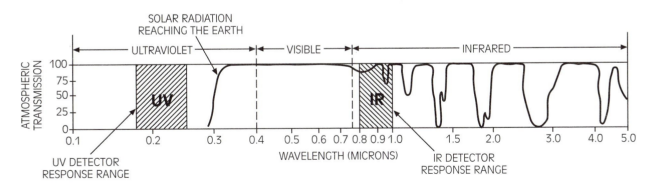

Figure 12-22 *Radiation spectrum.*

Although these detectors may be capable of sensing fires up to several hundred feet away, close proximity of the detector to the anticipated source of flame, careful aiming, and shielding or screening may be necessary to isolate the detector from unwanted signals such as lightning and welding arcs. A limitation to the use of UV detectors is fogging or loading of the lens, which can occur when these detectors are used in dusty or moist atmospheres, adversely affecting the detector's reaction time. Unless special arrangements are provided, such as air shields to divert dust away from the lens, UV detectors may be unsuitable for a dusty, moist, or greasy environment.

Infrared (IR) Flame Detectors

Infrared (IR) flame detectors use a photocell to search for the IR radiation spectra and are most effective for use at distances up to 50 feet. They respond to infrared radiation involving wavelengths between 8500 and 12,000 angstroms. A triple-band infrared flame detector—one that requires three abnormal conditions within the infrared spectrum to activate—is shown on **Figure 12-23**.

infrared (IR) flame detectors
devices designed to respond to infrared radiation involving wavelengths from 8500 to 12,000 angstroms

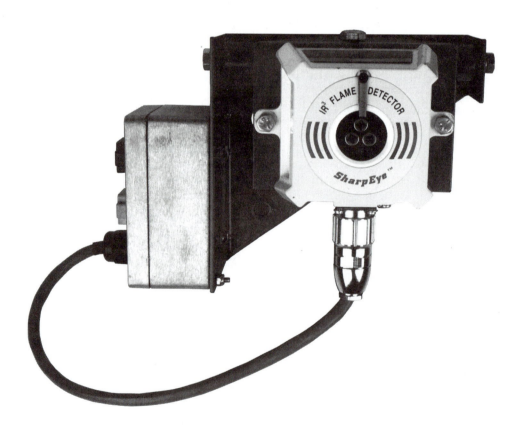

Figure 12-23 *Triple-band infrared flame detector. (Courtesy of Ansul, Inc.)*

Most infrared detectors have a built-in timing unit that can delay actuation by as much as 30 seconds. If infrared detectors are used to detect rapidly spreading fires, the distance between hazard and detector must be kept to a minimum, effectively screening out most sources of unwanted signals, and the time delay must be eliminated or minimized. For increased reliability, multiple-band infrared detectors, such as triple-band infrared detectors, may be specified.

Combination Ultraviolet/Infrared (UV/IR) Detectors

combination ultraviolet/infrared (UV/IR) flame detectors devices that contain sensors for both ultraviolet and infrared wavelengths, and the detector requires that abnormal values in both the UV and the IR spectra be seen before a signal is sent to the FACU

Although ultraviolet (UV) and infrared (IR) detectors can detect flaming radiation rapidly, false alarms can occur. In cases where actuation time is of lesser concern than false alarms, **combination ultraviolet/infrared (UV/IR) flame detectors** can be specified. A UV/IR detector contains sensors for both radiation wavelengths, and the detector requires that abnormal values in both the UV and the IR spectra be seen before a signal is sent to the FACU. This process is known as the "and-gate," based on the principle of logic that involves two statements in which both statements must be true for the logical presumption to be valid.

Selecting a Flame Detection System

Although UV and IR flame detectors are approved for both indoor and outdoor use, care should be taken when using them outdoors to minimize unwanted signals. Sources of flickering light, such as lights mounted on operating or vibrating machinery, may be perceived as being a fire by the detector and should be screened or eliminated. The following method is recommended to select a flame-detection system:

1. Know the anticipated fuel and the characteristics of the combustion of that fuel. Choose a detector capable of reacting to the wavelengths encountered in a fire involving the fuel in question. For example, an ultraviolet detector responds to a hydrogen fire, but an infrared detector operating in the 4.4-micron range does not. The radiation emission spectra of an anticipated fire must be matched precisely to the capabilities of a detector.

2. Survey the hazard area, the process, and the adjoining areas to determine sources of ambient non-fire signals. For example, if arc welding is expected within the view of a detector, ultraviolet detectors must be shielded from view by properly aiming the detectors away from such sources, or by shielding the sources from the detector. Another source of false signals is sunlight reflected into a room from the windshields of moving vehicles. A film or coating can be added to the window glass to filter and minimize the probability of receiving such signals.

3. Determine what elements could interfere with optimum operation of the detector. Dust or ice could severely hamper the detector's ability to function in accordance with its design parameters.

4. Ascertain the speed with which the detector is required to respond. The speed of detection is dependent on the combustion characteristics of the fuel and the proximity of personnel to the point of ignition.

5. Know the gaseous products of combustion for the fuel. Detection may be enhanced by specifying a detector that searches for specific gases, as discussed later in this chapter.

Radiation Spectrum for Flame Detectors

Detection ranges are shown on the radiation spectrum graph in Figure 12-22. Note how Figure 12-23 shows an intersection between the infrared detector sensitivity (between 0.8 and 1.0 microns) and the graph of solar radiation that reaches the Earth.

One can surmise by this intersection that an infrared detector would be receptive to, and adversely affected by, sunlight radiation. This sensitivity strongly affects the manner in which infrared detectors are used, as shown in **Figure 12-24**. Ultraviolet detectors are not affected by sunlight, according to the graph, and can be located without regard to the sun factor. High-intensity lights and high-temperature objects affect infrared detectors with no effect on ultraviolet detectors. However, arc welding, lightning, x-rays, and gamma rays

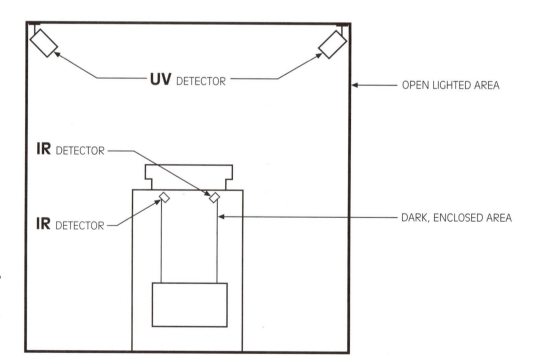

Figure 12-24

Example of ultraviolet (UV) and infrared (IR) flame detector placement.

affect ultraviolet detectors with no effect on infrared detectors. Because the two types of detectors are searching for different types of radiation and are affected by different interfering media, their applications are generally quite different.

Ultraviolet detectors, being insensitive to sunlight, are well suited for surveillance of an entire area or an area requiring a wide angle of vision. In contrast, infrared detectors are better suited for close surveillance of a specific contact point or supervision of an enclosed vessel.

Inverse Square Law for Flame Detectors

Ultraviolet detectors are subject to the inverse square laws of optics:

$$\frac{S_1}{(d_1)^2} = \frac{S_2}{(d_2)^2}$$

where d is the distance from the fire to the detector and S is the size of the fire, in square feet or meters. From this famous law of physics, one can see that sight distance is a powerful function of the equation. Increasing or reducing the sight distance dramatically affects the radiation received by the detector. The wisest use of ultraviolet detectors is to provide detection of the entire area by overlapping the viewing areas, or cones of vision, and to supplement the area detection system with additional detectors that view critical points closely. Reliability increases dramatically when critical areas are viewed by an overlapping cone of vision from different detectors in the same detection zone.

Because of their sensitivity to sunlight, infrared detectors are best located in enclosed, partially enclosed, or shielded spaces. Although infrared detectors adhere to the inverse square law, the hazard of receiving unwanted sunlight signals affects their performance at longer distances negatively. Shields are available to restrict the cone of vision of an infrared detector to prevent unwanted actuation, and downward positioning reduces dust buildup on the lens.

Spark/Ember Detectors

spark/ember detectors
specialized flame detectors with a photodiode that senses very small amounts of radiant energy, emitted before the full flaming stage

A **spark/ember detector** is a specialized type of flame detector that has a photodiode, which senses very small amounts of radiant energy, emitted before the full flaming stage. These detectors may be beneficial for explosion-suppression systems.

multi-sensor detectors
a class of detector that increases reliability by combining heat and smoke sensing, heat and carbon monoxide sensing, heat/co/ smoke sensing, or ionization plus photoelectric sensing

MULTI-SENSOR DETECTORS

Designers requiring a high level of reliability may specify one of a wide selection of **multi-sensor detectors** in combination with flame detectors. These detectors increase reliability by combining heat and smoke sensing, heat and carbon monoxide sensing, heat/co/smoke sensing, or ionization plus photoelectric sensing. Multi-sensor detectors contain algorithms for discerning sources of nuisance alarms. Analog versions send the multi-sensor values back to the control unit, where software algorithms are applied to determine the validity of the alarm.

GAS-SENSING DETECTORS

gas-sensing detectors
devices that detect the release of a flammable gas before it reaches its ignitable concentration

A gas-sensing detector can be chosen to detect the release of a flammable gas before it reaches its ignitable concentration. **Gas-sensing detectors** also can be selected to detect specific gases associated with combustion, such as carbon monoxide—the primary killer gas in fires.

PRESSURE-SENSING FIRE DETECTORS

pressure-sensing fire detectors
devices that consist of a plate that depresses when the enclosure pressure increases to a predetermined level and activates a signal to the FACU

In tightly enclosed process machinery or small rooms or enclosures where personnel are not present, a fire increases the pressure within the enclosure. **Pressure detectors** consist of a plate that depresses when the enclosure pressure increases to a predetermined level. Contacts close upon depression of the plate, and a signal is sent to the FACU. The FACU initiates discharge of an explosion suppression system, opens relief vents, or initiates discharge of inerting gas or suppressing agent into the enclosure.

FIRE PROTECTION SYSTEM FLOW AND SUPERVISORY SWITCHES

Fire protection initiating devices include the following:
- sprinkler system waterflow switches
- fire suppression system releasing switches
- fire suppression system supervisory switches

Sprinkler System Waterflow Switches

Sprinkler system waterflow switches sense the movement of water in a wet pipe sprinkler system, caused when water flows through sprinkler pipes after a heat-sensitive automatic sprinkler actuates. Waterflow switches come in two varieties—the vane-type paddle waterflow switch and the water pressure

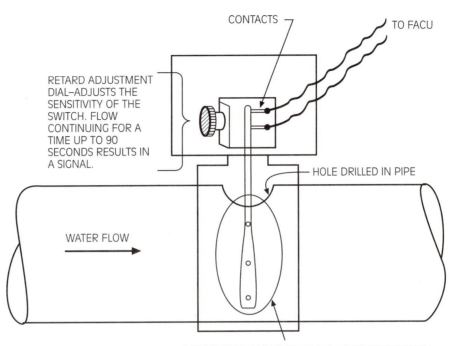

CONTACTS

TO FACU

RETARD ADJUSTMENT DIAL—ADJUSTS THE SENSITIVITY OF THE SWITCH. FLOW CONTINUING FOR A TIME UP TO 90 SECONDS RESULTS IN A SIGNAL.

HOLE DRILLED IN PIPE

WATER FLOW

POLYETHYLENE VANE IS ROLLED UP TO FIT THROUGH HOLE, THEN UNFURLS. THE VANE MOVES IN THE DIRECTION OF WATER FLOW, CLOSING THE CONTACTS AND SENDING A SIGNAL TO THE FACU.

Figure 12-25   *Vane-type waterflow switch.*

switch. These switches send a signal to the FACU, which can initiate an alarm, initiate smoke control systems, notify the fire department, or perform other functions.

The vane-type waterflow switch shown in **Figure 12-25** uses a polyethylene paddle or vane to sense water movement in the sprinkler pipe. Vane-type switches are not permitted to be installed on dry pipe sprinkler systems whose pipes normally are not filled with water because the initial force of water hitting the delicate vane could dislodge the vane and block waterflow to one or more automatic sprinklers.

The water pressure switch shown in **Figure 12-26** uses water pressure on a plate to close contacts and send a signal to the FACU. Pressure switches normally are installed on the valve trim of an automatic sprinkler system alarm valve or dry pipe valve.

Because water pressure within a sprinkler system may fluctuate, a retard function is included, which delays sending an alarm signal up to 90 seconds. A designer does not want a waterflow switch to be so sensitive that it will send false signals in fluctuating conditions but wants to ensure that the fastest, most reliable signal is sent. Field adjustment is needed to ensure that the switch will

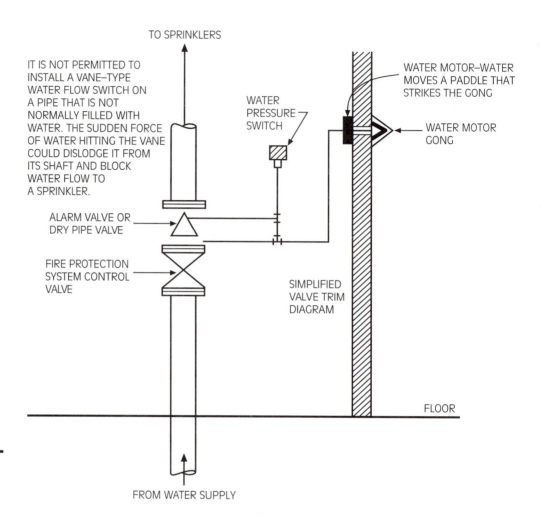

TO SPRINKLERS

IT IS NOT PERMITTED TO
INSTALL A VANE–TYPE
WATER FLOW SWITCH ON
A PIPE THAT IS NOT
NORMALLY FILLED WITH
WATER. THE SUDDEN FORCE
OF WATER HITTING THE VANE
COULD DISLODGE IT FROM
ITS SHAFT AND BLOCK
WATER FLOW TO
A SPRINKLER.

WATER
PRESSURE
SWITCH

WATER MOTOR–WATER
MOVES A PADDLE THAT
STRIKES THE GONG

WATER MOTOR
GONG

ALARM VALVE OR
DRY PIPE VALVE

FIRE PROTECTION
SYSTEM CONTROL
VALVE

SIMPLIFIED
VALVE TRIM
DIAGRAM

FLOOR

FROM WATER SUPPLY

Figure 12-26 *Water pressure switch installation.*

activate reliably on the flow of a single automatic sprinkler. Most modern retard mechanisms are set in the 15 sec. to 30 sec. range.

For further information on automatic sprinkler systems, please refer to *Design of Water-Based Fire Protection Systems*, also published by Delmar Learning.

Fire Suppression System Releasing Switches

Many of the suppression systems discussed in this book require actuation by a fire detection system as shown on **Figure 12-27**. Some systems can be actuated by a signal received by a fire alarm control unit and relayed to a solenoid valve that can initiate mechanical action of a suppression system deluge valve, and others can be actuated by a normally closed control valve that is designed to be opened upon receiving a detection signal.

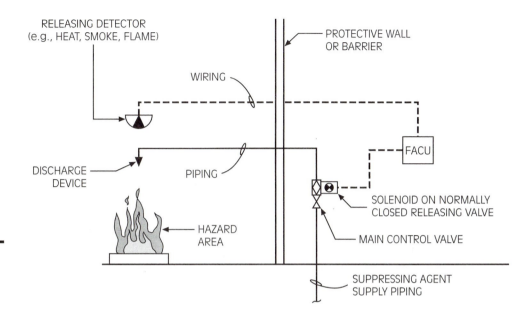

Figure 12-27   *Fire suppression releasing system schematic.*

Fire Suppression System Supervisory Switches

Many functions of fire suppression systems are required to be supervised. These include the following:

- *Water level for a water storage tank.* A high-level switch can shut down water filling, and a low-level switch can initiate a supervisory alarm signal or commence automatic water refill.

- *Water temperature or room temperature supervision.* A water temperature level on a water-based fire protection system that is too high could cause the heat-responsive elements on an automatic sprinkler or other heat-actuated device to actuate in the absence of fire. If the automatic sprinkler system water temperature drops below 32°F, it will freeze and render the suppression system inoperative or damage the piping system. Where water conditions could affect performance of the system, water temperature supervision often is mandated by law.

- *Control valve position switches.* A fire protection system is of no value if the control valve to the system is closed. NFPA requires that the control valve position be supervised in the open position. Although chains and locks are permitted, valve supervisory switches indicate control valve closure on a fire alarm control unit.

- *Air pressure switches.* Dry pipe automatic sprinkler systems and water pressure storage tanks require supervision of the air pressure to ensure proper functioning of the system. A drop in pressure is sensed by a low-pressure switch, initiating the starting of an air compressor. Excessive air pressure is sensed by a high-pressure switch, which stops the operation of the air compressor. If air pressure is not regulated by the high-pressure and low-pressure switches, a supervisory signal is sent to the FACU to initiate maintenance.

- *Fire pump supervisory switches, per NFPA 20.* Fire pump operation and control can be monitored by a fire alarm control unit, providing annunciation of valuable information to responding fire service personnel.

- *Power generator supervisory switches.* A backup power supply should be monitored by a fire alarm control unit.

SUMMARY

Initiating devices on a fire alarm system are watchdogs that supervise essential equipment and report fire conditions to the fire alarm control unit (FACU). For a system of initiating devices to be reliable, it must be protected, properly wired and supported, placed within the ceiling jet boundary, tested frequently, suitable for use in the enclosure conditions to which it is exposed, installed for specific points of hazard, selected correctly with respect to the anticipated fire, and zoned wisely.

Detectors can be chosen from a variety of heat-sensing, smoke-sensing, particle-sensing, flame-sensing, gas-sensing, or pressure-sensing fire detectors. It is important to carefully select a detector that precisely detects the byproducts of an anticipated fire. Because stratification can negate the effectiveness of a ceiling fire-detection system, prediction of stratification is essential in tall, large-volume spaces.

REVIEW QUESTIONS

1. What criteria are used to determine whether to specify the following manually operated devices:
   a. coded or non-coded manual stations
   b. presignal or general alarm manual stations
   c. single action or double-action manual stations
   d. breakglass or nonbreakglass manual stations

2. Why is knowledge of an anticipated ceiling jet profile of value to a fire alarm system designer?

3. List the available types of heat-sensing fire detectors, and explain the operating principle of each. Compare and contrast spot-type and line-type heat detectors.

4. Evaluate the criteria used to select each of the types of smoke detectors discussed in this chapter.

5. What is a "pressure sandwich," and how can a fire alarm system accomplish such a feat?

6. Using the radiation spectrum as a guide, determine:
   a. which flame detector is affected by sunlight.

b. which flame detector is unaffected by sunlight.

c. which flame detector is less prone to false alarms.

7. If a flame detector is to be installed in a dusty atmosphere, what can be done to ensure that the lens is kept clean?

8. What are sources of the false alarms with flame detectors, and how can they be avoided?

9. A wet pipe sprinkler system is connected to an underground water pressure tank. What supervisory functions should be monitored by a fire alarm system?

DISCUSSION QUESTIONS

1. What does NFPA 72 mean in reference to the use of "proper engineering judgment?" What reliability methodology would be used by a fire alarm system designer employing such judgment?

2. Define stratification, and explain how this phenomenon is predicted. How can stratification be defeated? How are fires detected where stratification is predicted to occur?

ACTIVITIES

1. Interview a fire service official. What fire alarm system requirements exist in the fire service jurisdiction to whom the official reports?

2. Visit a fire detection and alarm design and installation firm. Inspect fire alarm initiating devices, and ask for help in determining their function.

3. Inspect a facility and determine the type of initiating devices installed. Try to ascertain the criteria used for selection of the initiating device.

Chapter 13

FIRE ALARM SYSTEM NOTIFICATION APPLIANCES

Learning Objectives

Upon completion of this chapter, you should be able to:

- Discuss the four groups of notification signals, explain why the temporal-coded signal is now the national standard signal, and outline research conducted with respect to recognition of the temporal-coded signal.
- List the available types of audible notification appliances.
- Evaluate public- and private-mode audibility requirements for fire alarm systems and explain why the two modes differ.
- Determine the effectiveness of an audible notification appliance, given ambient sound levels, door or wall attenuation, and inverse square law losses.
- Compare wall-mounted and ceiling-mounted visible notification appliance requirements.
- Locate visible notification appliances in a room, corridor, or sleeping room.
- Use the multiple-square layout to optimize visual notification appliance location.
- Explain the function of an annunciator panel.
- Determine the reasons why tactile notification appliances may be necessary, and discuss how they can be used effectively.
- Compare NFPA 72, ANSI, UL, and ADA requirements for visible notification appliances.
- Discuss the conditions that make strobe synchronization necessary.

The initiating devices reviewed in Chapter 12 monitor conditions that match predetermined criteria for which the devices are rated and selected. Once the design criteria are met, a signal is sent to the fire alarm control unit for interpretation. If the control unit determines that the signal is valid, a signal is sent to all notification appliances on the fire alarm system, or to one or more preselected groups of notification appliances in the vicinity of the detection signal received in accordance with the fire protection plan for a building.

Audible notification signals fall into four groups:

1. *Non-coded.* A signal that conveys one bit of information, such as a constantly sounding bell, for occupant notification,

2. *Coded.* A signal with more than one bit of information that provides specific information relative to location of the fire or provides notification to a specific individual or group of individuals. An example of a signal that identifies location could be a bell that sounds twice, pauses briefly, sounds three times, then pauses before recycling. Such a signal could be designed to identify the second floor, zone three. Voice communication fire alarm systems can notify specific individuals in code, such as "Dr. Red, report to the third floor nursing station 3," which could be intended to notify all responsible nursing personnel on zone 3 of the third floor to commence orderly evacuation of that zone without creating panic among patients. Responsible personnel are trained to recognize such signals.

3. *March time.* A coded signal of 120 beats per minute, with 1/4 second pulse on, 1/4 second pulse off. NFPA 72 has replaced this signal with a requirement for the temporal-coded signal.

4. *Temporal-coded.* A coded signal, illustrated in Figure 11-5, which sounds in a pattern of three half-second signals on, each separated by one half-second off. Each cycle pattern is separated by 1½ seconds. This signal became the national standard on July 1, 1997, for new fire alarm audible notification systems but is not retroactive, meaning that it does not apply to fire alarm systems installed before that date. This means that a building that has had many additions during the past decades, such as a hospital or school, could have differing notification signals in different wings of the building unless a designer has taken steps to standardize the signal for the building as the temporal signal.

Guylene Proulx of the National Research Council of Canada found low public recognition of the temporal-3 evaluation signal.[1] While 98% of respondents correctly recognized and identified a car horn, only 6% correctly identified a temporal-3 evaluation signal.

In some cases, providing specific verbal information or instructions in addition to the temporal signal may be valuable in increasing the probability of recognition. A voice communication system could supplement an audible temporal-coded notification signal, in which the temporal signal is interrupted periodically by spoken instructions.

Notification appliances may be selected from the following list, based on an evaluation of effectiveness for the occupants of the specific building being designed:

- audible notification appliances
- exit-marking audible notification appliances
- visible notification appliances
- audible/visible notification appliances
- textual audible and visible notification appliances
- annunciation notification appliances
- tactile notification appliances

AUDIBLE NOTIFICATION APPLIANCES

Household fire alarm systems, as discussed in Chapter 11, primarily use audible notification appliances that notify occupants by creating a sound of sufficient decibel level to alert and rouse individuals to leave the building. Commercial, industrial, and other public fire alarm systems employ a wide variety of audible notification appliances.

Audible appliances include

- buzzers, as shown in **Figure 13-1**
- horns, as shown in **Figure 13-2**
- sirens, as shown in **Figure 13-3**
- speakers, as shown in **Figure 13-4**

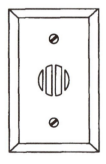

Figure 13-1 *Fire alarm buzzers. (Courtesy of Simplex Grinnell.)*

BUZZER WITH SINGLE GANG PLATE
(a)

BUZZER LESS MOUNTING PLATE
(b)

Figure 13-2 *Fire alarm electronic horns. (Courtesy of Simplex Grinnell.)*

ELECTRONIC HORN

Figure 13-3 *Fire alarm sirens. (Courtesy of Simplex Grinnell.)*

SINGLE PROJECTOR SIREN
(a)

SINGLE PROJECTOR
HEAVY DUTY SIREN
(b)

**RECTANGULAR WALL MOUNT SPEAKERS ARE AVAILABLE
AS RED WITH WHITE "FIRE" LETTERING AND
WHITE WITH RED "FIRE" LETTERING**

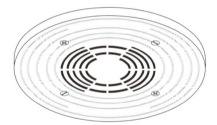

Figure 13-4 *Fire
alarm speakers.
(Courtesy of
Simplex Grinnell.)*

ROUND SPEAKERS ARE AVAILABLE IN OFF-WHITE (NO LETTERING)

Audible Notification Appliance Audibility Requirements

Audible notification appliances are designed for use in either the public mode or
the private mode.

NFPA

NFPA 72, National Fire
Alarm Code (2007)

Public-Mode Audibility Requirements Public-mode requirements are established by
NFPA 72 to provide a general evacuation alarm for occupants or inhabitants of a
building. The following requirements apply to each separate public area (notification zone), defined by NFPA 72, National Fire Alarm Code (2007) as an area
separated from other areas on a floor by a 2-hour fire wall. Public-mode requirements address not only the minimum decibel sound level required for initiation
of egress, but also maximum sound levels. The safe duration for exposure to high
sound levels is shown in **Table 13-1**.

Public-mode audibility requirements addressed by NFPA 72 (2007) include:

Table 13-1 *Permissible noise exposures.*

Duration (hours)	L_A (dBA)
8	90
6	92
4	95
3	97
2	100
1.5	102
1	105
0.5	110
0.25	115
0.125 (7.5 minutes)	120

Source: NFPA 72 (2007), Figure A.7.4.1.2 (OSHA 29 CFR 1910.5, Table G-16, Occupational Noise Exposure). Reprinted with permission from NFPA 72, *National Fire Alarm Code®*, Copyright © 2007, National Fire Protection Association, Quincy, MA 02169. This reprinted material is not the complete and official position of the National Fire Protection Association on the referenced subject which is represented only by the standard in its entirety.

- 15 dB above the average measured ambient sound level measured at 5 ft (1.5 m) above the floor. **Table 13-2** shows sound levels for various occupancies and may be used for the design of fire alarm systems in new buildings, but accurately measuring sound levels is the most reliable means of determining the ambient sound level in existing buildings.

- 5 dB above the maximum measured sound level, having a duration of at least 60 sec., measured at 5 ft (1.5 m) above the floor.

- 75 dB minimum for sleeping rooms, or 15 dB above average sound levels and 5 dB above the maximum sound level.

- visible notification is required if the ambient sound level exceeds 105 dB, such as might be encountered in a steel mill or a power plant.

- fire alarm systems whose fire alarm control units are designed to electronically stop or reduce ambient noise are permitted when approved by the authority having jurisdiction (AHJ).

Private-Mode Audibility Requirements Private-mode requirements apply to audible or visible signaling, affecting persons concerned directly with the implementation and direction of emergency action in the area protected by the fire alarm system. Examples of private-mode signaling applications include: hospital nursing stations, supervising station personnel or other trained personnel in constant responsible charge of receiving and responding to initial fire alarm signals, including control room operators, switchboard operators, and security officers. Because personnel are trained to expect and interpret the signal and are assumed to be awake and alert, private-mode requirements are less stringent than requirements for the public mode.

Table 13-2 *Average ambient sound level according to location.*

Location	Average Ambient Sound Level (dBA)
Business occupancies	55
Educational occupancies	45
Industrial occupancies	80
Institutional occupancies	50
Mercantile occupancies	40
Mechanical rooms	85
Piers and water-surrounded structures	40
Places of assembly	55
Residential occupancies	35
Storage occupancies	30
Thoroughfares, high-density urban	70
Thoroughfares, medium-density urban	55
Thoroughfares, rural and suburban	40
Tower occupancies	35
Underground structures and windowless buildings	40
Vehicles and vessels	50

Source: NFPA 72 (2007), Figure A.7.4.2. Reprinted with permission from NFPA 72, *National Fire Alarm Code®*, Copyright © 2007, National Fire Protection Association, Quincy, MA 02169. This reprinted material is not the complete and official position of the National Fire Protection Association on the referenced subject which is represented only by the standard in its entirety.

Requirements for private-mode notification include:

- 10 dB minimum above the average ambient sound level measured at 5 ft (1.5 m) above the floor.
- 5 dB maximum above the maximum sound level having a duration of 60 sec, measured at 5 ft (1.5 m) above the floor.
- audible alarms in elevator cars are permitted to conform to private audibility requirements.
- fire alarm systems designed to stop or reduce ambient noise are permitted when approved by the AHJ.

Audible Notification Appliance Intelligibility

The traditional design concept for audible notification was to make a loud noise from a small number of speakers. Intelligibility research demonstrates that more speakers, spaced more closely, will increase the probability of a voice message being heard and understood. NFPA 72 requires that voice communications systems be intelligible, and NFPA 72 Annex material provides advice on attaining this goal.

Audible Notification Appliance Location

NFPA 72 (2007) requires that wall-mounted audible notification appliances, where subject to mechanical damage, be protected from physical injury and mounted at a height no less than

- 90 inches (2.3 m) minimum above the floor, and
- 6 inches (150 mm) minimum below the ceiling

Ambient sound levels in existing buildings should be measured for the most accurate results. For the design of fire alarm systems in new buildings, some typical sound levels are given in **Table 13-3**, and some of the more common methods of sound attenuation, or sound reduction, through various doors and walls, are listed in **Table 13-4**. (The minimum sound levels required for the public and private modes are given previously in the chapter.)

Sound pressure declines in intensity as the distance from the ear to the notification appliance increases, in accordance with the **inverse square law**. This law states that as the distance from the ear to the notification appliance doubles, sound is reduced by 6 dB along the centerline of the notification appliance. This principle is illustrated in **Figure 13-5**.

The ability of a person to perceive sound pressure also reduces as a person moves away from the centerline of the notification appliance. Each appliance has its unique loss characteristics, and polar charts may be obtained for a specific appliance used. This loss could be offset by reflections of sound from floors, walls, and ceilings, or exacerbated by sound-attenuating wall or ceiling linings,

inverse square law
a law stating that as the distance from the ear to the notification appliance doubles, sound is reduced by 6 dB along the centerline of the notification appliance

Table 13-3 *Typical sound levels. (Courtesy of Simplex Grinnell.)*

Whisper @ 5 Feet	18 dB
Home	40 dB
Office	55 dB
Conversation @ 3 Feet	70 dB
Auto @ 15 Feet	70 dB
Subway	102 dB
Pain Threshold	130 dB

Table 13-4 *Sound attenuation: Items that reduce sound levels. (Courtesy of Simplex Grinnell.)*

Material	Sound Measurement
Open door	−5 dB
Hollow core door	−10 dB
Solid core door	−20 dB
Gasket door	−24 dB
Stud wall	−41 dB

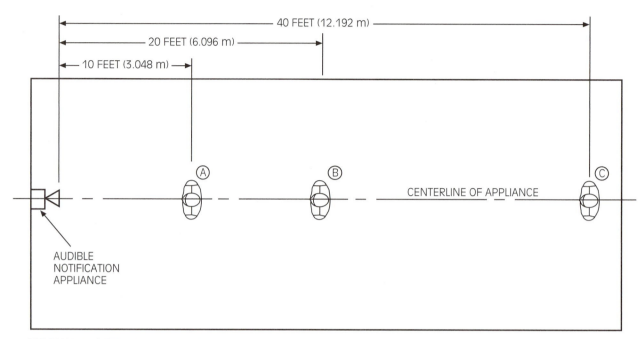

PERSON "A" 10 FEET (3.048 m) FROM APPLIANCE
• HEARS RATED SOUND LEVEL OF APPLIANCE
• EXAMPLE: 96 dB AT 10' (3.048 m)

PERSON "B" AT 20 FEET FROM (6.096 m) APPLIANCE (TWICE THE DISTANCE FROM NOTIFICATION APPLIANCE AS PERSON "A")
• HEARS 6 dB LESS THAN PERSON "A"
• EXAMPLE: 90 dB AT 20' (6.096 m)

PERSON "C" AT 40 FEET (12.192 m) FROM APPLIANCE (TWICE THE DISTANCE FROM NOTIFICATION APPLIANCE AS PERSON "B")
• HEARS 6 dB LESS THAN PERSON "B"
• EXAMPLE: 84 dB AT 40' (12.192 m)

Figure 13-5 *Illustration of the inverse square law; sound pressure reduces 6 dB every time the distance from the audible notification appliance doubles.*

such as cork. A polar chart showing typical off-centerline losses is shown in **Figure 13-6**.

The challenges in designing audible notification systems are to obtain accurate ambient noise levels, to consider sound attenuation, to calculate inverse square losses along the sound distribution centerline, and to calculate off-centerline losses, taking into account wall linings that may enhance or attenuate audibility, to ensure that the minimum audibility requirements of NFPA 72 are met.

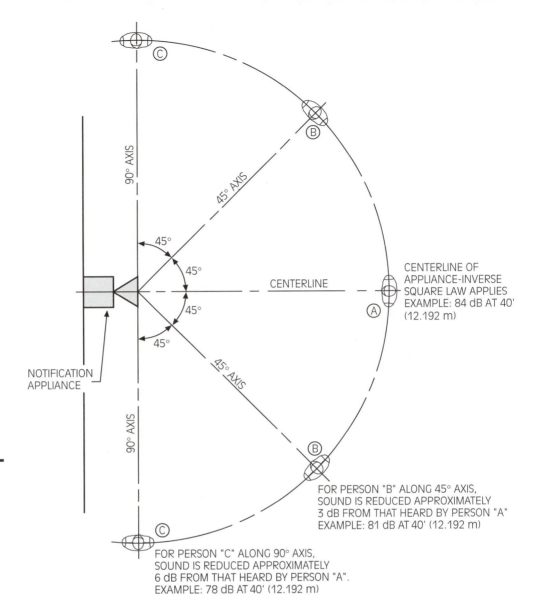

Figure 13-6 *Off-centerline sound losses; persons "B" and "C" are at the same lineal distance from the notification appliance as person "A."*

EXAMPLE 13-1

Calculation of Audible Notification Appliance Audibility

An audible notification appliance is mounted on a corridor ceiling, rated at 100 dB at 10 feet. A person is in a sleeping room with an ambient noise level of 60 dB, at a distance of 80 feet from the notification appliance, as shown in **Figure 13-7**. A

403

solid-core door separates the sleeping room from the corridor. Neglecting the elevation differential between the audible appliance and the ear, how many decibels will the person hear if the door is closed?

Is an additional audible notification appliance required in the sleeping room?

Solution

The audible notification appliance is rated at 100 dB at 10 feet. From this level, the inverse square law tells us that 6 dB are lost at the 20-foot distance, another 6 dB lost at the 40-foot distance, and another 6 dB lost at the 80-foot distance. Table 13-4 indicates that a closed solid-core door attenuates the sound by a value of 20 dB. The ambient noise level was given as 60 dB. We can calculate what the person hears:

100 dB	(appliance rated at 10 ft)
−6 dB	(loss at 20-ft centerline distance)
−6 dB	(loss at 40-ft centerline distance)
−6 dB	(loss at 80-ft centerline distance)
−20 dB	(loss through closed solid-core door)
−60 db	(ambient sound level)
2 dB	(sound heard by a person in sleeping room)

The 2 dB in the sleeping room at 80 feet obtained in Example 13-1 indicates that even under the best conditions, the notification appliance is not likely to be heard unless the door is open, and the appliance is certainly not capable of rousing a sleeping individual. In this case, an additional notification appliance in the sleeping room is clearly justified because the door is likely to be closed and

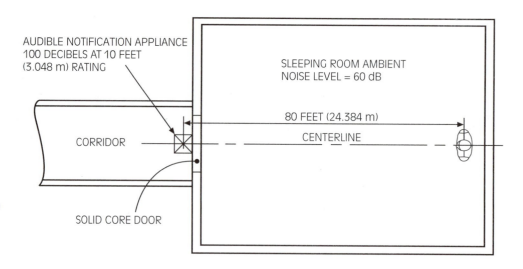

Figure 13-7
Illustration for Example 13-1.

the appliance must be capable of providing a minimum of 15 dB over the net ambient sound level. The minimum acceptable sound level for sleeping rooms is 75 dB. Subtracting the ambient sound level from 75 dB leaves 15 dB, so notification appliances in the sleeping room will be required to be located and rated to account for inverse square losses in the room.

"EXIT MARKING" AUDIBLE NOTIFICATION APPLIANCES

"exit marking" audible notification appliances
installed at the entrance of all building exits and areas of refuge, and emit distinct sound pressure levels capable of directing occupants to the exits

A new feature in NFPA 72 (2007) consists of requirements related to **"exit marking" audible notification appliances**, sometimes referred to as "directional sound" audible appliances. These appliances are installed at the entrance of all building exits and areas of refuge, and emit distinct sound pressure levels capable of directing occupants to the exits. The appliances differ from standard audible appliances, which use broadband frequencies in the 20 Hz to 20 kHz range.

VISIBLE NOTIFICATION APPLIANCES

candela (cd)
the standard unit of light intensity measurement

blackbody
an ideal body that would absorb all incident radiation and reflect none

Visible notification appliances, shown in **Figure 13-8**, use the **candela (cd)** as the standard unit of light intensity measurement. The candela unit once was referred to as a unit of candle power. A candela is defined scientifically as 1/60 of the luminous intensity per square centimeter of a blackbody radiator operating at the temperature of freezing platinum. A **blackbody** is an ideal body that would absorb all incident radiation and reflect none. The candela is inspired by the light intensity of a candle, the essential primary lighting fixture of bygone days. In a manner similar to the use of the horsepower as a measure of power production, our past continues to define the more sophisticated innovations of the present.

As with audible appliances, NFPA 72 (2007) gives requirements for the public mode and for the private mode, as outlined in the passages that follow.

Public-Mode Visibility Requirements

In a manner similar to audible appliances, it has been determined by testing that a flashing light is more likely to be noticed and more likely to inspire a person to leave an area than is a constantly shining visible appliance. A flashing or strobe effect, therefore, is desired. NFPA 72 (2007) refers to effective light intensity or the equivalent light intensity that would be perceived by a flashing light as compared to a constantly illuminated light. A flashing light with an effective intensity of 15-cd has the same apparent brightness as a 15 cd device burning steadily.

Because a flash rate that is too slow may be perceived as a constant light, and because a flash rate that is too rapid may lead to problems for people with

VISIBLE NOTIFICATION APPLIANCES FOR WALL MOUNTING

**ADDRESSABLE STROBES ARE AVAILABLE IN RED WITH
WHITE LETTERING AND WHITE WITH RED LETTERING**

(a)

VISIBLE NOTIFICATION APPLIANCES FOR CEILING MOUNTING

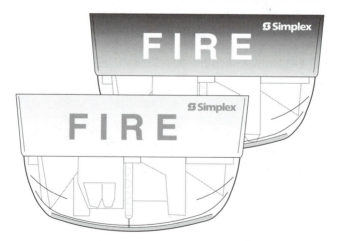

Figure 13-8   *Visible
notification
appliances.
(Courtesy of
Simplex Grinnell.)*

**CEILING MOUNT STROBES ARE AVAILABLE IN RED WITH
WHITE LETTERING AND WHITE WITH RED LETTERING**

(b)

MOUNTING WITH OPTIONAL GUARD

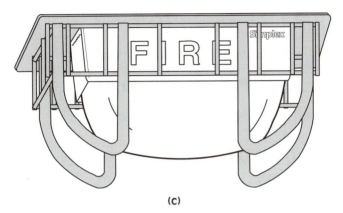

Figure 13-8
(continued)

(C)

photo-sensitive epilepsy, the Americans with Disabilities Act Accessibility Guidelines (ADAAG) and NFPA 72 have declared some boundaries relative to visible notification appliance flashing rates and attributes:

- Minimum flash rate is 1 flash per second.
- Maximum flash rate is 2 flashes per second.
- Maximum pulse duration of 0.2 seconds.
- Lens color is to be white or clear.
- Flash intensity is 1000 cd maximum.

ADAAG Visible Appliance Requirements

Visible appliance location requirements are based on NFPA 72 (2007) requirements. The ADAAG requirements are intended to provide notification for individuals with various physical or sensory limitations. ADAAG requirements apply when disabled occupants are likely to be located in a building, or when ADAAG is required by law for a building or a jurisdiction. In the past, ADAAG requirements have differed from NFPA 72 requirements for visible notification appliance intensity, and confusion has resulted when designing visible appliances for buildings where both requirements apply.

The Draft Final "Americans with Disabilities Act and Architectural Barriers Act Accessibility Guidelines", dated April 2, 2002, was published in the *Federal Register,* July 23, 2004. As of the printing of this book, the Justice Department was still performing its review and had not officially adopted the new guidelines. The new guidelines with regard to visible appliances will require visible notification appliances to be installed in accordance with requirements of NFPA 72 (1999 and 2002 editions). This is an effort by ADAAG and NFPA to coordinate requirements.

Private-Mode Visibility Requirements

NFPA 72 provides advice for audibility requirements in the private mode but no specific advice for visible appliances in the private mode, because trained and alert supervisory persons normally do not require high candela effective intensity ratings to commence their notification responsibilities. The designer must obtain the approval of the AHJ to establish private-mode visibility criteria.

Wall-Mounted Visible Notification Appliance Location

NFPA 72 (2007) requires that the bottom of a wall-mounted visible notification appliance lens be between 80 inches (2.0 m) and 96 inches (2.4 m) above the floor and spaced in accordance with **Figure 13-9**, with the maximum distance between appliances not to exceed 100 feet (30.4 m). As can be seen, the minimum required effective intensity is 15 cd for a 20 foot by 20 foot spacing in accordance with NFPA 72 (2007), and that 635 cd per appliance is the maximum candela rating listed in Figure 13-9.

The maximum intensity per appliance as shown in Figure 13-9 is 635 cd, but caution is urged that a temporary blinding effect may result with the use of high-intensity visible notification appliances. Designers of fire alarm systems use candela ratings that are significantly lower than the maximum, and for the 130 ft × 130 ft (39.6 m × 39.6 m) room in Figure 13-9, are likely to use four or more strobes per room. Most fire alarm equipment vendors do not manufacture high-candela appliances.

UL 1971, *Standard for Signaling Devices for the Hearing Impaired*, requires that wall- and ceiling-mounted visible appliances provide a polar distribution of light, and that their plots become part of the listing of the appliance. With that in mind, it is important to recognize that a strobe listed for wall mounting is not appropriate for ceiling mounting.

Ceiling-Mounted Visible Notification Appliance Location

Table 13-5 shows spacing for visible notification appliances mounted to ceilings with heights ranging from 10 feet (3.05 m) to 30 feet (9.14 m). In rooms with ceiling heights higher than 30 feet (9.14 m) it is required that the notification appliance be mounted on a suspension pole that lowers the visual notification appliance to an elevation no higher than 30 feet (9.14 m) above the floor.

Visible Appliance Spacing in Rooms That Are Not Square

The example in **Figure 13-10** shows a correct and incorrect installation for an irregular room with dimensions 40 ft (12.2 m) long by 22 ft (6.7 m) wide. The correct placement is a visible appliance mounted at the halfway distance on the 40-foot-long (12.2 m) wall, not the 22-foot (6.7 m) wall, and the appliance would

Maximum Room Size		Minimum Required Light Output (Effective Intensity, cd)		
m	ft	One Light per Room	Two Lights per Room (Located on Opposite Walls)	Four Lights per Room (One Light per Wall)
6.10 × 6.10	20 × 20	15	NA	NA
8.53 × 8.53	28 × 28	30	Unknown	NA
9.14 × 9.14	30 × 30	34	15	NA
12.2 × 12.2	40 × 40	60	30	15
13.7 × 13.7	45 × 45	75	Unknown	19
15.2 × 15.2	50 × 50	94	60	30
16.5 × 16.5	54 × 54	110	Unknown	30
16.8 × 16.8	55 × 55	115	Unknown	28
18.3 × 18.3	60 × 60	135	95	30
19.2 × 19.2	63 × 63	150	Unknown	37
20.7 × 20.7	68 × 68	177	Unknown	43
21.3 × 21.3	70 × 70	184	95	60
24.4 × 24.4	80 × 80	240	135	60
27.4 × 27.4	90 × 90	304	185	95
30.5 × 30.5	100 × 100	375	240	95
33.5 × 33.5	110 × 110	455	240	135
36.6 × 36.6	120 × 120	540	305	135
39.6 × 39.6	130 × 130	635	375	185

NA: Not allowable.

Figure 13-9 *Room spacing for wall-mounted visible appliances.*

Source: NFPA 72 (2007), Table 7.5.4.3.1(a) and Figure 7.5.4.3.1. Reprinted with permission from NFPA 72, *National Fire Alarm Code*®, Copyright © 2007, National Fire Protection Association, Quincy, MA 02169. This reprinted material is not the complete and official position of the National Fire Protection Association on the referenced subject which is represented only by the standard in its entirety.

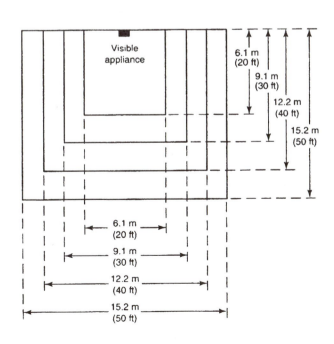

Table 13-5   *Room spacing for ceiling-mounted visible appliances.*

Maximum Room Size		Maximum Ceiling Height		Minimum Required Light Output (Effective Intensity); One Light (cd)
m	ft	m	ft	
6.1 × 6.1	20 × 20	3.0	10	15
9.1 × 9.1	30 × 30	3.0	10	30
12.2 × 12.2	40 × 40	3.0	10	60
13.4 × 13.4	44 × 44	3.0	10	75
15.2 × 15.2	50 × 50	3.0	10	95
16.2 × 16.2	53 × 53	3.0	10	110
16.8 × 16.8	55 × 55	3.0	10	115
18.0 × 18.0	59 × 59	3.0	10	135
19.2 × 19.2	63 × 63	3.0	10	150
20.7 × 20.7	68 × 68	3.0	10	177
21.3 × 21.3	70 × 70	3.0	10	185
6.1 × 6.1	20 × 20	6.1	20	30
9.1 × 9.1	30 × 30	6.1	20	45
13.4 × 13.4	44 × 44	6.1	20	75
14.0 × 14.0	46 × 46	6.1	20	80
15.2 × 15.2	50 × 50	6.1	20	95
16.2 × 16.2	53 × 53	6.1	20	110
16.8 × 16.8	55 × 55	6.1	20	115
18.0 × 18.0	59 × 59	6.1	20	135
19.2 × 19.2	63 × 63	6.1	20	150
20.7 × 20.7	68 × 68	6.1	20	177
21.3 × 21.3	70 × 70	6.1	20	185
6.1 × 6.1	20 × 20	9.1	30	55
9.1 × 9.1	30 × 30	9.1	30	75
15.2 × 15.2	50 × 50	9.1	30	95
16.2 × 16.2	53 × 53	9.1	30	110
16.8 × 16.8	55 × 55	9.1	30	115
18.0 × 18.0	59 × 59	9.1	30	135
19.2 × 19.2	63 × 63	9.1	30	150
20.7 × 20.7	68 × 68	9.1	30	177
21.3 × 21.3	70 × 70	9.1	30	185

be a 60 cd wall-mounted appliance per Figure 13-9, based upon the assumption of a 40-foot (12.2 m) by 40-foot (12.2 m) square room, as determined by Figure 13-10.

Multiple Square Layout of Visible Notification Appliances

Not all rooms are square, and not all rooms are configured in such a way that Figure 13-9 or Table 13-5 can be applied directly, with one visible appliance positioned in the room. Certainly one can see from Figure 13-9 that a 20 foot by

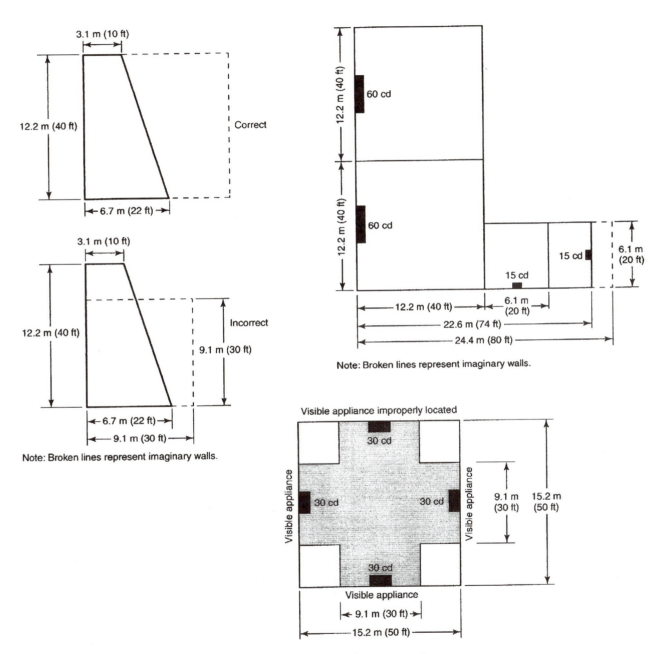

Figure 13-10 *Correct and incorrect positioning of visible notification appliances.*

Source: NFPA 72 (2007), Figure 7.5.4.3(a)(b)(d). Reprinted with permission from NFPA 72, *National Fire Alarm Code®*, Copyright © 2007, National Fire Protection Association, Quincy, MA 02169. This reprinted material is not the complete and official position of the National Fire Protection Association on the referenced subject which is represented only by the standard in its entirety.

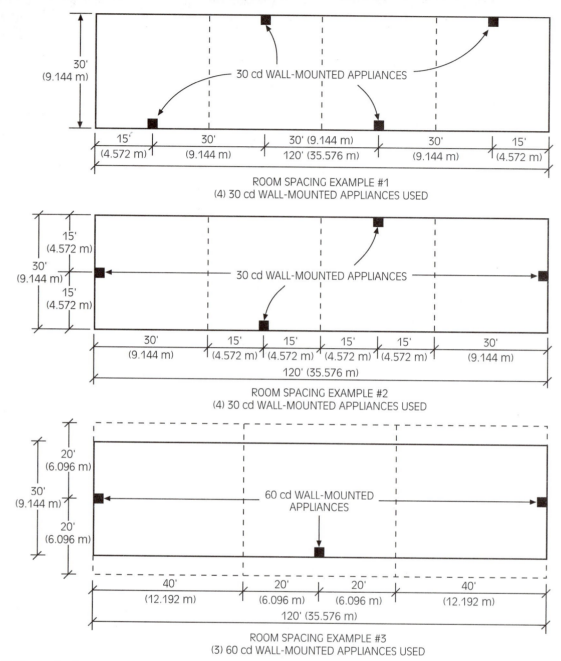

Figure 13-11 *Multiple square layout of visible notification appliances showing the same room designed three different ways; other arrangements are also possible.*

Source: NFPA 72 (2007), Figure A.7.5.4.4. Reprinted with permission from NFPA 72, *National Fire Alarm Code*®, Copyright © 2007, National Fire Protection Association, Quincy, MA 02169. This reprinted material is not the complete and official position of the National Fire Protection Association on the referenced subject which is represented only by the standard in its entirety.

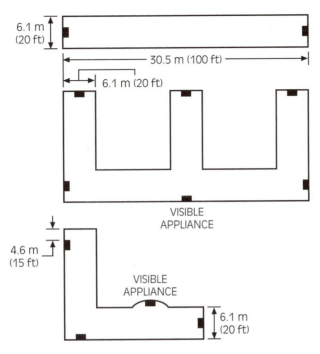

Figure 13-11
(continued)

CORRIDOR SPACING FOR VISIBLE APPLIANCES.

20 foot (6.1 m × 6.1 m) square room can be provided notification by one centrally located 15-cd wall-mounted visible notification appliance. One could imagine a room with width and length divisible by 20 ft (6.1 m) with visible notification appliances spaced to cover 20 ft (6.1 m) squares. **Figure 13-11** shows a 30 ft by 120 ft (9.144 m × 35.576 m) rectangular room with appliances positioned using three different methods using multiple squares of notification patterns from Figure 13-9. NFPA 72 (2007) requires that visible notification appliances spaced closer than 55 feet (16.764 m) apart must flash in synchronization. One can conclude from this multiple squares analysis that there is more than one way to position visible notification appliances in a room.

Optimization of Visual Notification Coverage

Figure 13-12 shows two methods for placement of four wall-mounted 30-cd visible notification appliance spacing allocated in a 100 ft by 100 ft (30.4 m × 30.4 m) room, one correct and one that is incorrect. The incorrect spacing allocation shows a failure to properly analyze the multiple square layout of the room, leaving the corners improperly covered by the visible appliances.

The coverage provided by visible notification appliances can be optimized by offsetting the appliances as shown in Figure 13-12. Note that Figure 13-9

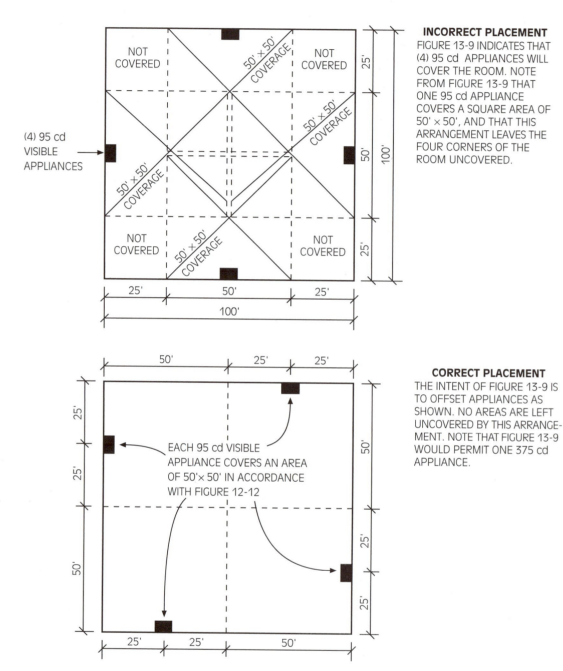

INCORRECT PLACEMENT
FIGURE 13-9 INDICATES THAT (4) 95 cd APPLIANCES WILL COVER THE ROOM. NOTE FROM FIGURE 13-9 THAT ONE 95 cd APPLIANCE COVERS A SQUARE AREA OF 50' × 50', AND THAT THIS ARRANGEMENT LEAVES THE FOUR CORNERS OF THE ROOM UNCOVERED.

CORRECT PLACEMENT
THE INTENT OF FIGURE 13-9 IS TO OFFSET APPLIANCES AS SHOWN. NO AREAS ARE LEFT UNCOVERED BY THIS ARRANGEMENT. NOTE THAT FIGURE 13-9 WOULD PERMIT ONE 375 cd APPLIANCE.

Figure 13-12 *Optimization of visible notification coverage.*

would permit replacement of the four 95-cd appliances with one 375-cd appliance, but most manufacturers of notification appliances do not manufacture an appliance with a luminous intensity as high as 375-cd.

Placement of Visible Appliances in Corridors

For corridors that do not exceed 20 feet (6.1 m) in width, **Table 13-6** may be used. If corridor width exceeds 20 feet (6.1 m), the table in Figure 13-9 is used because most room applications rely on reflected light, whereas corridors use direct (non-reflected) light.

When spacing visible appliances in a corridor in accordance with Table 13-6, the following rules apply:

- Maximum corridor width is 20 feet (6.1 m).
- Corridors greater than 20 feet (6.1 m) in width are spaced using NFPA 72 requirements for room spacing.
- Minimum effective visible appliance intensity is 15 cd.
- Visible appliances may not be spaced further than 15 feet (4.57 m) from the end of the corridor.
- Appliances may not be spaced further than 100 feet (30.4 m) apart.
- An obstruction or change in direction in the corridor necessitates that portions of the corridor on either side of the obstruction or change in direction must be considered as separate corridors when locating visible notification appliances.

Figure 13-13 illustrates some of the corridor rules that apply to spacing of visible appliances.

Table 13-6 *Corridor spacing allocation for wall-mounted visible appliances.*

Corridor Length		Minimum Number of 15-cd Visible Appliances Required
(ft)	(m)	
0–30	0–9.14	1
31–130	9.45–39.6	2
131–230	39.93–70	3
231–330	70.4–100.6	4
331–430	100.9–131.1	5
431–530	131.4–161.5	6

Note: Use this table for corridors up to 20 feet (6.1 m) in width; for corridors exceeding 20 feet (6.1 m) in width, use Figure 13-9.

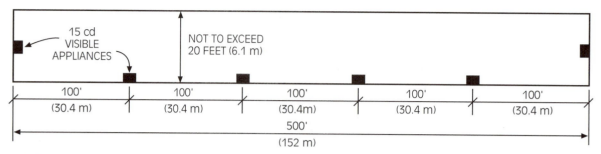

TABLE 13-6 STATES THAT SIX 15 cd APPLIANCES ARE USED IN A CORRIDOR OF LENGTH FROM 431 FT. TO 530 FT. A 500 FT. CORRIDOR CAN BE SPACED AS SHOWN ABOVE.

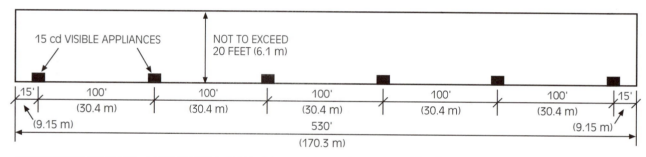

NOTE THAT WHEN SIX 15 cd APPLIANCES ARE USED IN A CORRIDOR BETWEEN 500 FT. (170.3 m) AND 530 FT. (161.5 m) LONG, APPLIANCES CANNOT BE LOCATED AT THE ENDS OF THE CORRIDOR WITHOUT EXCEEDING 100 FT. (30.4 m) BETWEEN APPLIANCES. APPLIANCES ARE PERMITTED TO BE LOCATED UP TO 15 FT. (9.15 m) FROM THE END OF A CORRIDOR.

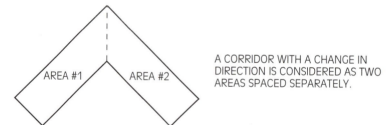

A CORRIDOR WITH A CHANGE IN DIRECTION IS CONSIDERED AS TWO AREAS SPACED SEPARATELY.

Figure 13-13   *Corridor spacing of visible appliances.*

Spacing of Visible Appliances in Sleeping Areas

NFPA 72 (2007) requires that visible appliances in sleeping rooms with no room dimension exceeding 16 feet (4.87 m) be spaced in accordance with **Table 13-7**. If a sleeping room has a dimension exceeding 16 feet (4.87 m), the visible appliance must be located within 16 feet (4.87 m) of the pillow. In addition to visible notification appliances, combination smoke detectors or smoke alarms shall be installed.

Table 13-7 *Effective intensity requirements for sleeping area visible notification appliances.*

Distance from Ceiling to Top of Lens		Intensity (cd)
mm	in.	
≥610	≥24	110
<610	<24	177

Source: NFPA 72 (2007), Table 7.5.4.6.2. Reprinted with permission from NFPA 72, *National Fire Alarm Code®*, Copyright © 2007, National Fire Protection Association, Quincy, MA 02169. This reprinted material is not the complete and official position of the National Fire Protection Association on the referenced subject which is represented only by the standard in its entirety.

Combination Audible/Visible Notification Appliances

combination audible/visible notification appliance
possesses both audible and visible notification components

A **combination audible/visible notification appliance**, as shown in **Figure 13-14**, possesses both audible and visible notification components. The visible notification placement rules overrule the audible notification appliance placement rules; therefore, appliances are spaced in accordance with the visible requirements listed in Figure 13-9 and Table 13-5. Most fire alarm system designs specify both audible and visible appliances, and many specify combination audible/visible appliances in buildings, because of concern for individuals who lack either the ability to hear or the ability to see. Audible appliances can be heard outside of the room where they are installed, but visible appliances must be in the same room as the individual being notified. Combination audible/visible notification appliances will increase the probability of people being notified who lack either sight or hearing.

Synchronization of Visible Appliances

When more than one unsynchronized visible appliance is positioned in a room, the viewer can perceive more than the maximum flash rate of two flashes per second in the room, either by direct viewing or from reflected-light viewing. Also, it has been found that unsynchronized flashes may prompt epileptic seizures in individuals who are prone to photosensitive epilepsy.

NFPA 72 (2007) provides synchronization as one option in visible notification appliance design:

1. A single visible appliance in a room requires no synchronization.
2. Two visible appliances on opposite walls require no synchronization.
3. Two groups of visible notification appliances, where visible appliances of each group are synchronized, in the same room or adjacent space within the

**SPEAKER/VISIBLE APPLIANCES; SPEAKER WITH STROBE
SELECTABLE AS FREE-RUN OR SYNCHRONIZED**

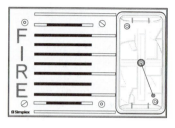

(a)

S/V WITH OPTIONAL GUARD

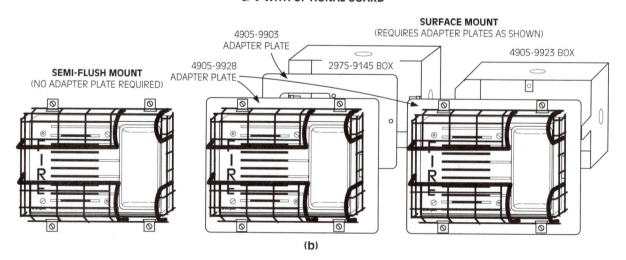

(b)

Figure 13-14 *Audible/visible notification appliances with speakers are available for horizontal or vertical mounting. (Courtesy of Simplex Grinnell.)*

STROBE SYNCHRONIZATION CIRCUIT DETAILS

- SYNCHRONIZING MODULE CONTROLS SYNCHRONIZED STROBES AND ARE AVAILABLE AS STYLE Y (CLASS B) OR STYLE Z (CLASS A)
- THE SAME CIRCUIT CAN POWER SYNCHRONIZED STROBES, UNSYNCHRONIZED STROBES, AND 24 VDC AUDIBLE APPLIANCES, IF REQUIRED

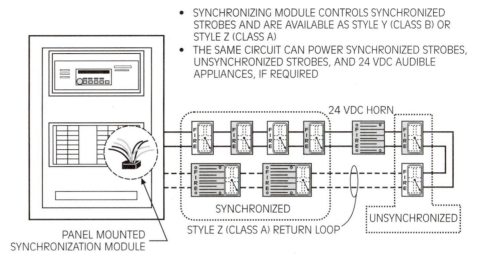

24 VDC HORN

SYNCHRONIZED

UNSYNCHRONIZED

PANEL MOUNTED
SYNCHRONIZATION MODULE

STYLE Z (CLASS A) RETURN LOOP

Figure 13-15 *Strobe synchronization circuit details. (Courtesy of Simplex Grinnell.)*

field of view, are permitted. This is effectively saying that no more than two flash *rates* (not necessarily two visible *appliances*) are permitted. For example, an acceptable option would be four strobes in one room with strobe #1 and strobe #3 synchronized with each other and strobe #2 and strobe #4 synchronized with each other, but not synchronized with #1 and #3.

A methodology for strobe synchronization is shown in **Figure 13-15**. A synchronization module is installed in the FACU, and the strobe circuit is connected to the module. Other methods of strobe synchronization are available as well. Most modern fire alarm control units are capable of synchronizing strobe flash rates where required.

TEXTUAL AUDIBLE AND VISIBLE NOTIFICATION APPLIANCES

An example of a textual audible notification appliance is a loudspeaker that provides verbal fire alarm notification from emergency fire service-operated microphones, telephones, or prerecorded messages from a voice alarm system. Audible requirements are the same as for other audible notification appliances.

A textual visible notification appliance provides alphanumeric fire alarm notification or instructions, usually on an annunciator, fire alarm control unit, or a panel remote from the main FACU. Textual notification information can be provided by LEDs on an alphanumeric display window.

Textual appliances can provide more precise information than can be provided by other audible and visible appliances and may provide notification for persons with hearing disabilities. Central station operators, security guards, and fire watch personnel often are provided with textual appliances to record the location of the detection signals for rapid fire service response and postfire analysis.

ANNUNCIATION NOTIFICATION APPLIANCES

annunciator panel

a device that provides visible notification of the location of an initiating device by zone and floor

A form of visible notification is the **annunciator panel**. Annunciator panels provide visual notification of the location of an initiating device by zone and floor, as shown in **Figure 13-16**. They can also indicate elevator status, fan operation status, HVAC damper closure status, fire pump status, status of the speakers on a voice system, status of firefighters' phones, and a wide range of detection, notification, and trouble conditions. The annunciator should be placed in a fire control room or at the main entrance to a building so fire service

Figure 13-16

Annunciator panel display; a typical annunciator provides indication for zone, floor, and type of device. Note: The panel shown is displaying smoke detector and water flow switch activation on the 6th floor, zone 2. (Courtesy of Simplex Grinnell.)

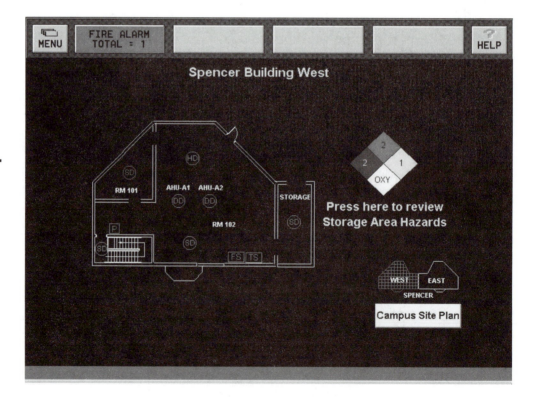

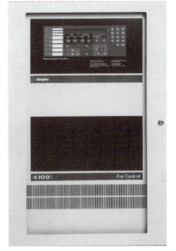

Figure 13-16
(continued)

personnel can identify the fire location and implement suppression strategy quickly.

NFPA 72 Annex material addresses a standardized firefighter interface that uses standardized symbols, locations of switching on interface panels, and a commonality of appearance of annunciator panels. NEMA (National Electrical Manufacturers Association) Publication 5830-2005, "Fire Service Annunciator and Interface," provides requirements.

TACTILE NOTIFICATION APPLIANCES

Audible appliances are effective for those who can hear, visible appliances are effective for those who can see, and combination audible/visible appliances are effective for those who possess either the senses of hearing or sight. Tactile notification appliances are for individuals who can neither see nor hear. An example of a tactile notification appliance is a vibrating device that could be installed in a pillow to rouse sleeping individuals or on a belt to notify individuals who are awake.

When one specifies tactile notification appliances, training must be provided so that occupants are capable of finding their way out of the building on their own. If this is not possible, a method must be provided for the notification of supervisory personnel who would be responsible for assisting occupants to safety.

Bed shakers are also an available alternative for the rousing of sleeping individuals. Some tactile appliances, such as vibrating belt pagers, and variable-speed ceiling fans, cannot be readily supervised, and are therefore considered supplemental notification appliances.

SUMMARY

Just as initiating devices must be selected carefully to be congruent with the properties of the anticipated fire, notification appliances must be selected carefully to be congruent with the needs of the building occupants, the supervisory or security personnel responsible for interpreting the alarm, and the fire service personnel who would be responding to the alarm.

Notification appliances must be designed to accommodate persons with disabilities and must be designed both to arouse a sleeping person and to spur a person who is awake into leaving the building or relocating to a fire-safe area within the building. To accomplish this task, this chapter has outlined minimum NFPA 72 (2007) requirements for notification by audible, visible, and tactile methods.

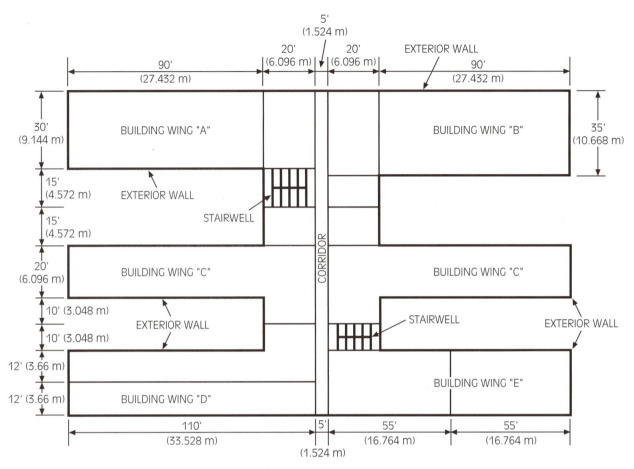

Figure 13-17   *Illustration for review question 9: Dimension locations of all audible appliances, visual appliances, and audible/visual appliances in each room.*

REVIEW QUESTIONS

1. A mechanical room, 20 feet by 20 feet, has a measured ambient sound level of 55 dB. Specify a decibel rating for an audible notification appliance in the room.

2. A steel mill has a measured average ambient sound level of 120 dB. Specify notification appliances for the mill.

3. A shoe store, 40 feet by 40 feet, has a measured average ambient sound level of 40 dB. Specify a decibel rating for audible notification appliances in the store.

4. A notification appliance is rated at 75 dB at 10 feet. Determine the decibel level perceived by an individual 40 feet from the appliance, on the other side of a stud wall, along the centerline of the appliance. The ambient sound level is 30 dB.

5. An audible notification appliance emits 95 dB at 10 feet. What dB rating would be perceived by a person 80 feet from the appliance, at a 45° angle from the appliance?

6. Visible notification appliances are specified for a 90 foot × 90 foot room. Determine candela requirements and sketch appliance locations for the following situations:

   a. one wall-mounted appliance

   b. two wall-mounted appliances

   c. four wall-mounted appliances

   d. two appliances mounted to a 20-foot ceiling

7. A rectangular room is 30 feet long and 15 feet wide. Determine the candela rating if one visible appliance is used, and sketch the appliance location.

8. A corridor is 430 feet long. Specify the number of 15-cd visible appliances required, and sketch locations with all appliances dimensioned.

9. Draw and dimension the locations of the following appliances for all rooms on the plan shown in **Figure 13-17**:

   a. audible appliances

   b. visible appliances

   c. audible/visible appliances

DISCUSSION QUESTIONS

1. What are the responsibilities of fire protection professionals relative to the needs of occupants with disabilities?

2. What are the responsibilities of fire protection professionals relative to requirements of the fire service?

ACTIVITIES

1. Visit a school or office building and determine the location and candela or decibel rating of all visible and audible notification appliances by marking their locations and ratings on a plan. Evaluate the locations and ratings with respect to NFPA 72 (2007).

2. With an experienced fire alarm system designer, survey a new or existing building and plan for the installation of notification appliances in accordance with NFPA 72. Determine if the Americans with Disabilitise Act applies to the building survey, and determine the effect of local requirements on the building.

3. Obtain the most recent edition of NFPA 101, *Code for Safety to Life in Buildings and Structures,* and list requirements that apply to notification appliances.

4. Determine the building code (such as the National Building Code, BOCA, or the Uniform Building Code) that applies to your area, and list requirements that apply to notification appliances.

NOTE

1. Construction Innovation Vol. 6, No. 4, Fall 2001.

Chapter 14

FIRE DETECTOR PLACEMENT

Learning Objectives

Upon completion of this chapter, you should be able to:

- Describe how detectors are listed by Underwriters Laboratories (UL).
- Explain why detectors are not permitted to be positioned too close to the corners of a room where the ceiling meets the wall.
- Determine the spacing of detectors in a room with a smooth flat ceiling.
- Determine locations of detectors in a room with unusually configured walls.
- Space detectors in a corridor.
- Discuss rules for placement of detectors on sloped ceilings.
- Determine the locations of detectors on ceilings with solid joists or beams.
- Explain why detectors are not permitted to be located too close to supply air ducts.
- Describe the effect of floor-mounted partitions on a ceiling jet.
- Determine the optimum locations for beam detectors, flame detectors, gas detectors, and duct detectors.

Spacing of detectors is a direct function of the listing of the detector. A recognized laboratory, such as Underwriters Laboratories (UL) or Factory Mutual (FM), tests detectors and lists them for their capability to meet the performance objectives established for each type of detector. NFPA 72, National Fire Alarm Code (2007) provides numerous detector spacing requirements for fire alarm system designers.

HEAT DETECTOR LISTINGS

Heat detectors are listed by UL as follows:
- Heat detector type
    fixed temperature
    rate-of-rise
    rate-of-rise/fixed temperature
    rate compensated
    rate-compensated/fixed-temperature
    analog-addressable
- Temperature rating

Ambient Ceiling Temperature	Detector Temperature
up to 100°F	135° to 165°F devices to be installed
100°F–150°F	175° to 225°F devices to be installed
150°F–225°F	250° to 300°F devices to be installed
225°F–300°F	325° to 360°F devices to be installed

- Listed smooth ceiling spacing between detectors
- Listed smooth ceiling spacing to wall or partition
- Contact arrangement
    N.O.—normally open contacts, circuit closes on detection
    N.C.—normally closed contacts, circuit opens on detection
- Detector model number

SMOKE DETECTOR LISTINGS

Smoke detectors are listed by UL under category UROX as follows:
1. APPLICATION
   - open area protection
   - releasing service

- open-air protection/releasing service
- duct detection
  - detector mounts inside duct
  - sampling tubes—detector mounts on outside of duct and tubes penetrate into the duct
- special application—a separate listing is maintained for detectors that are used for specific purposes, such as suitable for high air velocities

2. DETECTOR TYPE
- photoelectric
- ionization
- combination photoelectric/ionization
- projected beam
- air-sampling

3. COMPATIBILITY RESTRICTIONS

4. INSTALLATION CRITERIA
- standards: NFPA 72
- spacings: NFPA 72
- environmental considerations
- effect of velocity
- stability test—determines suitability of detector after installation

SPOT-TYPE DETECTOR SPACING

A fire alarm system designer must recognize that spot-type heat and smoke detector installations must conform to some basic rules prescribed in NFPA 72.

Corner Placement for Spot-Type Detectors: The 4-Inch Rule

Ceiling jets do not travel at precise 90° angles when encountering a wall or obstruction. **Figure 14-1** shows an example of a trash can fire against a wall. Note that the velocity of the plume creates dead air space in the corner, within which a detector may not be effective. NFPA 72 requires that a ceiling spot detector be placed no closer than 4 inches to a wall and that a wall-mounted spot detector be placed no closer than 4 inches or farther than 12 inches below the ceiling.

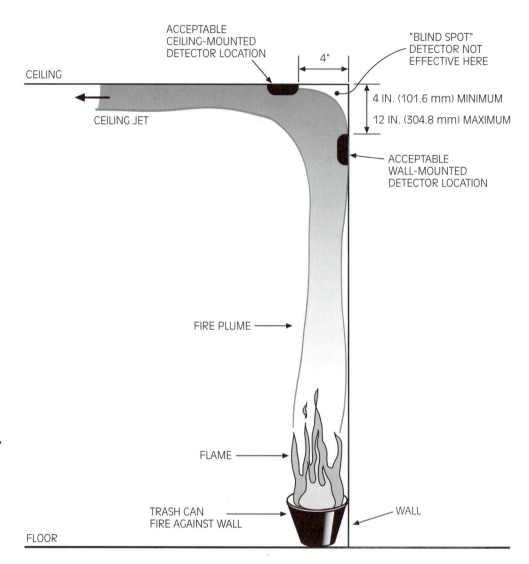

Figure 14-1 *Ceiling jet dynamics for corner fires; NFPA 72 prohibits spot detectors being placed within the "blind spot".*

Smooth Ceiling Spacing of Spot-Type Detectors

Underwriters Laboratories and Factory Mutual provide a listed maximum smooth-ceiling spacing (S_{max}) that is permissible for spot-type heat detectors and a listed maximum distance from a wall or partition. Smoke detectors have the manufacturer's recommended spacings. The ideal spacing condition (S_{max}) exists when the ceiling is smooth, when the detectors are spaced at their maximum listed or recommended spacing, and when the room length and room width are both exactly divisible by the maximum listed spacing, as shown in **Figure 14-2**.

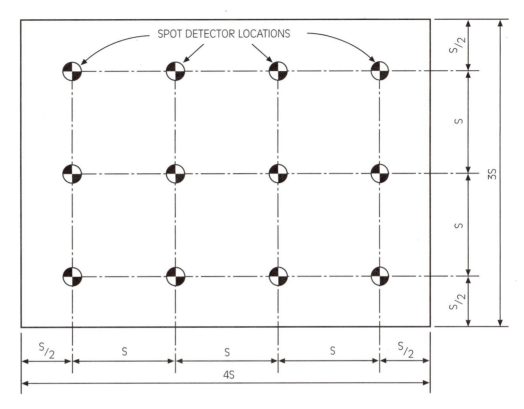

Figure 14-2 *Ideal maximum spacing for spot detectors; applies when ceiling is smooth, detectors are spaced at maximum allowable listed spacing, and room dimensions are exactly divisible by the listed spacing.*

EXAMPLE 14-1

Ceiling Spacing of Spot-Type Detectors—Ideal Spacing

A room is 200 feet long and 150 feet wide, and the 8-foot-high ceiling is smooth and flat. A spot heat detector has been selected that is listed for a 50-foot maximum spacing between detectors and a 25-foot maximum distance from a wall. Draw the locations of the heat detectors on a plan, and show the dimensions between detectors and to the walls.

Solution

The solution is shown in **Figure 14-3**. Note that the room length, 200 feet, and the room width, 150 feet, are both divisible by the maximum detector spacing (S_{max}) of 50 feet, so the ideal maximum spacing of detectors can be obtained.

Rarely does a designer encounter a situation in which the room length and the room width both are exactly divisible by the maximum detector spacing. Let us consider a more realistic example.

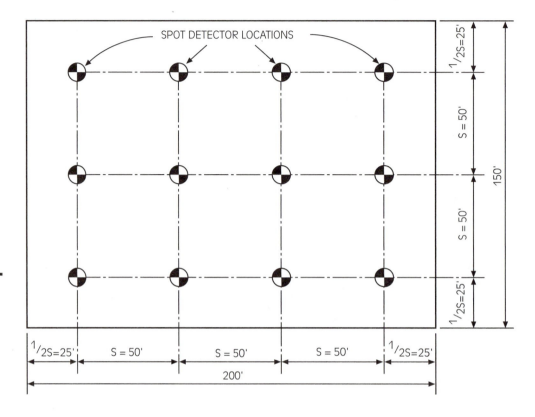

SPOT DETECTOR LOCATIONS

$^1/_2S=25'$

S = 50'

150'

S = 50'

$^1/_2S=25'$

$^1/_2S=25'$ S = 50' S = 50' S = 50' $^1/_2S=25'$

200'

Figure 14-3 *Solution to Example 14-1: Spot detectors are laid out using square areas as shown, with detectors lined up in rows.*

EXAMPLE 14-2

Ceiling Spacing of Spot-Type Detectors—Spacing Less than Listed Spacing

A room is 227 feet long and 166 feet wide, and the ceiling is smooth and flat. A spot heat detector has been selected that is listed for a 50-foot maximum spacing (S_{max}) between detectors and a 25-foot maximum distance from a wall. Space the detectors for this room and draw the detector layout.

Solution

As can be seen, the room dimensions are not exactly divisible by the maximum listed spacing; therefore, detectors will be spaced at less than their listed spacing. To determine the number of detectors per row on a smooth ceiling, divide the room dimension by the maximum listed spacing, (S_{max}):

$$\text{Number of detectors per row} = \frac{\text{room dimension}}{S_{max}}$$

Let us look at the number of detectors along the length of the room:

$$\text{Number of detectors along length} = \frac{227 \text{ ft}}{50 \text{ ft}} = 4.54$$

A fractional detector does not exist, so we always round up to the nearest whole detector:

Number of detectors along length = 5

Spacing between detectors (S_{actual}) is determined by dividing the room dimension by the number of detectors:

$$S_{\text{actual}} = \frac{\text{room dimension}}{\text{number of detectors per row}}$$

For Example 14-2, the room length is 227 feet and the number of detectors along the length is 5:

$$\text{Length } S_{\text{actual}} = \frac{227 \text{ ft}}{5 \text{ detectors}}$$

$$= 45.4 \text{ ft}$$

Distance to the wall is $(1/2) \times (S_{\text{actual}})$. In this example, S_{actual} is 45.4 ft.

$$\text{Length} = (1/2) \times (S_{\text{actual}})$$
$$= (1/2) \times (45.4 \text{ ft})$$
$$= 22.7 \text{ ft}$$

The room width is computed similarly:

$$\text{Number of detectors along width} = \frac{166 \text{ ft}}{50 \text{ ft}} = 4.$$

$$\text{The width is } S_{\text{actual}} = \frac{166 \text{ ft}}{4 \text{ detectors}} = 41.5 \text{ ft}$$

$$\text{Distance to wall is } (1/2) \times (S_{\text{actual}}) = (1/2)(41.5 \text{ ft})$$
$$= 20.75 \text{ ft}$$

The solution to this problem is shown in **Figure 14-4**. As computed, five detectors are spaced along the length at 45.4 feet between detectors and four detectors are spaced along the width at 41.5 feet between detectors, for a total of 20 detectors in the room.

Note that because S_{max} in Example 14-2 is 50 feet, detectors are located at less than their listed spacing. If the location of detectors conflicts with lighting fixtures or other ceiling-mounted equipment, the flexibility exists to move detectors, with the following limitations:

- S_{max} may not be exceeded for any detector in the room.
- The maximum distance from a detector to any wall may not exceed the listed distance to a wall.
- No detector can be mounted closer than 4 inches to a wall, per Figure 14-1.

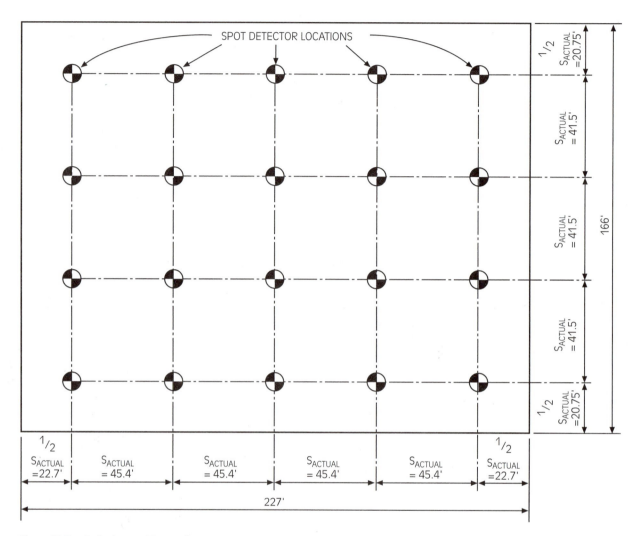

SPOT DETECTOR LOCATIONS

$^1/_2$ S_{ACTUAL} =20.75'

S_{ACTUAL} = 41.5'

S_{ACTUAL} = 41.5'

S_{ACTUAL} = 41.5'

$^1/_2$ S_{ACTUAL} =20.75'

166'

$^1/_2$ S_{ACTUAL} =22.7'

S_{ACTUAL} = 45.4'

S_{ACTUAL} = 45.4'

S_{ACTUAL} = 45.4'

S_{ACTUAL} = 45.4'

$^1/_2$ S_{ACTUAL} =22.7'

227'

Figure 14-4   *Solution to Example 14-2.*

Spot-Type Detector Spacing for Unusual Wall Configurations

In Example 14-2, the room is rectangular and the smooth ceiling spacing is relatively straightforward. Where should detectors be positioned in an unusually shaped room, such as one shaped in a five-pointed star or an odd-shaped polygon?

NFPA 72 (2007) permits detectors to be positioned on a smooth ceiling at a distance up to $(0.7) \times (S_{max})$ to a corner of a room. To understand this rule, consider a detector as illustrated in **Figure 14-5**. The detector is in the exact center

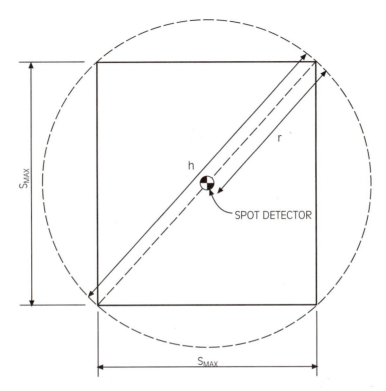

Figure 14-5
Illustration of .7 S_{max} rule, or "circle of coverage".

of a square with sides equal to S_{max} in length, which represents the maximum ideal square spacing for a spot detector. Note that the detector also can be the exact center of a circle whose perimeter intersects the four corners of the square.

The diagonal of the square, h, as shown in Figure 14-5, intersects the detector position and is calculated using the Pythagorean theorem from algebra:

$$h^2 = (S_{max})^2 + (S_{max})^2$$

$$h = \sqrt{(S_{max})^2 + (S_{max})^2}$$

$$= \sqrt{2 S_{max}^{\,2}}$$

$$= (1.4) \times (S_{max})$$

The distance from the detector to the corner of the square, r, as shown in Figure 14-5, is 1/2 h.

$$r = (1/2) \times (h)$$

$$= (1/2) \times (1.4) \times (S_{max})$$

$$= (0.7) \times (S_{max})$$

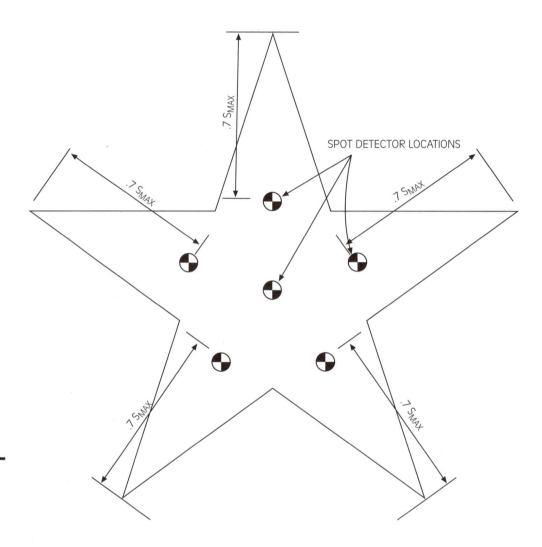

Figure 14-6 *Spot detector spacing in rooms of unusual configuration.*

We have proved that the distance from a detector to the corner of a square room is 0.7 S_{max}. Let us use this result to solve some challenging detector spacing problems. First, consider a room shaped like a five-pointed star, as shown in **Figure 14-6**. For rooms of unusual configuration, detectors can be spaced up to .7 S_{max} from the star points, with the spacing in the center not to exceed S_{max} between detectors.

Spot-Type Detector Spacing in Corridors

In Figure 14-5, we investigated a detector with a square area of coverage, S_{max} by S_{max} and a circle whose radius r intersects the corners of the square. Using

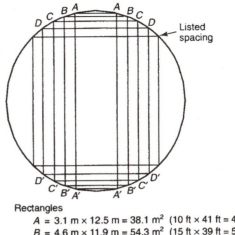

Rectangles
$A = 3.1\ m \times 12.5\ m = 38.1\ m^2$ ($10\ ft \times 41\ ft = 410\ ft^2$)
$B = 4.6\ m \times 11.9\ m = 54.3\ m^2$ ($15\ ft \times 39\ ft = 585\ ft^2$)
$C = 6.1\ m \times 11.3\ m = 68.8\ m^2$ ($20\ ft \times 37\ ft = 740\ ft^2$)
$D = 7.6\ m \times 10.4\ m = 78.9\ m^2$ ($25\ ft \times 34\ ft = 850\ ft^2$)
Listed spacing for = $9.1\ m \times 9.1\ m = 83.6\ m^2$ ($30\ ft \times 30\ ft = 900\ ft^2$)
heat detectors only

Note: Smoke detectors are not listed for spacing. Use manufacturer's
coverage recommendations and this figure.

Figure 14-7   *Heat detector spacing in rectangular areas.*

the $.7\ S_{max}$ rule we derived previously, we can space detectors in long narrow spaces, such as corridors. NFPA 72 (2007) provides an example, shown in **Figure 14-7**, where the listed spacing (S_{max}) is 30 feet and which shows rectangular room arrangements permissible in accordance with NFPA 72 (2007).

Note that in Figure 14-7 the spot detector is in the center of a square, identified as "listed spacing" where $S_{max} = 30$ feet (9.1 m). As we determined previously, the radius of the circle is $.7\ S_{max} = 21$ feet (6.4 m).

Notice that using a radius of 21 feet (6.4 m), we can create rectangles resembling corridors within the circle. Note that Figure 14-7 illustrates that, for a detector whose S_{max} is 30 feet (9.1 m) a corridor that is 10 feet wide is permitted to have spot detectors spaced up to 41 feet (12.5 m) apart.

Other spacings can be handled similarly. For a detector whose S_{max} is 50 feet (50.24 m), .7 S_{max} is 35 feet (10.668 m), and detectors are permitted to be spaced in 10-foot-wide corridors up to about 68 feet (20.72 m) apart.

Figure 14-7 applies specifically to heat detectors with a listed spacing of 30 feet (9.1 m). The same concepts, however, can be used for smoke detectors.

Spot-Type Detector Spacing for Smooth Peaked Ceilings

Figure 14-8 illustrates how detectors are spaced per NFPA 72 (2007) for a smooth peaked ceiling. Note two important points from this figure:

1. Detector spacing is determined by the vertical projections of the detectors to the floor, as shown in Figure 14-8. They are not spaced along the angle of the roof.
2. A detector must be placed within 3 feet (900 mm) of the roof peak. For fires where the ceiling jet moves up the sloped roof to the peak at a rapid velocity, the detector at the peak may actuate before the detectors along the sloped roof.

Spot-Type Detector Spacing for Smooth Shed Ceilings

Figure 14-9 illustrates how detectors are spaced for a smooth shed ceiling. As with peaked ceilings, the detectors are spaced by their vertical projections on the floor, and a detector must be positioned within 3 feet (900 mm) of the peak.

Spot-Type Detector Spacing for Solid Joist Construction

An unfinished basement in a residence has exposed wood joists on the ceiling as shown in **Figure 14-10**. These joists, usually spaced 12 inches (304.8 mm) to 16 inches (406.4 mm) apart, affect the ceiling jet by rendering it more turbulent and slowing it down. Detectors must be installed with reduced spacing to accommodate the altered ceiling jet characteristics associated with such construction.

Factors affecting spacing of spot-type heat detectors on solid joist construction, in accordance with NFPA 72 (2007), include the following:

- If the joists are 4 inches (100 mm) or less in depth, consider the ceiling smooth, mount the heat detectors to the bottom of the joists, and space the detectors at S_{max}.
- If the joists exceed 4 inches (100 mm) in depth, mount the heat detectors to the bottom of the joists and space them at 1/2 S_{max} when measured to right angles to the joists, to account for turbulence in the ceiling jet.

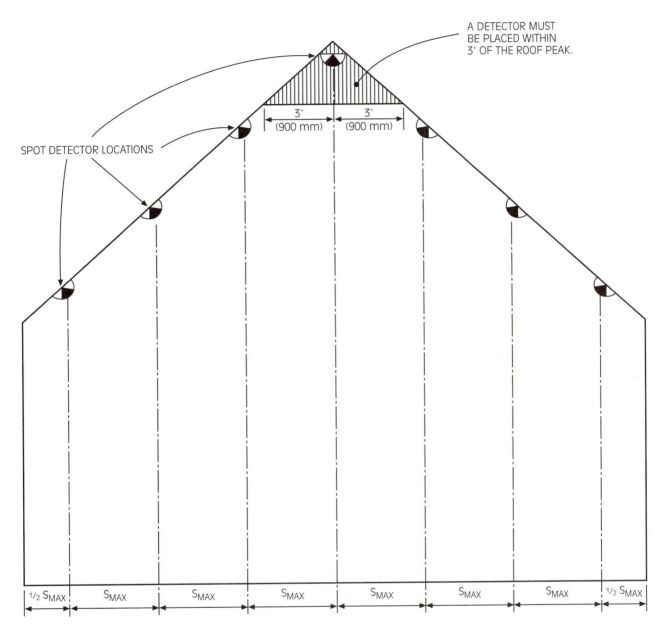

Figure 14-8   *Sectional view of a peaked ceiling; a detector must be placed within 3 feet (900 mm) of the roof peak, and spacing is measured along the floor, not along the slope.*

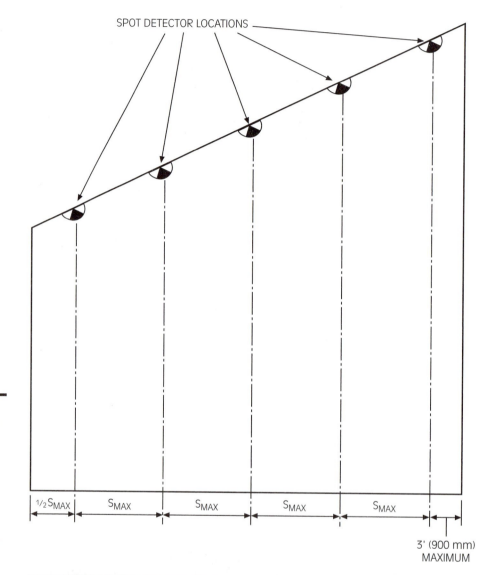

SPOT DETECTOR LOCATIONS

$\frac{1}{2}S_{MAX}$ S_{MAX} S_{MAX} S_{MAX} S_{MAX}

3' (900 mm)
MAXIMUM

Figure 14-9

Sectional view of a shed ceiling; a detector must be placed within 3 feet (900 mm) of the roof peak, and spacing is measured along the floor, not along the slope.

SPOT-TYPE HEAT DETECTOR SPACING FOR BEAMED CONSTRUCTION—BODY OF NFPA 72

NFPA 72 provides requirements in the body of the standard, which are requirements in jurisdictions referencing NFPA 72. Those requirements are as follows:

- If the beam depth is less than or equal to 4 inches (100 mm), assume a smooth ceiling and space detectors at S_{max}.
- If the beam depth is greater than 4 inches (100 mm), space detectors at 2/3 S_{max} at right angles to the direction of the beam.

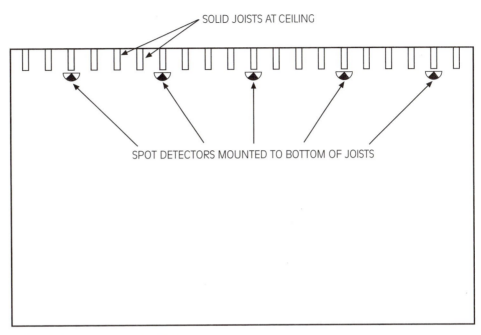

SOLID JOISTS AT CEILING

SPOT DETECTORS MOUNTED TO BOTTOM OF JOISTS

Figure 14-10 *Spot detector placement on solid joist ceilings.*

• HEAT DETECTORS – SPACING MAY NOT EXCEED $(1/2)$ (S_{MAX})

- If beam depth is less than than 12 inches (300 mm) and and the distance between beams is less than 8 feet (2.4 m) on center, detectors are permitted to be placed on the bottom of beams. If this statement is not true, detectors are required to be placed in each beam pocket.
- If waffle or pan-type ceilings have beams or solid joists no greater than 24 inches (600 mm) deep, and no greater than 12 feet (3.66 m) center-to-center spacing, smooth ceiling spacing shall be permitted.

Spot-Type Heat Detector Spacing for Beamed Construction—Annex of NFPA 72

The Annex of NFPA 72 provides nonmandatory advice for designers that addresses beam depth, beam spacing, and ceiling height. Note that requirements in the body of the standard do not consider ceiling height as a factor in placement or spacing of detectors. Designers may compare the two detector placement approaches (body of NFPA 72 and Annex of NFPA 72) when spacing detectors in beamed construction, using the most conservative or demanding placement methodology. The Annex recommendations, as shown in **Figure 14-11**, are as follows:

- If the beam depth divided by the ceiling height (d/h) is greater than 0.10 *and* if the distance between beams divided by the ceiling height (w/h) is greater than 0.40, detectors are placed between each beam, within the beam pocket.

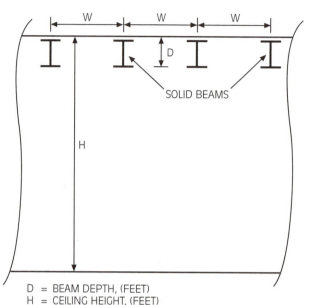

HEAT DETECTOR SPACING RULES-BODY OF STANDARD

#1. IF D ≤ 4" (100 mm) ASSUME SMOOTH CEILING AND SPACE DETECTORS AT S_{MAX}

#2. IF D > 4" (100 mm) SPACE DETECTORS AT (2/3) S_{MAX}

#3. IF D < 12" (300 mm) AND W < 8 FEET (2.4 m), DETECTORS ARE PERMITTED TO BE INSTALLED ON BOTTOM OF BEAMS.

HEAT DETECTOR SPACING RULES-ANNEX OF STANDARD

#4. IF D/H > 0.10 AND W/H > 0.40 PLACE DETECTORS BETWEEN EACH BEAM

#5. IF D/H < 0.10 OR W/H < 0.40 PLACE DETECTORS ON THE BOTTOMS OF THE BEAMS, SPACED PER RULE #1 AND RULE #2 ABOVE.

D = BEAM DEPTH, (FEET)
H = CEILING HEIGHT, (FEET)
W = DISTANCE BETWEEN BEAMS, (FEET)

Figure 14-11 *Heat detector placement for beamed construction, using the body of NFPA 72, and Annex of NFPA 72.*

- If the beam depth divided by the ceiling height (d/h) is less than 0.10 *or* if the distance between beams divided by the ceiling height (w/h) is less than 0.40, detectors are permitted to be placed on the bottoms of the beams.

- Though not stated specifically in the Annex, if either ratio is *equal* to the specified value, designers may consider either option. Many designers choose the most conservative of the two options in such cases.

EXAMPLE 14-3

Detector Placement for Beamed Construction

A room is 10 feet 6 inches high from floor to ceiling, with 30-inch-deep beams spaced at 39 inches on center. A heat detector whose S_{max} is 50 feet is specified. Determine location and permissible spacing for the detectors, using the Annex of NFPA 72.

Solution

In solving this problem, make sure that units are uniform for all calculations. In this case, we will use units of feet:

$$d/h = \frac{2.5 \text{ ft}}{10.5 \text{ ft}} = 0.24$$

$$w/h = \frac{3.25 \text{ ft}}{10.5 \text{ ft}} = 0.31$$

Note that d, w, and h must be in the same unit for the formula to work. In this case, because w/h is less than 0.40, we are permitted to install detectors on the bottoms of the beams using the annex formulae. We can space the detectors up to 2/3 S_{max} at the bottoms of the beams. Detectors are placed within a beam pocket when smoke is highly likely to collect within the pockets. When plume and ceiling jet velocities are very high, or where air movement within the room is unpredictable, detectors on the bottoms of beams may be more effective. The performance objective of NFPA 72 requirements is to place the detectors where heat is most likely to collect and actuate a detector.

Effect of Ceiling Height on Spot-Type Heat Detectors

As discussed in Chapter 11, air entrainment into the fire plume increases with ceiling height, and the temperature of the plume therefore decreases with ceiling height. NFPA 72 (2007) recognizes this and requires reduced spacing for ceiling heights from 10 to 30 feet (3.05 to 9.14 m), as **Table 14-1** shows.

SPACING OF LINE-TYPE HEAT DETECTORS

Line-type heat detectors predominately provide detection for linear hazards, such as coal conveyors and cable trays. They also are permitted to protect rooms, as shown in **Figure 14-12**.

Table 14-1   *Heat detector spacing reduction Based on ceiling height.*

Ceiling Height Above		Up to and Including		Multiply Listed Spacing by
m	ft	m	ft	
0	0	3.05	10	1.00
3.05	10	3.66	12	0.91
3.66	12	4.27	14	0.84
4.27	14	4.88	16	0.77
4.88	16	5.49	18	0.71
5.49	18	6.10	20	0.64
6.10	20	6.71	22	0.58
6.71	22	7.32	24	0.52
7.32	24	7.93	26	0.46
7.93	26	8.54	28	0.40
8.54	28	9.14	30	0.34

Source: NFPA 72 (2007), Table 5.6.5.5. Reprinted with permission from NFPA 72, *National Fire Alarm Code*®, Copyright © 2007, National Fire Protection Association, Quincy, MA 02169. This reprinted material is not the complete and official position of the National Fire Protection Association on the referenced subject which is represented only by the standard in its entirety.

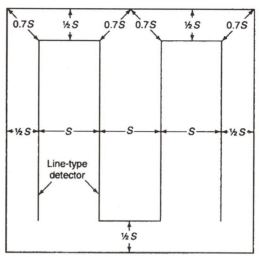

S = Space between detectors

Figure 14-12   *Line-type detectors—spacing layouts, smooth ceiling.*

Source: NFPA 72 (2007), Figure A.5.6.5.1(b). Reprinted with permission from NFPA 72, *National Fire Alarm Code*®, Copyright © 2007, National Fire Protection Association, Quincy, MA 02169. This reprinted material is not the complete and official position of the National Fire Protection Association on the referenced subject which is represented only by the standard in its entirety.

Line-type detectors have spacing requirements that are different from spot detectors:

- Line-type detectors on a ceiling must be positioned not more than 20 inches from the ceiling.
- Line-type detectors are mounted to the bottom of solid joists.
- Line-type detectors are mounted to the bottoms of beams, where beams are less than 12 inches deep and less than 8 feet apart on center.

SPOT-TYPE SMOKE DETECTOR SPACING

Figure 14-13 illustrates NFPA 72 (2007) requirements for smoke detectors:

- The 4-inch (100 mm) rule, shown in Figure 14-1 applies.
- 30 ft (9.1 m) spacing shall be permitted unless the listed spacing differs.

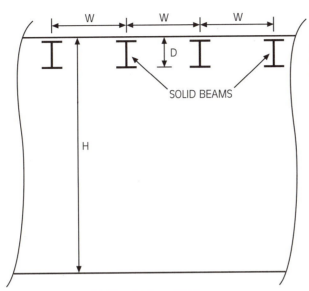

NFPA 72 (2007) SMOKE DETECTOR SPACING RULES FOR BEAMED CONSTRUCTION APPLY:

- FOR LEVEL CEILINGS WITH BEAM DEPTHS OF LESS THAN 10 PERCENT OF THE CEILING HEIGHT (0.1 H), SMOOTH CEILING SPACING.

- FOR LEVEL CEILINGS WITH BEAM DEPTHS EQUAL TO OR GREATER THAN 10 PERCENT OF THE CEILING HEIGHT (0.1 H) AND BEAM SPACING EQUAL TO OR GREATER THAN 40 PERCENT OF THE CEILING HEIGHT (0.4 H), SPOT-TYPE DETECTORS SHALL BE LOCATED ON THE CEILING IN EACH BEAM POCKET.

D = BEAM DEPTH, (FEET) (METERS)
H = CEILING HEIGHT, (FEET) (METERS)
W = DISTANCE BETWEEN BEAMS, (FEET) (METERS)

Figure 14-13 *Smoke detector spacing for beamed construction.*

- The 0.7 rule, illustrated in Figure 14-5, applies.
- For beam depths of less than 10% of the ceiling height (0.1 H), smooth ceiling spacing, as shown in Figure 14-13, shall be permitted.
- For beam depths equal to or greater than 10% of the ceiling height (0.1 H) and beam spacing equal to or greater than 40% of the ceiling height (0.4 H), spot-type detectors located on the ceiling in each beam pocket, as shown in Figure 14-13, shall be permitted.
- For waffle or pan-type ceilings with beams or solid joists no greater than 24 inch (600 mm) deep and no greater than 12 ft (3.66 m) center-to-center spacing, smooth ceiling spacing with detectors on ceilings or on bottoms of beams shall be permitted.
- For corridors 15 ft (4.5 m) in width or less having ceiling beams or solid joists perpendicular to the corridor length, smooth ceiling spacing with detectors located on ceilings, sidewalls, or bottom of beams shall be permitted.
- For rooms of 900 ft.^{2} (84 m^2) area or less, only one smoke detector shall be required.
- For sloped ceilings with beams running parallel to (up) the slope, the spacing for level beamed ceilings shall be used, and the ceiling height shall be taken as the average height over slope. For slopes greater than 10 degrees

the detectors located at one-half the spacing from the low end shall not be required, and spacing shall be measured along a horizontal projection of the ceilings.

- For sloped ceilings with beams running perpendicular to (across) the slope, the spacing for level beamed ceilings shall be used and the ceiling height shall be taken as the average height over slope.

- For sloped ceilings with solid joists, the detectors shall be located on the bottom of the joist.

Spacing of Spot-Type Smoke Detectors near Air Vents

The effectiveness of smoke detectors is influenced greatly by the movement of air in the space. NFPA 72 (2007) requires that smoke detectors not be placed directly in the air stream of a wall-mounted supply or return-air diffuser, or closer than 3 feet from a ceiling-mounted supply or return-air diffuser, as illustrated in **Figure 14-14**. This separation distance recognizes the turbulence surrounding HVAC diffusers and also the propensity for dust to be drawn into smoke detectors.

NFPA 72 (2007) requires that detectors be located so they can sense smoke as it is being drawn toward return air diffusers, but recent research from the National Institute for Standards and Technology (NIST) indicates that turbulence near return-air diffusers may retard smoke detector activation in cases of high-velocity air movement.

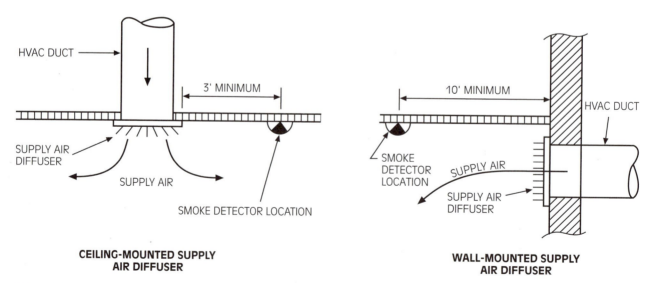

Figure 14-14 *Positioning smoke detectors near supply air diffusers. (Note: If supply air velocities are very high, distances should be increased. Separation distances apply also to return air diffusers.)*

Although no air supply vent placement requirements currently apply to heat detectors, a conservative approach would be to observe the same placement criteria as required for smoke detectors.

Spacing of Spot-Type Smoke Detectors near Floor-Mounted Partitions

Floor-mounted partitions that extend closer than 18 inches (457.2 mm) to the ceiling are likely to break up the ceiling jet, as shown in **Figure 14-15**. NFPA 72 (2007) requires the designer to consider a reduction of spacing. The conservative

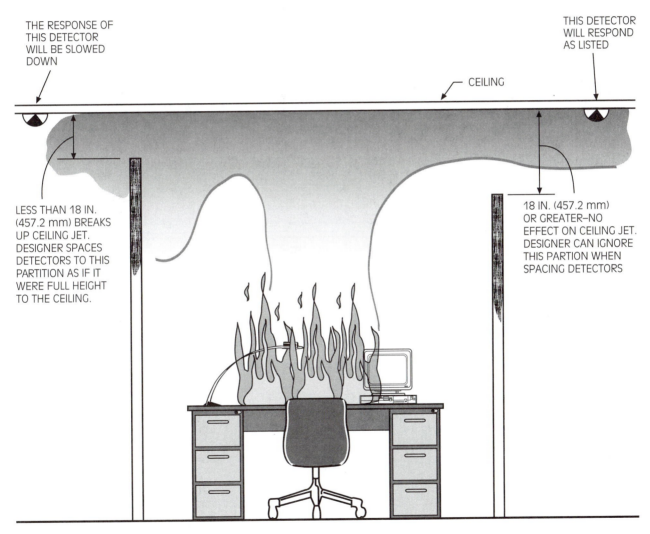

THE RESPONSE OF THIS DETECTOR WILL BE SLOWED DOWN

THIS DETECTOR WILL RESPOND AS LISTED

CEILING

LESS THAN 18 IN. (457.2 mm) BREAKS UP CEILING JET. DESIGNER SPACES DETECTORS TO THIS PARTITION AS IF IT WERE FULL HEIGHT TO THE CEILING.

18 IN. (457.2 mm) OR GREATER–NO EFFECT ON CEILING JET. DESIGNER CAN IGNORE THIS PARTION WHEN SPACING DETECTORS

Figure 14-15   *Floor-mounted partitions and detector spacing configuration.*

approach for such cases would be to space detectors as if the partitions were full-height walls. The dilemma involved with such an arrangement is that when movable or portable partitions are used and if numerous partition relocations are performed regularly, many detectors would be rendered ineffective. Designers of new buildings can avoid the dilemma by ensuring that the clearance between the ceiling and the top of the portable partition exceeds 18 inches in such situations.

Spot-Type Smoke Detection for Raised Floors and Suspended Ceilings

Figure 14-16 illustrates an area with a raised floor and a suspended ceiling. Suspended ceilings are commonly installed in educational, office, mercantile, and assembly occupancies. Raised floors are installed to conceal wires and cables in spaces such as control rooms and computer rooms.

NFPA 72 (2007) requires smoke detection in both spaces (above and below the suspended ceiling), regardless of whether the construction is combustible or noncombustible. For areas above ceilings, detection is permitted to be omitted

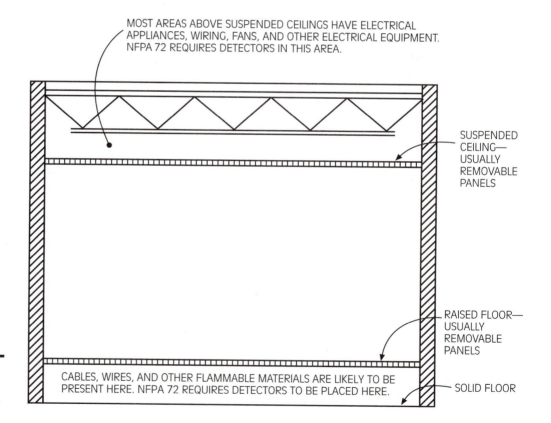

Figure 14-16 *Raised floors and suspended ceilings.*

HVAC plenum

a space that is kept under negative pressure by fans, allowing stale air to enter the space through grilles on the ceiling

if the same space is used as an **HVAC plenum**. An HVAC plenum is a space that is kept under negative pressure by fans, allowing stale air to enter the space through grilles on the ceiling. The HVAC system exhausts the stale air to the outside.

For plenums and raised floor areas, NFPA 72 (2007) requires that the detectors be mounted in a position that does not allow dust to be introduced into the detector. Detectors used for this application must be listed for the environment selected, with spacing based on the air-flow anticipated. Positioning for under-floor areas is illustrated in **Figure 14-17.**

Spot-Type Smoke Detectors in High Air Movement Areas

When air in a room is used for the cooling of items such as telecommunications and Internet switching equipment, it will have an effect on the performance of

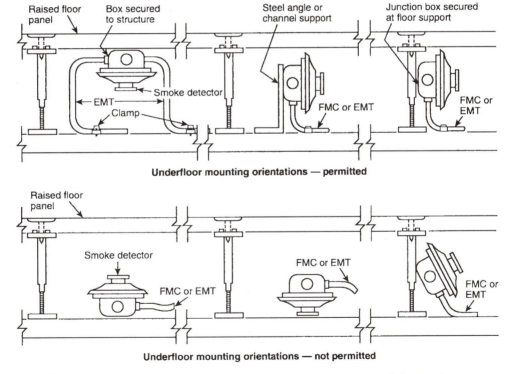

Figure 14-17 *Mounting installations permitted (top) and not permitted (bottom).*

Source: NFPA 72 (2007), Figure A.5.7.3.2.2. Reprinted with permission from NFPA 72, *National Fire Alarm Code®*, Copyright © 2007, National Fire Protection Association, Quincy, MA 02169. This reprinted material is not the complete and official position of the National Fire Protection Association on the referenced subject which is represented only by the standard in its entirety.

air change

the cycle of air movement from HVAC unit to underfloor through equipment and back to the HVAC unit

smoke detectors. In such rooms, HVAC handling equipment is installed to allow rapid circulation of air to underfloor areas, and through equipment for cooling, draw it back to the HVAC unit for re-cooling. The cycle of air movement from the HVAC unit to underfloor through equipment and back to the HVAC unit is called an **air change**. The measured or rated rate of air change is in minutes, or air changes per minute. This rate can be easily converted to air changes per hour.

If an item of equipment in this arrangement is in a state of smoldering combustion, the air moving through the equipment is being cooled by the HVAC unit and is moving rapidly, which makes for a challenging smoke detection scenario. NFPA 72 recognizes this challenge and provides information for designers relative to placement of smoke detectors as a function of air changes per hour. NFPA 72 requires that smoke detectors not be placed in the direct air stream of HVAC supply registers and stipulates further that the spacing of smoke detectors in high air-flow areas be reduced in accordance with **Figure 14-18**, which decreases detector spacing as the number of air changes per hour increases.

To determine air changes per hour, the designer must ascertain the flow rate of the HVAC unit in cubic feet per minute (CFM). Assuming supply air CFM is equal to return air CFM:

$$\text{Air changes/hour} = [(H)/(V)] \times (60 \text{ minutes}/1 \text{ hour})$$

H = HVAC supply air/return air-flow, in cubic feet per minute (CFM)
V = room volume served by the HVAC unit, in cubic feet. The total room volume is divided by the number of HVAC units to obtain V for this equation.
H/V = air changes/minute

Spot-Type Smoke Detectors for Smoke Control Systems

A smoke control system is a system designed to perform one or more of the following functions:

- Remove and exhaust smoke from zones where smoke has been detected, and prevent distribution of the smoke to occupied areas.
- Provide pressurization of key areas of egress, such as stairwells, to prevent smoke from migrating to those areas.
- Isolate smoke spread to certain zones by initiating the closing of doors and duct dampers.

Detectors installed within ducts must be listed specifically for this purpose and are not permitted to replace general open-area protection. When they detect smoke within the ducts, they signal shutdown of air-moving equipment to prevent recirculation of smoke.

Smoke Detector Spacing Based on Air Movement

Minutes per Air Change	Air Changes per Hour	Spacing per Detector	
		m^2	ft^2
1	60	11.61	125
2	30	23.23	250
3	20	34.84	375
4	15	46.45	500
5	12	58.06	625
6	10	69.68	750
7	8.6	81.29	875
8	7.5	83.61	900
9	6.7	83.61	900
10	6	83.61	900

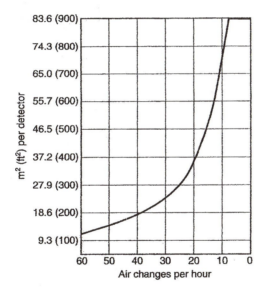

Figure 14-18   *Smoke detector placement in high air-movement areas.*

Source: NFPA 72 (2007), Table 5.7.5.3.3 and Figure 5.7.5.3.3. Reprinted with permission from NFPA 72, *National Fire Alarm Code*®, Copyright © 2007, National Fire Protection Association, Quincy, MA 02169. This reprinted material is not the complete and official position of the National Fire Protection Association on the referenced subject which is represented only by the standard in its entirety.

Duct detectors are manufactured in two basic varieties—the air-sampling duct detector, shown in **Figure 14-19**, and the in-duct detector, shown in **Figure 14-20**. Sampling detectors are used commonly where air velocity exceeds 400 feet (121.92 m) per minute, and in-duct detectors usually are used where air velocity is less than 400 feet (121.92 m) per minute.

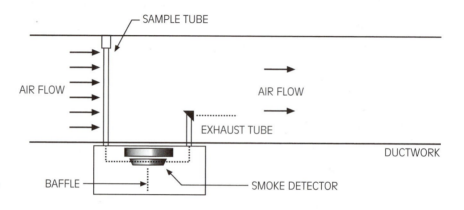

**TYPICAL DUCT DETECTOR APPLICATION
(GREATER THAN 400 FPM)**

Figure 14-19 *Air-sampling duct detector. (Courtesy of Simplex Grinnell.)*

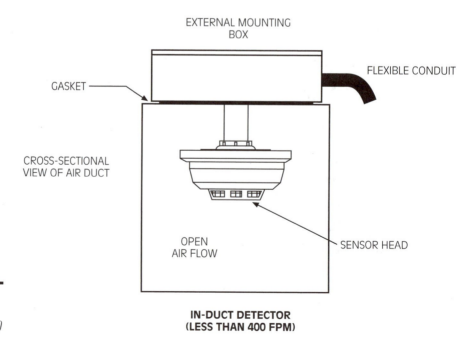

**IN-DUCT DETECTOR
(LESS THAN 400 FPM)**

Figure 14-20 *In-duct detector. (Courtesy of Simplex Grinnell.)*

One duct detector centered in the duct opening is permitted to be installed for ducts up to 36 inches wide, and two detectors are required for ducts up to 72 inches (182.88 cm) wide with one additional detector required for each 24 inches (60.96 cm) above 72 inches (182.88 cm) in width.

PROJECTED BEAM SMOKE DETECTOR SPACING

Beam detectors work on the following principles, as **Figure 14-21** illustrates.

- For most applications, a beam detector consists of two components—a sender and a receiver.
- NFPA 72 (2007) requires that the sender and receiver be mounted firmly to avoid movement of the projected beam and possible unwanted alarms.
- NFPA 72 (2007) requires that the projected beam remain unobstructed by storage, ceiling-mounted equipment, or other objects that would result in a trouble condition on the FACU and prohibit the detectors from sensing smoke.

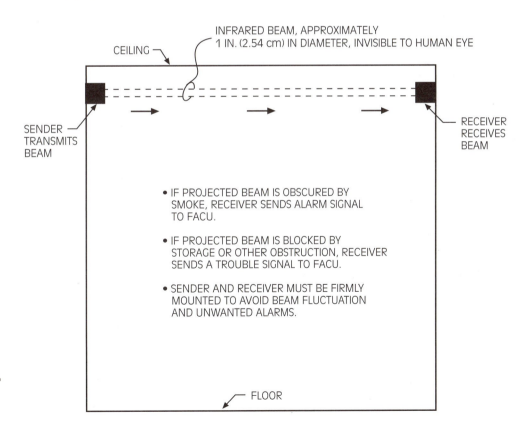

INFRARED BEAM, APPROXIMATELY
1 IN. (2.54 cm) IN DIAMETER, INVISIBLE TO HUMAN EYE

CEILING

SENDER
TRANSMITS
BEAM

RECEIVER
RECEIVES
BEAM

- IF PROJECTED BEAM IS OBSCURED BY SMOKE, RECEIVER SENDS ALARM SIGNAL TO FACU.

- IF PROJECTED BEAM IS BLOCKED BY STORAGE OR OTHER OBSTRUCTION, RECEIVER SENDS A TROUBLE SIGNAL TO FACU.

- SENDER AND RECEIVER MUST BE FIRMLY MOUNTED TO AVOID BEAM FLUCTUATION AND UNWANTED ALARMS.

FLOOR

Figure 14-21 *Beam detector principles.*

- Smoke obscures the projected beam and reduces transmission of the infrared beam to the receiver. The receiver sends a signal to the FACU when transmission is reduced to a predetermined level.
- Beam detectors are for indoor use, and are to be mounted to stable surfaces.
- Mirrors are permitted to be used with projected beams, in accordance with the manufacturer's instructions, provided that the mirrors are mounted to stable surfaces.
- Spacing for smooth ceilings is to be in accordance with the manufacturer's instructions. **Figure 14-22** illustrates recommendations of a manufacturer's,

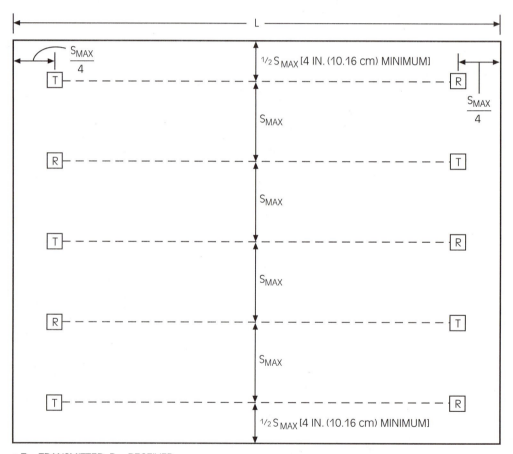

- T = TRANSMITTER, R = RECEIVER
- S_{MAX} VARIES BY MANUFACTURER [USUALLY 45 FEET (13.716 m) MAXIMUM RECOMMENDED RANGE]
- STAGGER SENDERS AND RECEIVERS AS SHOWN
- L VARIES BY MANUFACTURER [USUALLY 330 TO 350 FEET (100.584 TO 106.68 m)]

Figure 14-22 *Beam detector spacing; S_{max} applies in either the vertical plane (for tall buildings) or in the horizontal plane (for smooth flat ceilings).*

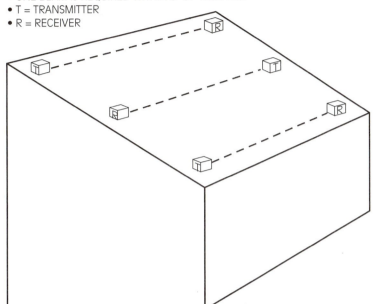

- PROJECTED BEAMS MUST RUN PARALLEL TO THE PEAK AS SHOWN
- ONE BEAM IS REQUIRED WITHIN 3' OF THE PEAK
- T = TRANSMITTER
- R = RECEIVER

Figure 14-23 *Sloped ceiling installation for beam detectors.*

such as Detection Systems, Inc. More detail is found in their booklet *Principles of Long-Range Beam Smoke Detection.*

- For beamed construction, some manufacturers require that beam detectors be mounted in each beam pocket if the beam depth exceeds 18 inches and the support beam spacing is greater than 8 feet.
- For solid joists less than 8 inches in depth, some manufacturers require that the ceiling be considered as a smooth ceiling.
- NFPA 72 requires that beam detectors on sloped ceilings be installed parallel to the peak. One line of beam detectors must be positioned within 3 feet (900 mm) of the peak as shown in **Figure 14-23**.

FLAME DETECTOR PLACEMENT

A flame detector can perceive a fire that is viewed within its cone of vision, as shown in **Figure 14-24**. Its ability to sense a fire is dependent upon the position of the fire within the cone of vision, the distance from the fire to the detector, the size of the fire, and matching the spectral response of the detector to the spectral emissions of the fire to be detected. Each flame detector is listed for a specific

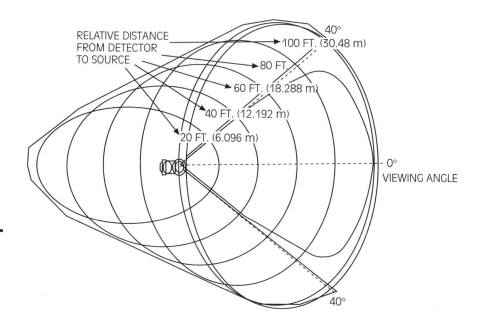

RELATIVE DISTANCE FROM DETECTOR TO SOURCE

40°
100 FT. (30.48 m)
80 FT.
60 FT. (18.288 m)
40 FT. (12.192 m)
20 FT. (6.096 m)

0°
VIEWING ANGLE

40°

Figure 14-24 *Cone of vision for a flame detector. (Courtesy of Simplex Grinnell.)*

viewing angle that defines the cone of vision. The detector in Figure 14-24 has a viewing angle of 40°. NFPA 72 (2007) requires that there be no interference that would block the detector from seeing a fire within its cone of vision, and that for area protection, sufficient detectors be placed such that the cones of vision overlap and all portions of the room are provided with detection. Placement of flame detectors usually requires considerable field experience and consultation with the manufacturer.

GAS-SENSING DETECTOR PLACEMENT

Gas-sensing detectors may sense a potentially hazardous condition in advance of ignition, or may sense gaseous combustion byproducts. Gas-sensing detectors are to be spaced in accordance with NFPA 72 (2007) and their listing. NFPA 72 has replaced detailed spacing requirements on gas-sensing fire detectors with more general requirements permitting a wide variety of gas-sensing detectors to be installed in accordance with manufacturers' recommendations and installation instructions corresponding to the detector's listing. Spacing rules for gas-sensing detectors include the following:

- Solid joists or beams less than 8 inches (20.32 cm) in depth are considered to be smooth ceiling construction.

- Spacing of gas detectors is required to be reduced when solid joists exceed 8 inches in depth.

- If beams exceed 8 inches (20.32 cm) in depth, detector spacing is required to be reduced.
- If beams exceed 18 inches (45.72 cm) in depth and are spaced at more than 8 feet (2.438 m) on center, detectors must be installed in each beam pocket.
- Sloped ceiling provisions used for spot detectors apply also to gas detectors.
- Partition provisions that apply to smoke detectors apply also to gas detectors.
- Positioning of detectors near supply air diffusers is the same as for smoke detectors.
- Gas-sensing detectors must be selected with proper consideration given to the ambient conditions. Nonfire gases may be produced that could trigger a false alarm. In such cases, the designer specifies whether receipt of a gas-sensing signal triggers a trouble indication, a general alarm, or an actuation of the suppression system.

AN ENGINEERING APPROACH TO HEAT DETECTOR PLACEMENT

The best way to ensure proper detection in a building is to conduct fire testing in numerous areas of the building to ascertain the most appropriate detector and the most advantageous detector locations. This usually is not possible and is performed rarely. Fire protection engineers use data obtained under test and laboratory conditions as a reference "design fire" in determining the characteristics of the fire that are used as the basis for spacing detectors and specifying suppression systems. Annex B of NFPA 72 (2007) provides a hand calculation methodology for spacing detectors, using characteristics obtained from design fires.

Calculated Heat Detector Placement per NFPA 72 Annex B (2007 Edition)

Degreed fire protection engineers perform calculations on fire growth in specific spaces and create computer models that simulate detection based on the results of these calculations. NFPA 72 provides a convenient formulaic structure for fire protection engineers in performing such calculations, called the "Fire Detection Design and Analysis Worksheet," in Annex B of NFPA 72, as shown in **Figure 14-25**.

This worksheet provides a structure for fire growth calculations, detector characteristics, design goals, and detector performance with respect to the fire characteristics, fire growth, and room attributes. An example is worked in Annex B of NFPA 72 to assist fire protection engineers.

Fire Detection Design and Analysis Worksheet [28]
Design Example

1.	Determine ambient temperature (T_a) ceiling height or height above fuel (H).	$T_a =$ _____ °C + 273 = _____ K $H =$ _____ m
2.	Determine the fire growth characteristic (α or t_g) for the expected design fire.	$\alpha =$ _____ kW/sec^2 $t_g =$ _____ sec
3a.	Define the characteristics of the detectors.	$T_s =$ _____ °C + 273 = _____ K RTI = _____ m$^{1/2}$sec$^{1/2}$ $\dfrac{dT_d}{dt} =$ _____ °C/min $\tau_0 =$ _____ sec
3b. or	*Design* — Establish system goals (t_{CR} or Q_{CR}) and make a first estimate of the distance (r) from the fire to the detector.	$t_{CR} =$ _____ sec $r =$ _____ m $Q_{CR} =$ _____ kW
3b.	*Analysis* — Determine spacing of existing detectors and make a first estimate of the response time or the fire size at detector response ($Q = \alpha t^2$).	$r =$ _____ *1.41 = _____ = S (m) $Q =$ _____ kW $t_d =$ _____ sec
4.	Using equation B.21, calculate the nondimensional time (t^*_{2f}) at which the initial heat front reaches the detector.	$t^*_{2f} = 0.861\left(1 + \dfrac{r}{H}\right)$ $t^*_{2f} =$
5.	Calculate the factor A defined by the relationship for A in equation B.20.	$A = \dfrac{g}{C_p T_a \rho_o}$ $A =$
6.	Use the required response time (t_{CR}) along with the relationship for t^*_p in equation B.19 and $p = 2$ to calculate the corresponding value of t^*_2.	$t^*_2 = \dfrac{t_{CR}}{A^{-1/(3+p)}\, \alpha^{-1/(3+p)} H^{4/(3+p)}}$ $t^*_2 =$
7.	If $t^*_2 > t^*_{2f}$, continue to step 8. If not, try a new detector position (r) and return to step 4.	
8.	Calculate the ratio $\dfrac{u}{u^*_2}$ using the relationship for U^*_p in equation B.17.	$\dfrac{u}{u^*_2} = A^{1/(3+p)}\, \alpha^{1/(3+p)} H^{(p-1)/(3+p)}$ $\dfrac{u}{u^*_2} =$
9.	Calculate the ratio $\dfrac{\Delta T}{\Delta T^*_2}$ using the relationship for ΔT^*_p in equation B.18.	$\dfrac{\Delta T}{\Delta T^*_2} = A^{2/(3+p)}(T_a/g)\, \alpha^{2/(3+p)} H^{-(5-p)/(3+p)}$ $\dfrac{\Delta T}{\Delta T^*_2} =$
10.	Use the relationship for ΔT^*_2 in equation B.23 to calculate ΔT^*_2.	$\Delta T^*_2 = \left[\dfrac{t^*_2 - t^*_{2f}}{(0.146 + 0.242r/H)}\right]^{4/3}$ $\Delta T^*_2 =$
11.	Use the relationship for $\dfrac{u^*_2}{(\Delta T^*_2)^{1/2}}$ in equation B.24 to calculate the ratio $\dfrac{u^*_2}{(\Delta T^*_2)^{1/2}}$.	$\dfrac{u^*_2}{(\Delta T^*_2)^{1/2}} = 0.59\left(\dfrac{r}{H}\right)^{-0.63}$ $\dfrac{u^*_2}{(\Delta T^*_2)^{1/2}} =$
12.	Use the relationships for Y and D in equations B.27 and B.28 to calculate Y.	$Y = \left(\dfrac{3}{4}\right)\left(\dfrac{u}{u^*_2}\right)^{1/2}\left[\dfrac{u^*_2}{(\Delta T^*_2)^{1/2}}\right]^{1/2}\left(\dfrac{\Delta T^*_2}{RTI}\right)\left(\dfrac{t}{t^*_2}\right)D$ $Y =$
13.	*Fixed Temperature HD* — Use the relationship for $T_d(t) - T_d(0)$ in equation B.25 to calculate the resulting temperature of the detector $T_d(t)$.	$T_d(t) = \left(\dfrac{\Delta T}{\Delta T^*_2}\right)\Delta T^*_2\left[1 - \dfrac{(1-e^{-Y})}{Y}\right] + T_d(0)$ $T_d(t) =$
14.	*Rate of Rise HD* — Use the relationship for $\dfrac{dT_d(t)}{dt}$ in equation B.26.	$dT_d = \left[\left(\dfrac{4}{3}\right)\left(\dfrac{\Delta T}{\Delta T^*_2}\right)(\Delta T^*_2)^{1/4}\dfrac{(1-e^{-Y})}{[(t/t^*_2)D]}\right]dt$ $dT_d =$

15.	If: 1. $T_d > T_s$ 2. $T_d < T_s$ 3. $T_d = T$	**Repeat Procedure Using** **Design** 1. a larger r 2. a smaller r 3. $s = 1.41 \times r =$ _____ m	**Analysis** 1. a larger t_r 2. a smaller t_r 3. $t_r =$ _____ sec

Figure 14-25 *Fire detection design and analysis worksheet.*

Source: NFPA 72 (2007), Figure B.3.3.4.4. Reprinted with permission from NFPA 72, *National Fire Alarm Code*®, Copyright © 2007, National Fire Protection Association, Quincy, MA 02169. This reprinted material is not the complete and official position of the National Fire Protection Association on the referenced subject which is represented only by the standard in its entirety.

SUMMARY

The placement of detectors must take into consideration the type of fire, dynamics of the ceiling jet, attributes of the detector, and properties of the ceiling or surface onto which the detectors are mounted. NFPA 72 provides minimum requirements for the spacing of detectors, but designers must carefully analyze the anticipated fire and the space requiring detection. An engineered approach to positioning detectors is provided by Annex B of NFPA 72, based on fire tests conducted under controlled conditions.

Heat detectors, smoke detectors, beam detectors, and gas detectors have specific requirements that are based on testing and listing of the detectors.

REVIEW QUESTIONS

1. Space rate-of-rise heat detectors in a room 964 feet long and 677 feet wide, with a smooth ceiling, using detectors that have an S_{max} of 45 feet.

2. Space smoke detectors in a corridor that is 1542 feet long and 6 feet wide, using a detector with an S_{max} of 50 feet.

3. A room is 15 feet high, with 13-inch-deep solid beams spaced at 4 feet 3 inches on center. Determine whether heat detectors are mounted to the bottoms of beams or whether they must be mounted within each beam pocket. Determine whether reduced spacing is required.

4. Smoke detectors are specified for a room that is 10 feet high, with 19-inch beams spaced 98 inches on center. Determine whether reduced spacing is required, and determine whether detectors are to be mounted to the bottoms of the beams or within each beam pocket.

5. A line-type detection system has an S_{max} of 50 feet. Sketch the detection system in a room with a smooth flat ceiling that is 900 feet long by 300 feet wide, showing dimensions for all portions of the system.

6. An atrium has 42 stories, each 10 feet high. Design a beam detection system for the atrium, assuming that the atrium is 250 feet wide and S_{max} for the detector is 45 feet.

7. Determine the number of air-sampling duct detectors required for a 97-inch-wide duct.

DISCUSSION QUESTIONS

1. Under what circumstances would you provide a beam-detection fire alarm system in lieu of a spot-detection fire alarm system?

2. Cite specific reasons for specifying a listed fire alarm component in lieu of one that is not listed.

ACTIVITIES

1. Visit a laboratory facility where fire testing is being performed. For the test observed, provide answers to the following questions:

   a. Does the combustion observed burn cleanly with little smoke or with a large volume of smoke?

   b. What is the maximum temperature of the combustion observed?

   c. What fire growth rate is recorded?

2. Visit a company that designs and installs detection systems. Is NFPA 72 used directly in all cases, or is an engineered approach used for detector spacing?

3. Visit a very tall building with an atrium.

   a. Is beam detection used? If so, what is the spacing?

   b. Are detectors located at the roof? If so, what are their ratings or characteristics?

   c. Evaluate the items that could burn on the atrium floor. Do you think a ceiling-mounted detection system would be effective? Why or why not?

Chapter 15

FIRE ALARM CIRCUIT DESIGN AND FIRE ALARM CONTROL UNITS

Learning Objectives

Upon completion of this chapter, you should be able to:

■ List the types of initiating, notification, and signaling line circuits, and discuss the differences.

■ Explain the function of Class A and Class B circuits.

■ Demonstrate understanding of the function of an end-of-line resistor.

■ Compare and contrast the effect of an open conductor, a grounded circuit, and a wire-to-wire short.

■ Explain the function and advantages of an alarm verification feature.

■ Evaluate the differences between hardwired and multiplex fire alarm systems.

■ Draw a riser diagram, a fire alarm system plan view, or a schematic fire alarm system diagram for a system.

■ Calculate the required battery capacity of a fire alarm system.

■ Select a fire alarm circuit based upon predetermined requirements for grounded conductors, wire-to-wire shorts, open conductors, or number of devices on a circuit.

Fire alarm circuits connect initiating devices and notification appliances, discussed in previous chapters, to the fire alarm control unit (FACU). NFPA 72, *National Fire Alarm Code* (2007), recognizes the following three types of circuits, with each type of circuit having several classifications and styles:

1. Initiating device circuits (IDC)
2. Notification appliance circuits (NAC)
3. Signaling line circuits (SLC)

INITIATING DEVICE CIRCUITS

initiating device circuit (IDC)

a circuit to which automatic or manual initiating devices are connected to the fire alarm control unit (FACU), where the signal received by the FACU identifies an alarm condition on the circuit but does not identify the specific device actuated

An **initiating device circuit (IDC)** is defined by NFPA 72 (2007) as a circuit to which automatic or manual initiating devices are connected to the FACU, where the signal received by the FACU identifies an alarm condition on the circuit but does not identify the specific device actuated.

Initiating Device Circuit Classifications

Initiating device circuits have two classifications: Class A and Class B. These two classifications have existed for many years and continue to be preferred by many designers. Before the 2007 edition, NFPA 72 made reference to five styles of IDCs (three styles of Class B circuits and two styles of Class A circuits), but in the 2007 edition, these references were simplified to Class A and Class B. NFPA 72 (2007) does not define when to use Class A or Class B, but Class B (the two-wire circuit) is less expensive to install, and Class A (the four-wire circuit) can be specified when greater reliability is desired.

Class B Initiating Device Circuits

end-of-line resistor (ELR)

a device that reduces the current of the circuit by creating resistance for the incoming current, allowing less current to leave the resistor than the amount entering it

Class B initiating device circuits, as shown in **Figure 15-1** have two wires connecting normally open initiating devices to an IDC module. An **end-of-line resistor (ELR)** reduces the current of the circuit by creating resistance for the incoming current, allowing less current to leave the resistor than the amount that came in to the resistor.

The operation of a Class B IDC in the normal (non-alarm) condition (as shown in Figure 15-1) is as follows:

1. Current flows from the positive terminal on the IDC module.
2. Current bypasses all initiating devices, because the contacts remain open in the normal condition.
3. Current flows through the ELR.
4. ELR reduces the current.

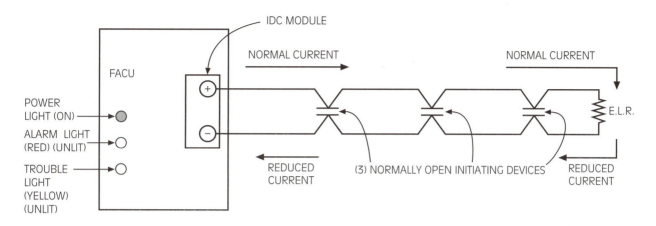

NORMAL CONDITION—CLASS B CIRCUIT

- INITIATING DEVICES REMAIN OPEN AND CURRENT GOES THROUGH THE ELR (END OF LINE RESISTOR) FROM THE POSITIVE TERMINAL (+) TO THE NEGATIVE TERMINAL (−)
- THE ELR REDUCES THE CURRENT RECEIVED BY THE IDC MODULE AT THE NEGATIVE TERMINAL (−)

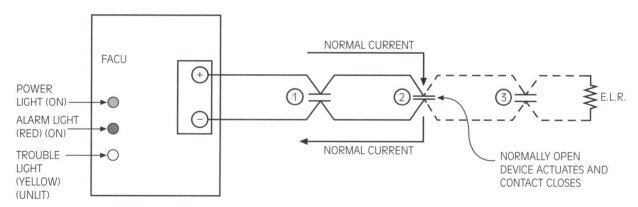

ALARM CONDITION—CLASS B CIRCUIT

- AN INITIATING DEVICE ACTUATES AND THE NORMALLY OPEN CONTACT CLOSES. NO CURRENT FLOWS TO THE ELR.
- CURRENT RECEIVED BY THE IDC MODULE INCREASES, AND THE FACU GOES INTO AN ALARM CONDITION. THE PANEL CAN'T IDENTIFY WHICH DEVICE ACTUATED.
- IF DEVICE 1 OR 3 WERE TO GO INTO ALARM, THE PANEL WOULD NOT BE ABLE TO DISCERN THAT ADDITIONAL DEVICES ARE IN ALARM. ONLY ONE ALARM SIGNAL IS CARRIED TO THE PANEL.

Figure 15-1 *Operation of a Class B initiating device circuit.*

5. Reduced current is received by the negative terminal on the IDC module.

6. The IDC module recognizes this reduced current as the normal condition.

The operation of a Class B IDC in the alarm condition (as shown in Figure 15-1) is as follows:

1. A normally open initiating device on the circuit senses a fire automatically by a detection device or is actuated manually by a manual pull station, and the contact closes.

2. The normal current issued by the positive terminal on the IDC module flows through the initiating device and is returned to the negative terminal on the IDC module at the fire alarm control unit.

3. Once one initiating device closes, no current flows to devices past the actuated device and no current flows to the ELR.

4. The same current that left the positive terminal is returned to the negative terminal on the IDC module that previously received reduced current from the ELR.

5. The IDC module senses the increase in current, and the FACU initiates an alarm.

Modern fire alarm control units contain voltage comparators within their IDC modules, which check circuit integrity and analyze current flow differentials.

Initiating Device Circuit Problems (Faults)

Problems that can occur in a circuit include an open conductor, a ground fault condition, and a wire-to-wire short.

- An **open conductor** on a Class B circuit, shown in **Figure 15-2**, can be caused by a cut or broken wire or a loose terminal. Figure 15-2 shows that when an open conductor occurs between devices 1 and 2, devices 2 and 3 are incapable of initiating an alarm and a trouble signal appears on the FACU. Device 1 is capable of operating under the scenario shown in Figure 15-2.

- A **ground fault** can be caused by a bare wire or terminal in contact with a grounded junction box or some other source of grounding, such as a junction box filled with water. After receiving a grounded signal, the IDC module reports a trouble condition. The initiating devices are required to continue to operate in the trouble condition for a single grounded wire but need not operate if more than one fault has occurred. In the case of two simultaneous ground faults, the FACU will most likely identify an alarm condition.

- A **wire-to-wire short** causes an alarm. A wire-to wire short is a condition in which two bare wires touch each other, creating an incomplete circuit.

open conductor
an IDC Class B fault that can be caused by a cut or broken wire or a loose terminal

ground fault
an IDC fault that can be caused by a bare wire or terminal in contact with a grounded junction box or some other source of grounding

wire-to-wire short
a condition in which two bare wires touch each other, creating an incomplete circuit

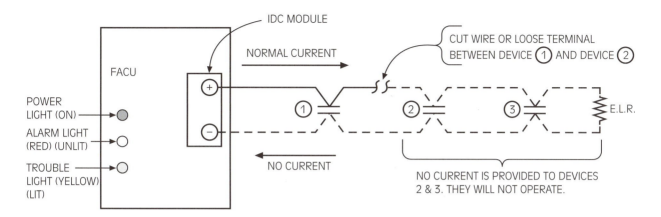

SINGLE OPEN CONDUCTOR CONDITION—CLASS B CIRCUIT

- A WIRE BREAK BETWEEN DEVICES #1 AND #2 DISABLES DEVICES #2 AND #3.
- NO CURRENT IS RECEIVED BY THE NEGATIVE TERMINAL ON THE IDC MODULE, AND A TROUBLE CONDITION IS REPORTED.
- DEVICE #1 WILL OPERATE AND REPORT AN ALARM, EVEN IN THE TROUBLE CONDITION.

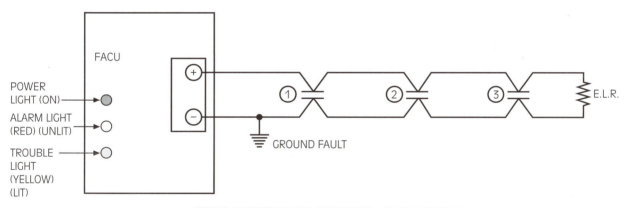

SINGLE GROUND FAULT CONDITION—CLASS B CIRCUIT

- A GROUND FAULT IS REPORTED AS A TROUBLE CONDITION AT THE FACU.
- CLASS B STYLE A LOSES ALARM RECEIPT CAPABILITY DURING A GROUND FAULT CONDITION.
- FOR CLASS B STYLE B ALL DEVICES ON THE SYSTEM FUNCTION IN THE TROUBLE MODE.
- IF BOTH A SINGLE OPEN CONDUCTOR AND A SINGLE GROUND FAULT OCCUR, A TROUBLE CONDITION IS REPORTED, AND DEVICES WILL NOT OPERATE.

Figure 15-2 *Class B circuit fault conditions.*

Class A Initiating Device Circuits

Class A circuits, shown in **Figure 15-3**, may provide more reliability and require more wire, more terminations, and more or larger conduit than Class B circuits. Class A circuits provide the ability to operate in a fault condition. They can be labor- and material-intensive.

The function of a Class A circuit in the normal operating condition occurs in the following sequence:

1. Current flows from the positive terminal (+) of the supervision circuit on the IDC module.

2. Current flows to the last initiating device and returns to the positive terminal (+) of circuit power on the IDC module.

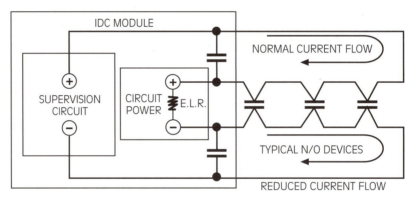

NORMAL OPERATION

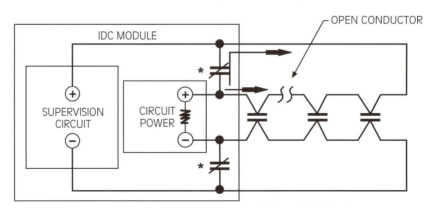

TROUBLE OPERATION WITH A SINGLE OPEN CONDUCTOR

*CONTACTS CLOSE ON OPEN CIRCUIT CONDITION

Figure 15-3

Operation of a Class A circuit. (Courtesy of Simplex Grinnell.)

3. Current then flows through an end-of-line resistor within circuit power on the IDC module.

4. Current drops as it flows through the end-of-line resistor (ELR).

5. Lowered current travels from the negative terminal (–) circuit power to the negative terminal (–) on the supervision circuit.

When one of the normally open devices closes, the ELR is bypassed. The IDC module senses this increase in current, and the FACU initiates an alarm. With a single open conductor, no current is returned to the negative terminal on the supervision circuit. Two sets of normally open contacts on the IDC module (noted with an asterisk in Figure 15-3) sense this loss of current and close simultaneously, and current is fed to all initiating devices on the circuit in both directions. A trouble signal is sent to the FACU, but all devices are capable of operation in the trouble condition on a Class A circuit.

Initiating Device Circuit Selection

NFPA 72 (2007) provides an excellent table, shown here as **Table 15-1**, for designers to use when selecting an IDC. This table has a wealth of information, and the reader is asked to confirm the following statements using Table 15-1.

- Class B circuits and Class A circuits have required alarm receipt capability during a single ground fault.
- A Class A circuit has required alarm receipt capability during a single open connector, whereas a Class B circuit does not.
- Class A circuits and Class B circuits provide indication at protected premises in the trouble condition.

A designer may be guided by determining what minimum class of circuit is required when a set of performance requirements relative to circuit reliability has been established.

Table 15-1   *Performance of initiating device circuits (IDCs).*

Class	B			A		
	Alm	Trbl	ARC	Alm	Trbl	ARC
Abnormal Condition	1	2	3	4	5	6
Single open	——	X	—	—	X	R
Single ground	—	X	R	—	X	R

Alm: Alarm. Trbl: Trouble. ARC: Alarm receipt capability during abnormal condition. R: Required capacity.
X: Indication required at protected premises and as required by Chapter 8.

Initiating Device Circuit Alarm Verification Features

An additional level of reliability for an IDC is the optional provision of an alarm verification sequence to avoid unwanted alarms resulting from temporary non-fire conditions from detectors, particularly smoke detectors. Verification is performed in the following sequence:

1. A detector sends an alarm signal to the FACU.
2. The FACU receives the signal and clears or resets the circuit.
3. Power to the IDC is reestablished after a predetermined time, such as 60 seconds.
4. The FACU rechecks the detectors on the IDC.
5. If the detector is still in alarm, the FACU activates the alarm functions.

This procedure can prevent nuisance alarm signals from sources such as a smoke detector reacting to dust blown from an HVAC duct that had been idle for an extended period, or from temporary power surges, or from a signal received from a smoke detector positioned above an ashtray in an elevator lobby.

NOTIFICATION APPLIANCE CIRCUITS

notification appliance circuits (NACs)

fire alarm circuits to which fire alarm notification appliances are connected

Previous reference to four styles of **notification appliance circuits (NACs)** was modified in the 2007 edition of NFPA 72 to simple reference to Class A and Class B circuits. Notification appliance circuits are fire alarm circuits to which fire alarm notification appliances are connected. Notification appliance circuits may be chosen from Class A circuits or Class B circuits, as **Table 15-2** illustrates. A number of conclusions can be drawn from examination of Table 15-2:

Table 15-2   *Notification appliance circuits (NACs).*

Class	B		A	
	Trouble Indication at Protected Premises	**Alarm Capability During Abnormal Conditions**	**Trouble Indication at Protected Premises**	**Alarm Capability During Abnormal Conditions**
Abnormal Condition	1	2	3	4
Single open	X	—	X	R
Single ground	X	R	X	R
Wire-to-wire short	X	—	X	—

X: Indication required at protected premises. R: Required capability.

- Alarm receipt capability is required for a Class A circuit in the presence of a single open connector, but not for a Class B circuit.
- Both Class A and Class B circuits require alarm receipt capability in the presence of a single ground fault.
- Both Class A and Class B circuits are required to provide trouble indication at the protected premises in the presence of a wire-to-wire short.

Class B Notification Appliance Circuits

Figure 15-4 illustrates one of the more commonly encountered notification appliance circuits, the Class B circuit, in normal operation. In the normal mode, Figure 15-4 shows three speakers, horns, bells or chimes connected to the supervision circuit, with current flowing from the positive (+) contact on the supervision circuit, through the end-of-line resistor, where current is reduced, to the negative (–) contact on the supervision circuit. The supervision circuit, noticing this reduced current, recognizes the circuit to be in the normal mode. The diodes, in series with the appliances, prevent current from flowing through the appliances in a normal, supervisory mode.

Figure 15-5 shows the trouble and alarm modes for the Class B notification appliance circuit. Note that when an initiating device alarm signal is received, the NAC module shifts from the supervisory circuit to signal power and that if an open conductor occurs, appliances on the fire alarm control unit side of the open conductor operate but appliances beyond the open conductor will not operate in an alarm condition.

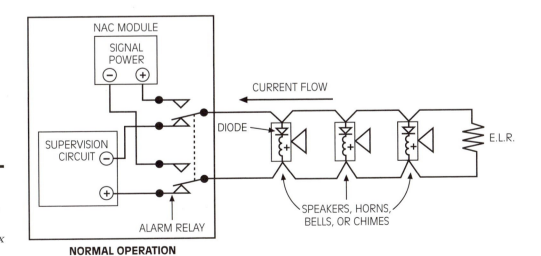

Figure 15-4   *Class B notification appliance circuit— normal operation. (Courtesy of Simplex Grinnell.)*

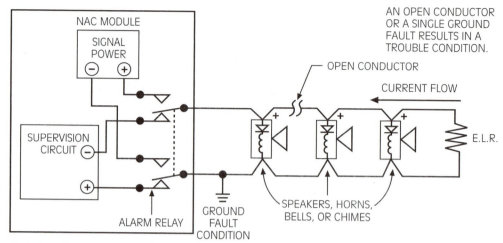

AN OPEN CONDUCTOR
OR A SINGLE GROUND
FAULT RESULTS IN A
TROUBLE CONDITION.

**CLASS B-TROUBLE OPERATION WITH A SINGLE OPEN
CONDUCTOR OR A SINGLE GROUND CONDITION**

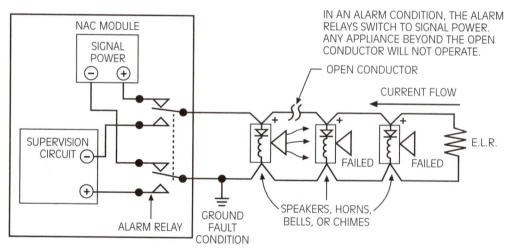

IN AN ALARM CONDITION, THE ALARM
RELAYS SWITCH TO SIGNAL POWER.
ANY APPLIANCE BEYOND THE OPEN
CONDUCTOR WILL NOT OPERATE.

Figure 15-5 *Class B
notification
appliance circuit in
trouble and in
alarm condition.
(Courtesy of Simplex
Grinnell.)*

**CLASS B-ALARM OPERATION WITH A SINGLE OPEN
CONDUCTOR OR A SINGLE GROUND CONDITION**

Class A Notification Appliance Circuits

The other classification of notification appliance circuit, the Class A circuit, is
shown in its normal mode in **Figure 15-6**. Note that the Class A notification
appliance circuit consists of four wires providing current to each notification
appliance from two directions.

When an initiating device alarm signal is received, the NAC module shifts
from the supervision circuit to the signal power circuit. Because the notification
appliances are receiving current from two directions, all appliances operate
when a single open conductor occurs, as shown in **Figure 15-7**.

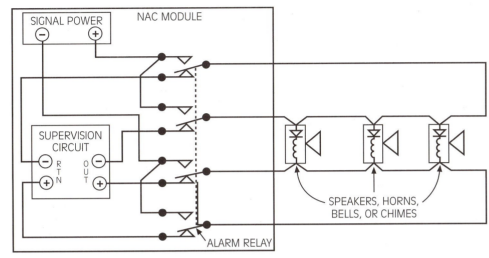

Figure 15-6 *Class A notification appliance circuit in normal operation. (Courtesy of Simplex Grinnell.)*

CLASS A-NORMAL OPERATION

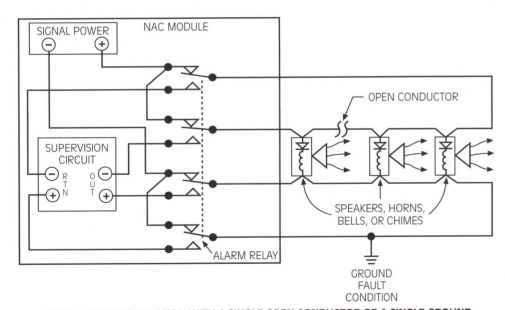

Figure 15-7 *Class A notification appliance circuit in alarm condition. (Courtesy of Simplex Grinnell.)*

CLASS A-ALARM OPERATION WITH A SINGLE OPEN CONDUCTOR OR A SINGLE GROUND

signaling line circuit (SLC)

a circuit over which multiple input and output signals of more than one fire alarm system, transmitter, or device is carried

SIGNALING LINE CIRCUITS

A **signaling line circuit (SLC)** is a circuit over which multiple input and output signals of more than one fire alarm system, transmitter, or device is carried. The SLC could be an output signal for remote notification from numerous fire alarm control units and/or an input signal to one or more devices or fire alarm control units. The

Table 15-3   *Performance of signaling line circuits (SLCs).*

Class	B			A			A		
Style	4			6			7		
	Alm	Trbl	ARC	Alm	Trbl	ARC	Alm	Trbl	ARC
Abnormal Condition	1	2	3	4	5	7	8	9	10
Single open	—	X	—	—	X	R	—	X	R
Single ground	—	X	R	—	X	R	—	X	R
Wire-to-wire short	—	X	—	—	X	—	—	X	R
Wire-to-wire short & open	—	X	—	—	X	—	—	X	—
Wire-to-wire short & ground	—	X	—	—	X	—	—	X	—
Open and ground	—	X	—	—	X	R	—	X	R
Loss of carrier (if used)/channel interface	—	X	—	—	X	—	—	X	—

Alm: Alarm. Trbl: Trouble. ARC: Alarm receipt capability during abnormal condition. R: Required capability.
X: Indication required at protected premises and as required by Chapter 8.

circuit to which addressable initiating devices, notification appliances, and circuit interfaces are connected to a FACU is designated as a signaling line circuit.

NFPA 72 (2007) provides performance and capacity guidelines for signaling line circuits, as shown in **Table 15-3**. Note that the circuits are Class A or Class B, with capabilities increasing with styles ranging from style 4, to style 6, to style 7. Many important and interesting findings can be derived from examining Table 15-3, such as:

- Class A styles 6 and 7 require alarm receipt capability in the presence of a single open conductor, whereas Class B circuits do not.
- All three styles require alarm receipt capability in the presence of a single ground fault.
- Class A style 7 requires alarm receipt capability in the presence of a wire-to-wire short, whereas the other two styles do not.

Figure 15-8 shows a Class B style 4 signaling line circuit in normal operation. The microprocessor individually interrogates or checks the status of each device or control unit on the signaling line circuit in sequence, from device 101 to 104, then continuously rechecks those devices in order. This process is referred to as **continuous software interrogation (CSI)**. The most commonly used SLC circuits are Class B style 4 (shown in Figure 15-8), Class A style 6 (shown in **Figure 15-9**), and Class A style 7. NFPA 72 (2007) was modified to recognize this.

Figure 15-9 illustrates a Class A style 6 signaling line circuit. Four wires provide a primary line and a secondary line to each device. The microprocessor switches continuously from the primary line to the secondary line, to provide

continuous software interrogation (CSI)

a process in which a microprocessor individually interrogates or checks the status of each device or control unit on the signaling line circuit in sequence, then continuously rechecks the devices in order

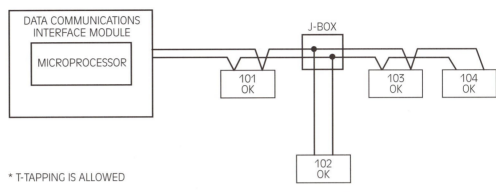

Figure 15-8
*Signaling line
circuit, Class B
style 4, in normal
operation. (Courtesy
of Simplex Grinnell.)*

* T-TAPPING IS ALLOWED

SIGNALING LINE CIRCUIT NORMAL OPERATION

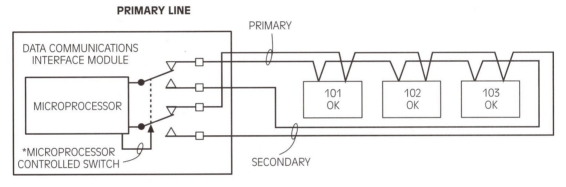

*PRIMARY LINE TRANSMISSION

SIGNALING LINE CIRCUIT NORMAL OPERATION

Figure 15-9   *Signaling line circuit, Class A style 6, in normal operation. (Courtesy of Simplex Grinnell.)*

supervision of the devices from two directions. The circuit continues to operate with a single open conductor. Addressable notification appliances, connected to an FACU, may be classified as SLCs.

HARDWIRED AND ADDRESSABLE (MULTIPLEX) FIRE ALARM SYSTEMS

Traditionally, fire alarm systems have notification appliances and initiating devices hardwired to zone modules on an FACU. With either Class A or Class B circuits, hardwired systems know which initiation zone is in alarm but cannot identify which specific device is in alarm.

Because it is becoming increasingly important for maintenance technicians to quickly identify trouble or alarm conditions, and because it it has become crucial for the fire service to know the exact location of a detector in alarm when responding to a fire call, an addressable (multiplex) system may be justified, and these systems are being specified increasingly.

An **addressable (multiplex) system** requires a central processing unit (CPU) or a computer and a software program that specifies and assigns the exact locations or addresses of each initiating device on the fire alarm system. The addressable system simultaneously or sequentially transmits a signal to each address, through wire or a radio, and the signal is returned and analyzed by the fire alarm control unit. If an alarm signal is received from a specific device, the notification circuits are activated and the address of the specific device in alarm is displayed on the fire alarm control unit.

The ability to identify the exact locations of the actuation of an initiating device can save the fire service significant time in locating and extinguishing a fire.

Multiplex systems are particularly valuable in large multizone buildings that undergo numerous renovations or changes, such as hospitals and some tenant-space office buildings. Relocating or rezoning detectors on a hardwired system may involve numerous electricians and can become expensive and time-consuming, whereas renovating or rezoning an addressable multiplex system simply involves a technician's reprogramming the CPU without having to relocate or rewire the detectors. Knowing the exact location of the origin of an alarm or trouble condition can reduce troubleshooting time and minimize system downtime.

addressable (multiplex) system

a system that requires a central processing unit (CPU) or a computer and a software program that specifies and assigns the exact locations or addresses of each initiating device on the fire alarm system

FIRE ALARM CONTROL UNITS

A fire alarm control unit (FACU), shown in **Figure 15-10**, is the nerve center for a fire alarm system and performs numerous functions for suppression systems and fire alarm systems, as described in previous chapters. The connection of devices to an FACU can be illustrated in three ways:

1. Fire alarm diagrams (riser and floor plan)
2. Fire alarm system plans
3. Schematic fire alarm system diagrams

Fire Alarm Riser Diagram

A fire alarm riser diagram, shown in **Figure 15-11**, is used commonly on fire alarm system contract drawings by fire alarm designers to illustrate the functions of a fire alarm system and the manner in which initiating and notification appliances are connected to an FACU. Note that the number of wires is shown at each juncture of the riser diagram.

FIRE ALARM CONTROLS NON-ADDRESSABLE FIRE ALARM CONTROL PANELS; TWO TO EIGHT ZONES

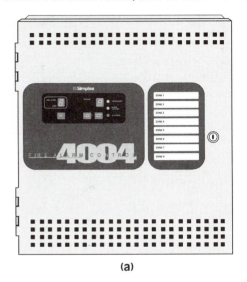

(a)

SUPPRESSION RELEASING PANEL FOR AUTOMATIC EXTINGUISHING, DELUGE AND PREACTION SPRINKLER CONTROL

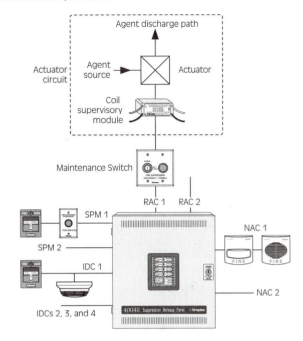

SUPPRESSION RELEASE PANEL ONE-LINE SYSTEM REFERENCE DRAWING

(c)

FIRE ALARM CONTROL PANEL FOR ADDRESSABLE DEVICES AND ADDRESSABLE SMOKE DETECTION

(b)

Figure 15-10 *Fire alarm control units. Counterclockwise from top left: (a) 2-unit cabinet, (b) 4-unit cabinet, (c) 6-unit cabinet with voice alarm. (Courtesy of Simplex Grinnell.)*

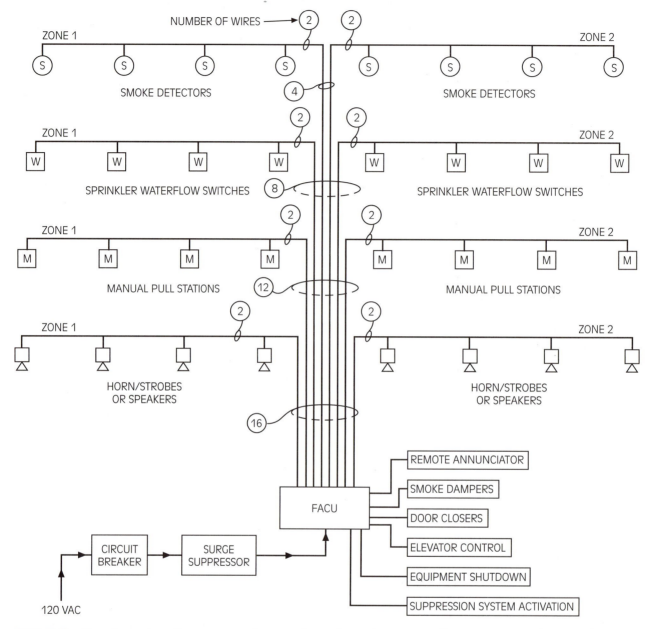

Figure 15-11 *Fire alarm riser diagram; two wires are shown for each circuit, a Class B system is shown. Note: Two zones are shown, with detection, sprinkler flow, manual pull stations, and horn-strobes located within each of the two zones. Clips are placed on the circuit breaker so it cannot be turned off inadvertently.*

Fire Alarm System Plan

A fire alarm system plan, shown in **Figure 15-12**, pictorially represents the location of all devices, dimensioned or shown to scale on a plan view of a building, with a diagram of the interconnection of the devices to the FACU shown. These plans are included in a set of fire alarm system contract drawings.

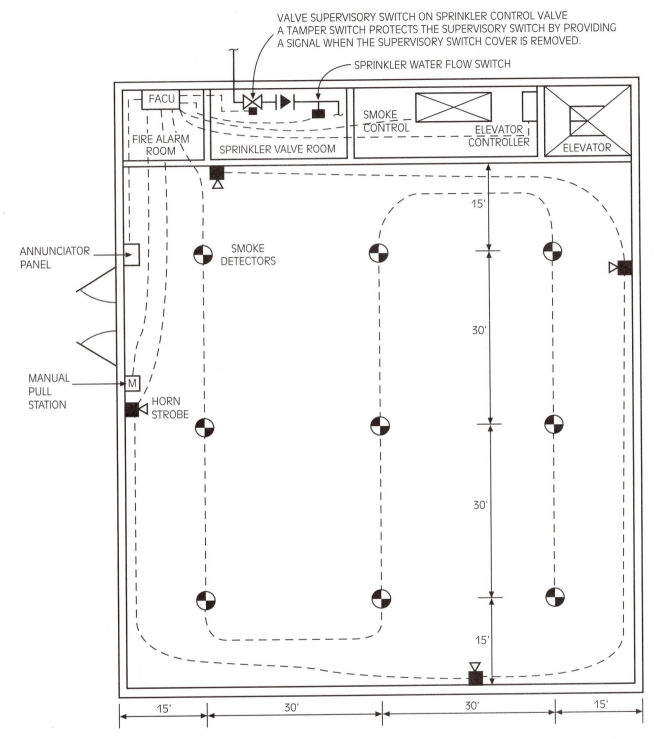

VALVE SUPERVISORY SWITCH ON SPRINKLER CONTROL VALVE
A TAMPER SWITCH PROTECTS THE SUPERVISORY SWITCH BY PROVIDING
A SIGNAL WHEN THE SUPERVISORY SWITCH COVER IS REMOVED.

SPRINKLER WATER FLOW SWITCH

FACU

FIRE ALARM
ROOM

SPRINKLER VALVE ROOM

SMOKE
CONTROL

ELEVATOR
CONTROLLER

ELEVATOR

ANNUNCIATOR
PANEL

SMOKE
DETECTORS

MANUAL
PULL
STATION

HORN
STROBE

15'

30'

30'

15'

15'   30'   30'   15'

Figure 15-12   *Fire alarm system plan. (NFPA 170 provides examples of symbols for use on fire alarm systems.)*

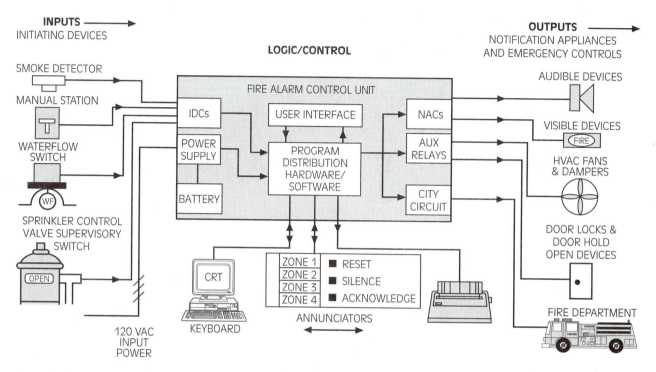

Figure 15-13   *Schematic fire alarm system diagram. (Courtesy of Simplex Grinnell.)*

Schematic Fire Alarm System Diagram

Another way of representing the central relationship of the FACU to the IDCs, NACs, and remote units is a schematic or pictorial diagram, shown in **Figure 15-13**. These diagrams also are commonly included in a set of fire alarm system contract drawings.

Calculating Fire Alarm Control Unit Battery Capacity

A fire alarm control unit is required to have a secondary power supply, and most fire alarm control units use standby batteries to perform this function. To calculate the capacity of a battery, the power requirements, in amp-hours, of each device or appliance on the system are added in the standby state, and the power requirements, in amp-hours, of each device operating in alarm are added in the alarm state.

EXAMPLE 15-1

Calculation of Battery Capacity

A fire alarm control unit is connected to the fire alarm equipment whose description and attributes are shown in the following list. Determine the minimum amp-hour rating of a battery that serves as the secondary power supply for this fire alarm system, assuming a requirement of 24 hours standby and 5 minutes in alarm, and assuming that no lamp tests are performed during the standby period. The power requirements for the fire alarm system in this example, which are not intended to replicate the attributes of any specific manufacturer, are as follows:

Description	Quantity	Amps in Supervision (per hour) (AMP-hours per unit)	Amps in Alarm (per hour) (AMP-hours per unit)
Control Unit	1	0.219	0.510
Output Modules	5	0.0065	0.04
Sirens	8	0.0	0.55
Supplementary Relay	2	0.0	0.045
Beam Smoke Detection Module	1	0.03	0.076
Gate Valve Module	1	0.020	0.22
Input Module	1	0.018	0.55
Horn	10	0.0	0.063
Strobe	6	0.0	0.025
Bells	8	0.0	0.063
Horn Strobe	14	0.0	0.50
Ionization Detectors	10	0.0001	0.08
Beam Detector	10	0.0013	0.06
Duct Detector	4	0.008	0.06

The values in the above list of components do not correspond to the power requirements of all commercially available devices and appliances. Refer to manufacturers' data for values appropriate to specific components. Note that in this list the sirens, supplementary relay, horns, strobes, bells, and horn-strobes draw no power in their standby state; they draw power only when they are operated in alarm. Ionization detectors, beam detectors, and duct detectors require power in standby to analyze smoke conditions.

Strobes are constant-power appliances. If voltage applied to strobes drops, the current they draw gets larger.

Solution

1. *Battery power requirement in supervision.* The amp-hours required in supervision is determined by multiplying the quantity of devices by the amps in supervision,

then multiplying that total by the standby time requirement. The total amp-hour requirement in standby is the sum of the amp-hours for each component.

Description	Quantity	Amps in Supervision (per unit)	Standby Time Requirement (hours)	Amp-Hours Required
Control Unit	1	0.219	24	5.256
Output Modules	5	0.0065	24	0.78
Sirens	8	0.0	24	0.0
Supplementary Relay	2	0.0	24	0.0
Beam Smoke Detection Module	1	0.03	24	0.72
Gate Valve Module	1	0.020	24	0.48
Input Module	1	0.018	24	0.432
Horn	10	0.0	24	0.0
Strobe	6	0.0	24	0.0
Bells	8	0.0	24	0.0
Horn-Strobe	14	0.0	24	0.0
Ionization Detectors	10	0.0001	24	0.024
Beam Detector	10	0.0013	24	0.312
Duct Detector	4	0.008	24	0.768
Total Amp-Hours in Standby				**8.772**

2. *Battery power requirement in alarm.* The amp-hours required in alarm is determined by multiplying the quantity of devices by the amps in alarm, then multiplying that total by the time requirement. In this case, 5 minutes is 5/60 hours, or 0.0833 hours. The total amp-hour requirement in alarm is the sum of the amp-hours for each component.

Description	Quantity	Amps in Alarm	Alarm Time Requirement (hours)	Amp-Hours Required
Control Unit	1	0.510	0.0833	0.042
Output Modules	5	0.04	0.0833	0.017
Sirens	8	0.55	0.0833	0.367
Supplementary Relay	2	0.045	0.0833	0.007
Beam Smoke Detection Module	1	0.076	0.0833	0.006
Gate Valve Module	1	0.22	0.0833	0.018
Input Module	1	0.55	0.0833	0.046
Horn	10	0.063	0.0833	0.052
Strobe	6	0.025	0.0833	0.012
Bells	8	0.063	0.0833	0.042
Horn-Strobe	14	0.50	0.0833	0.583
Ionization Detectors	10	0.08	0.0833	0.067
Beam Detectors	10	0.06	0.0833	0.050
Duct Detector	4	0.06	0.0833	0.020
Total Amp-Hours in Alarm				**1.329**

3. *Total battery power requirement.* The amp hours are determined by adding the power requirement in supervision to the power requirement in alarm.

Power requirement in supervision	8.772 amp-hours
Power requirement in alarm	1.329 amp-hours
Total minimum battery power requirement	10.101 amp-hours

A designer then selects the nearest commercially available battery that is larger than 10.101 amp-hours, using manufacturers' charts showing battery power availability curves.

VOLTAGE DROP ON NOTIFICATION APPLIANCE CIRCUITS

Current flowing through a wire can be compared to water as it flows through a pipe. They both start with an initial "pressure" or potential, then the "pressure" degrades as it proceeds to its destination. Just as a pipe exerts resistance on the flow of water, a wire exerts resistance to the flow of current. Similarly, as designers of automatic sprinkler systems perform calculations to ensure that each sprinkler is operating at its minimum performance objective, fire alarm designers perform voltage-drop calculations to ensure that each notification appliance performs as specified.

Standard Operating Voltage for Notification Appliances

To meet the minimum required voltage at each appliance, a higher initial voltage is required to be specified at the FACU. Designers must consider the worst-case scenario, when a NAC in a FACU is operating on batteries at the end of their useful life. UL 864, *Standard for Control Units for Fire Protection Signaling Systems*, defines this minimum as 20.4 V on a 24 VDC fire alarm system. On May 1, 2004, Underwriter's Laboratories (UL) initiated a standard operating voltage for notification appliances, within a range of 16 VDC to 33 VDC. The total maximum permissible voltage drop in a circuit is therefore 4.4 VDC (20.4 VDC minus 16 VDC).

Wire Size, Resistance, and Gauge

Designers use **Table 15-4**, from NFPA 70, *National Electric Code*, as a basis for determining wire size and calculating wire resistance.

For calculating wiring losses, the famous relation Ohm's Law is used:

$$V = I \times R$$

where:

V = Voltage
I = Current in amperes
R = Resistance in ohms

Table 15-4 *Conductor properties.*

Size (AWG or kcmil)	Area mm²	Area Circular mils	Stranding Quantity	Stranding Diameter mm	Stranding Diameter in.	Overall Diameter mm	Overall Diameter in.	Overall Area mm²	Overall Area in.²	Copper Uncoated ohm/km	Copper Uncoated ohm/kFT	Copper Coated ohm/km	Copper Coated ohm/kFT	Aluminum ohm/km	Aluminum ohm/kFT
18	0.823	1620	1	—	—	1.02	0.040	0.823	0.001	25.5	7.77	26.5	8.08	42.0	12.8
18	0.823	1620	7	0.39	0.015	1.16	0.046	1.06	0.002	26.1	7.95	27.7	8.45	42.8	13.1
16	1.31	2580	1	—	—	1.29	0.051	1.31	0.002	16.0	4.89	16.7	5.08	26.4	8.05
16	1.31	2580	7	0.49	0.019	1.46	0.058	1.68	0.003	16.4	4.99	17.3	5.29	26.9	8.21
14	2.08	4110	1	—	—	1.63	0.064	2.08	0.003	10.1	3.07	10.4	3.19	16.6	5.06
14	2.08	4110	7	0.62	0.024	1.85	0.073	2.68	0.004	10.3	3.14	10.7	3.26	16.9	5.17
12	3.31	6530	1	—	—	2.05	0.081	3.31	0.005	6.34	1.93	6.57	2.01	10.45	3.18
12	3.31	6530	7	0.78	0.030	2.32	0.092	4.25	0.006	6.50	1.98	6.73	2.05	10.69	3.25
10	5.261	10380	1	—	—	2.588	0.102	5.26	0.008	3.984	1.21	4.148	1.26	6.561	2.00
10	5.261	10380	7	0.98	0.038	2.95	0.116	6.76	0.011	4.070	1.24	4.226	1.29	6.679	2.04
8	8.367	16510	1	—	—	3.264	0.128	8.37	0.013	2.506	0.764	2.579	0.786	4.125	1.26
8	8.367	16510	7	1.23	0.049	3.71	0.146	10.76	0.017	2.551	0.778	2.653	0.809	4.204	1.28
6	13.30	26240	7	1.56	0.061	4.67	0.184	17.09	0.027	1.608	0.491	1.671	0.510	2.652	0.808
4	21.15	41740	7	1.96	0.077	5.89	0.232	27.19	0.042	1.010	0.308	1.053	0.321	1.666	0.508
3	26.67	52620	7	2.20	0.087	6.60	0.260	34.28	0.053	0.802	0.245	0.833	0.254	1.320	0.403
2	33.62	66360	7	2.47	0.097	7.42	0.292	43.23	0.067	0.634	0.194	0.661	0.201	1.045	0.319
1	42.41	83690	19	1.69	0.066	8.43	0.332	55.80	0.087	0.505	0.154	0.524	0.160	0.829	0.253
1/0	53.49	105600	19	1.89	0.074	9.45	0.372	70.41	0.109	0.399	0.122	0.415	0.127	0.660	0.201
2/0	67.43	133100	19	2.13	0.084	10.62	0.418	88.74	0.137	0.3170	0.0967	0.329	0.101	0.523	0.159
3/0	85.01	167800	19	2.39	0.094	11.94	0.470	111.9	0.173	0.2512	0.0766	0.2610	0.0797	0.413	0.126
4/0	107.2	211600	19	2.68	0.106	13.41	0.528	141.1	0.219	0.1996	0.0608	0.2050	0.0626	0.328	0.100
250	127	—	37	2.09	0.082	14.61	0.575	168	0.260	0.1687	0.0515	0.1753	0.0535	0.2778	0.0847
300	152	—	37	2.29	0.090	16.00	0.630	201	0.312	0.1409	0.0429	0.1463	0.0446	0.2318	0.0707
350	177	—	37	2.47	0.097	17.30	0.681	235	0.364	0.1205	0.0367	0.1252	0.0382	0.1984	0.0605
400	203	—	37	2.64	0.104	18.49	0.728	268	0.416	0.1053	0.0321	0.1084	0.0331	0.1737	0.0529
500	253	—	37	2.95	0.116	20.65	0.813	336	0.519	0.0845	0.0258	0.0869	0.0265	0.1391	0.0424
600	304	—	61	2.52	0.099	22.68	0.893	404	0.626	0.0704	0.0214	0.0732	0.0223	0.1159	0.0353
700	355	—	61	2.72	0.107	24.49	0.964	471	0.730	0.0603	0.0184	0.0622	0.0189	0.0994	0.0303
750	380	—	61	2.82	0.111	25.35	0.998	505	0.782	0.0563	0.0171	0.0579	0.0176	0.0927	0.0282
800	405	—	61	2.91	0.114	26.16	1.030	538	0.834	0.0528	0.0161	0.0544	0.0166	0.0868	0.0265
900	456	—	61	3.09	0.122	27.79	1.094	606	0.940	0.0470	0.0143	0.0481	0.0147	0.0770	0.0235
1000	507	—	61	3.25	0.128	29.26	1.152	673	1.042	0.0423	0.0129	0.0434	0.0132	0.0695	0.0212
1250	633	—	91	2.98	0.117	32.74	1.289	842	1.305	0.0338	0.0103	0.0347	0.0106	0.0554	0.0169
1500	760	—	91	3.26	0.128	35.86	1.412	1011	1.566	0.02814	0.00858	0.02814	0.00883	0.0464	0.0141
1750	887	—	127	2.98	0.117	38.76	1.526	1180	1.829	0.02410	0.00735	0.02410	0.00756	0.0397	0.0121
2000	1013	—	127	3.19	0.126	41.45	1.632	1349	2.092	0.02109	0.00643	0.02109	0.00662	0.0348	0.0106

Notes:

1. These resistance values are valid **only** for the parameters as given. Using conductors having coated strands, different stranding type, and, especially, other temperatures changes the resistance.

2. Formula for temperature change: $R_2 = R_1 [1 + \alpha (T_2 - 75)]$ where $\alpha_{cu} = 0.00323$, $\alpha_{AL} = 0.00330$ at 75°C.

3. Conductors with compact and compressed stranding have about 9 percent and 3 percent, respectively, smaller bare conductor diameters than those shown. See Table 5A for actual compact cable dimensions.

4. The IACS conductivities used: bare copper = 100%, aluminum = 61%.

5. Class B stranding is listed as well as solid for some sizes. Its overall diameter and area is that of its circumscribing circle.

FPN: The construction information is per NEMA WC8-1992 or ANSI/UL 1581-1998. The resistance is calculated per National Bureau of Standards Handbook 100, dated 1966, and Handbook 109, dated 1972.

To calculate losses in a fire alarm circuit, a designer first must determine wire gauge, calculated as follows:

$$V_{load} = (V_{terminals}) - [(I_{load}) \times (R_{conductor})]$$

where:

V_{load} = minimum operating voltage, 16 VDC
$V_{terminals}$ = 20.4 VDC for a 24-volt fire alarm system
I_{load} = total current draw of connected appliances
$R_{conductor}$ = total conductor resistance

By solving the above equation for $R_{conductor}$ and using Table 12-4, a designer can determine the minimum wire size.

Calculating Voltage Drop in Fire Alarm Wiring

A variety of methods can be used to calculate voltage drop. This chapter employs the lump-sum method, also commonly referred to as the engineering handbook method, described as follows:

$$\text{Total voltage drop} = (A \times L \times 21.6)/CM$$

where:

A = amperage (current) flowing through the fire alarm circuit
L = wire length, which is a one-way length only (e.g., do not double the length for a two-wire Class B circuit)
21.6 = a constant relating to the resistive characteristics of copper conductors
CM = the circular mill area of the conductor, as obtained from Table 15-4.

EXAMPLE 15-2

Calculation of Voltage Drop

A notification appliance circuit is wired to a 24 VDC FACU, as shown in **Figure 15-14**, and features:

three horns rated at 0.035 A @ 24 VDC
two horn/strobes rated at 0.105 A @ 24 VDC
one strobe rated at 0.070 A @ 24 VDC

The circuit features the lengths and gauges of wire shown in Figure 15-14. Determine whether the circuit will perform in accordance with NFPA 70, NFPA 72, and UL requirements.

Solution

Total voltage drop for the 16-gauge wire is determined as follows:

$$\text{Total voltage drop} = (A \times L \times 21.6)/CM$$

where:

A   = amperage (current) flowing through the fire alarm circuit = 0.105 A as specified for the 16-gauge wire supplying the most remote horn/strobe as shown in Figure 15-14.

L   = wire length, which is a one-way length only = 104 feet of 16-gauge wire, as shown in Figure 15-14.

21.6 = a constant relating to the resistive characteristics of copper conductors

CM   = circular mill area of the 16-gauge conductor, as obtained from Table 15-4 = 2580 mills.

Total voltage drop (16-gauge wire) = $(A \times L \times 21.6)/CM$

Total voltage drop (16-gauge wire) = $[(0.105 \text{ A}) \times (104 \text{ ft}) \times 21.6]/(2580 \text{ mills})$

Total voltage drop (16 gauge wire) = 0.091 VDC

Total voltage drop for the 14-gauge wire is determined as follows:

$$\text{Total voltage drop} = (A \times L \times 21.6)/CM$$

where:

A   = amperage (current) flowing through the fire alarm circuit

A   = as shown in Figure 15-14: (0.105 A) + (0.035 A) + (0.035 A) + (0.070 A) + (0.035 A) + (0.105 A) = 0.385 A

L   = wire length of 14-gauge wire, as shown in Figure 15-14, which is a one-way length only

L   = (64 ft) + (60 ft) + (45 ft) + (75 ft) + (63 ft) = 307 ft of 14-gauge wire.

21.6 = a constant relating to the resistive characteristics of copper conductors

CM   = the circular mill area of the 14-gauge conductor, as obtained from Figure 15-14 = 4110 mills

Total voltage drop (14-gauge wire)	= $(A \times L \times 21.6)/CM$
Total voltage drop (14-gauge wire)	= $[(0.385 \text{ A}) \times (307 \text{ ft}) \times 21.6]/(4110 \text{ mills})$
Total voltage drop (14-gauge wire)	= 0.621 VDC
Total voltage drop in circuit	= (drop in 16-gauge wire) + (drop in 16-gauge wire)
Total voltage drop in circuit	= (0.091 VDC) + (0.621 VDC)
Total voltage drop in circuit	= 0.712 VDC
Percentage of voltage drop in circuit = (Total voltage drop in circuit)/(24 VDC)
Percentage of voltage drop in circuit = (0.712)/(24 VDC) = 0.030 = 3% voltage drop

Circuit parameters:

16 VDC	= minimum required power to any appliance
4.4 VDC	= maximum permissible voltage drop in the circuit
20.4 VDC	= reduced battery potential at end of life

(20.4 VDC) − (0.712 VDC) = 19.688 VDC supplied in this example to the most remote appliance.

19.688 VDC exceeds the minimum requirement of 16 VDC; therefore, the circuit meets the requirements of NFPA 72, NFPA 70, and UL.

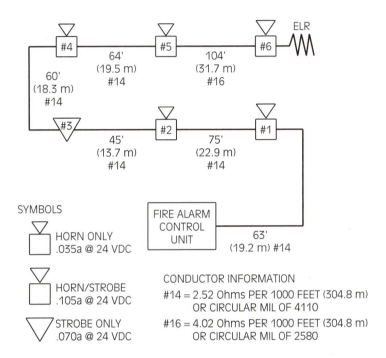

Figure 15-14
Illustration of Example 15-2.

SYMBOLS

HORN ONLY
.035a @ 24 VDC

HORN/STROBE
.105a @ 24 VDC

STROBE ONLY
.070a @ 24 VDC

CONDUCTOR INFORMATION
#14 = 2.52 Ohms PER 1000 FEET (304.8 m)
       OR CIRCULAR MIL OF 4110
#16 = 4.02 Ohms PER 1000 FEET (304.8 m)
       OR CIRCULAR MIL OF 2580

SUMMARY

Fire alarm circuits connect initiating devices and notification appliances to the fire alarm control unit in a logical manner that permits the system to meet predetermined performance objectives. Circuit logic, styles, and isolation can be chosen to increase the reliability of a fire alarm system.

The fire alarm control unit (FACU) is the nerve center of a fire alarm system, and system logic can be displayed in riser diagrams, plan views, and schematic diagrams. The battery backup for a fire alarm control unit is calculated by determining power requirements for all components in supervision and power requirements for all components in alarm and must consider battery degradation characteristics.

The calculation of voltage drop in a fire alarm circuit is an important responsibility for a fire alarm system designer because it provides evidence that notification appliances will work in accordance with their listing.

REVIEW QUESTIONS

1. Why does the power received by the IDC module on the FACU increase when an initiating device is actuated on a Class B circuit?

2. Select a class of IDC that provides all of the following functions:

   a. Provides alarm receipt capability for a single open connector.

   b. Provides an alarm during a wire-to-wire short.

   c. Provides alarm receipt capability for a single grounded wire.

3. Select a class of IDC that provides alarm receipt capability for a single open connector and also gives a trouble signal if encountering a wire-to-wire short.

4. Select a class of notification circuit that provides alarm receipt capability when encountereing a single open connector or a single grounded wire.

5. Draw a riser diagram showing the number of wires at each junction for a Class B hardwired fire alarm system consisting of the following components:

   a. one FACU.

   b. four zones of ionization detectors, with 10 detectors per zone.

   c. four zones of sprinkler waterflow switches, with two switches per zone.

   d. four zones of manual pull stations, with two stations per zone.

   e. four zones of horn-strobes stations, with six horn-strobes per zone.

6. Calculate the minimum size of battery required for the fire alarm system in Example 15-1, if an authority having jurisdiction requires a minimum of 30 hours in standby and 10 minutes in alarm.

7. Determine adequacy of the circuit shown in Figure 15-14 and in Example 15-2 if the 16-gauge wire is replaced with 14-gauge wire, and the 14-gauge wire shown on Figure 15-14 is replaced by 12-gauge wire.

DISCUSSION QUESTIONS

1. When specifing a fire alarm system, why is it important to show a detailed floor plan with all components?

2. Under what circumstances would you specify a Class A circuit in lieu of a Class B circuit?

ACTIVITIES

1. Survey the fire alarm system in a school or office building and determine the type of circuit chosen, then determine the logic used for zoning circuits. Draw a fire alarm riser diagram for the system. Design an annunciator panel for the system, displaying all system functions and capabilities.

2. Obtain a data book from a fire alarm system manufacturer, and determine the nearest available size of battery for the system in Example 15-1 and for the system in Review Question 5.

3. Using the data book described in Activity 2, develop a computerized spreadsheet for the calculation of battery capacity, referring to the actual power requirements listed for components in the data book. Use the spreadsheet to calculate the capacity of a battery for a fire alarm system, using the numbers of components listed in Example 15-1.

Appendix A

REFERENCE TABLES

Inches converted to decimals of a foot.

Inches		Decimal of a Foot	Inches		Decimal of a Foot
0	1/8	.010416	6	1/8	.510416
	1/4	.020833	(.50)	1/4	.520833
	3/8	.031250		3/8	.531250
	1/2	.041666		1/2	.541666
	5/8	.052083		5/8	.552083
	3/4	.062500		3/4	.562500
	7/8	.072916		7/8	.572916
1	1/8	.093750	7	1/8	.593750
(.083333)	1/4	.104166	(.583333)	1/4	.604166
	3/8	.114583		3/8	.614583
	1/2	.125000		1/2	.625000
	5/8	.135416		5/8	.635416
	3/4	.145833		3/4	.645833
	7/8	.156250		7/8	.656250
2	1/8	.177083	8	1/8	.677083
(.166666)	1/4	.187500	(.666666)	1/4	.687500
	3/8	.197916		3/8	.697916
	1/2	.208333		1/2	.708333
	5/8	.218750		5/8	.718750
	3/4	.229166		3/4	.729166
	7/8	.239583		7/8	.739583
3	1/8	.260416	9	1/8	.760416
(.250)	1/4	.270833	(.750)	1/4	.770833
	3/8	.281250		3/8	.781250
	1/2	.291666		1/2	.791666
	5/8	.302083		5/8	.802083
	3/4	.312500		3/4	.812500
	7/8	.322916		7/8	.822916
4	1/8	.343750	10	1/8	.843750
(.333333)	1/4	.354166	(.833333)	1/4	.854166
	3/8	.364583		3/8	.864583
	1/2	.375000		1/2	.875000
	5/8	.385416		5/8	.885416
	3/4	.395833		3/4	.895833
	7/8	.406250		7/8	.906250
5	1/8	.427083	11	1/8	.927083
(.416666)	1/4	.437500	(.916666)	1/4	.937500
	3/8	.447916		3/8	.947916
	1/2	.458333		1/2	.958333
	5/8	.468750		5/8	.968750
	3/4	.479166		3/4	.979166
	7/8	.489583		7/8	.989583

Decimal equivalents of fractions of an inch.

Inches	Decimal of an Inch	Inches	Decimal of an Inch
$1/64$	.015625	$33/64$	.515625
$1/32$	.03125	$17/32$	.53125
$3/64$	.046875	$35/64$	.546875
$1/16$	.0625	$9/16$	.5625
$5/64$	.078125	$37/64$	.578125
$3/32$	.09375	$19/32$	.59375
$7/64$	.109375	$39/64$	.609375
$1/8$	.125	$5/8$	.625
$9/64$	.140625	$41/64$	.640625
$5/32$	.15625	$21/32$	.65625
$11/64$	.171875	$43/64$	.671875
$3/16$	.1875	$11/16$	.6875
$13/64$	.203125	$45/64$	.703125
$7/32$	.21875	$23/32$	.71875
$15/64$	.234375	$47/64$	.734375
$1/4$	.25	$3/4$	.75
$17/64$	.265625	$49/64$	.765625
$9/32$	.28125	$25/32$	.78125
$19/64$	.296875	$51/64$	.796875
$5/16$	.3125	$13/16$	.8125
$21/64$	.328125	$53/64$	.828125
$1/3$	.333	$27/32$	.84375
$11/32$	.34375	$55/64$	.859375
$23/64$	.359375	$7/8$	.875
$3/8$	.375	$57/64$	.890625
$25/64$	.390625	$29/32$	.90625
$13/32$	.40625	$59/64$	.921875
$27/64$	.421875	$15/16$	.9375
$7/16$	.4375	$61/64$	.953125
$29/64$	.453125	$31/32$	.96875
$15/32$	.46875	$63/64$	.984375
$31/64$	.484375	1	1.
$1/2$	.5		

Minutes converted to decimals of a degree.

Min.	Deg.	Min.	Deg.	Min.	Deg.	Min.	Deg.	Min.	Deg.	Min.	Deg.
1	.0166	11	.1833	21	.3500	31	.5166	41	.6833	51	.8500
2	.0333	12	.2000	22	.3666	32	.5333	42	.7000	52	.8666
3	.0500	13	.2166	23	.3833	33	.5500	43	.7166	53	.8833
4	.0666	14	.2333	24	.4000	34	.5666	44	.7333	54	.9000
5	.0833	15	.2500	25	.4166	35	.5833	45	.7500	55	.9166
6	.1000	16	.2666	26	.4333	36	.6000	46	.7666	56	.9333
7	.1166	17	.2833	27	.4500	37	.6166	47	.7833	57	.9500
8	.1333	18	.3000	28	.4666	38	.6333	48	.8000	58	.9666
9	.1500	19	.3166	29	.4833	39	.6500	49	.8166	59	.9833
10	.1666	20	.3333	30	.5000	40	.6666	50	.8333	60	1.0000

Standard conversions.

To Change	To	Multiply By	To Change	To	Multiply by
Inches	Feet	0.0833	Inches of mercury	Inches of water	13.6
Inches	Millimeters	25.4	Inches of mercury	Feet of water	1.1333
Feet	Inches	12	Inches of mercury	Pounds per square inch	0.4914
Feet	Yards	0.3333	Ounces per square inch	Inches of mercury	0.127
Yards	Feet	3	Ounces per square inch	Inches of water	1.733
Square inches	Square feet	0.00694	Pounds per square inch	Inches of water	27.72
Square feet	Square inches	144	Pounds per square inch	Feet of water	2.310
Square feet	Square yards	0.11111	Pounds per square inch	Inches of mercury	2.04
Square yards	Square feet	9	Pounds per square inch	Atmospheres	0.0681
Cubic inches	Cubic feet	0.00058	Feet of water	Pounds per square inch	0.434
Cubic feet	Cubic inches	1728	Feet of water	Pounds per square foot	62.5
Cubic feet	Cubic yards	0.03703	Feet of water	Inches of mercury	0.8824
Cubic yards	Cubic feet	27	Atmospheres	Pounds per square inch	14.696
Cubic inches	Gallons	0.00433	Atmospheres	Inches of mercury	29.92
Cubic feet	Gallons	7.48	Atmospheres	Feet of water	34
Gallons	Cubic inches	231	Long tons	Pounds	2240
Gallons	Cubic feet	0.1337	Short tons	Pounds	2000
Gallons	Pounds of water	8.33	Short tons	Long tons	0.89285
Pounds of water	Gallons	0.12004			
Ounces	Pounds	0.0625			
Pounds	Ounces	16			
Inches of water	Pounds per square inch	0.0361			
Inches of water	Inches of mercury	0.0735			
Inches of water	Ounces per square inch	0.578			
Inches of water	Pounds per square foot	5.2			

Conversion factors for water.

1 U.S. gallon = 8.3356 pounds
1 U.S. gallon = 0.1337 cubic feet
1 U.S. gallon = 231 cubic inches
1 U.S. gallon = 0.83356 Imperial gallons
1 U.S. gallon = 3.7854 liters

1 Imperial gallon = 10.00 pounds
1 Imperial gallon = 0.16037 cubic feet
1 Imperial gallon = 277.12 cubic inches
1 Imperial gallon = 1.1997 U.S. gallons
1 Imperial gallon = 4.5413 liters

1 liter = 2.202 pounds
1 liter = 0.0353 cubic feet
1 liter = 61.023 cubic inches
1 liter = 0.2642 U.S. gallons
1 liter = 0.2202 Imperial gallons

1 cubic foot of water = 62.355 pounds
1 cubic foot of water = 1728.00 cubic inches
1 cubic foot of water = 7.4805 U.S. gallons
1 cubic foot of water = 6.2355 Imperial gallons
1 cubic foot of water = 28.317 liters

1 pound of water = .01604 cubic feet
1 pound of water = 27.712 cubic inches
1 pound of water = 0.11997 U.S. gallons
1 pound of water = 0.100 Imperial gallons
1 pound of water = 0.45413 liters

1 cubic inch of water = 0.0361 pounds
1 gallon of water = 8.33 pounds

1 inch of water = 0.0361 pounds per square inch
1 foot of water = 0.4334 pounds per square inch
1 pound per square inch = 2.310 feet of water
1 pound per square inch = 2.04 inches of mercury
1 atmosphere = 14.696 pounds per square inch

Geometric formula for area (A), volume (V), and circumference (C), radius (R), and diameter (D).

Circle

$A = 3.142 \times R^2$

$C = 3.142 \times D$

$R = \dfrac{D}{2}$

$D = 2 \times R$

Cone

$A_1 = 3.142 \times R \times S + 3.142 \times R^2$

$V = 1.047 \times R^2 \times H$

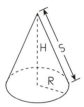

Elipse

$A = 3.142 \times A \times B$

$C = 6.283 \times \dfrac{\sqrt{A^2 + B^2}}{2}$

Cylinder

Area of curved surface =

$(2) \times (3.142) \times (R) \times (H)$

$V = 3.142 \times R^2 \times H$

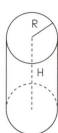

Parallelogram

$A = H \times L$

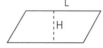

Rectangle

$A = W \times L$

Elliptical tank

$V = 3.142 \times A \times B \times H$

$A_1 = 6.283 \times \dfrac{\sqrt{A^2 + B^2}}{2} \times H + 6.283 \times A \times B$

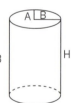

Sector of circle

$A = \dfrac{3.142 \times R^2 \times \alpha}{360}$

$L = .01745 \times R \times \alpha$

$\alpha = \dfrac{L}{.01745 \times R}$

$R = \dfrac{L}{.01745 \times \alpha}$

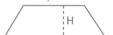

Rectangular solid

$A_1 = 2[W \times L + L \times H + H \times W]$

$V = W \times L \times H$

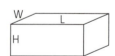

Trapezoid

$A = H \times \dfrac{L_1 + L_2}{2}$

Sphere

$A_1 = 12.56 \times R^2$

$V = 4.188 \times R^3$

Triangle

$A = \dfrac{W \times H}{2}$

$C^2 = H^2 + \left(\dfrac{W}{2}\right)^2$

For above figures when used as water containers:

Capacity in gallons $= \dfrac{V}{231}$ when V is in cubic feet

Capacity in gallons $= 7.48 \times V$ when V is in cubic feet

Sizes and capacity of steel pipe.

Nominal Pipe Size (inches)	Outside Diameter (inches)	Inside Diameter (inches)	Capacity of a One Foot Length of Pipe		Nominal Pipe Size (inches)	Outside Diameter (inches)	Inside Diameter (inches)	Capacity of a One Foot Length of Pipe	
			Cubic Feet	Gallons				Cubic Feet	Gallons
Schedule 40 Pipe					Schedule 80 Pipe				
½	0.840	0.622	0.0021	0.0158	½	0.840	0.546	0.0016	0.012
¾	1.050	0.824	0.0037	0.0276	¾	1.050	0.742	0.0030	0.022
1	1.315	1.049	0.0060	0.0449	1	1.315	0.957	0.0050	0.037
1¼	1.660	1.380	0.0104	0.0774	1¼	1.660	1.278	0.0089	0.066
1½	1.900	1.610	0.0142	0.106	1½	1.900	1.500	0.0123	0.092
2	2.375	2.067	0.0233	0.174	2	2.375	1.939	0.0205	0.153
2½	2.875	2.469	0.0332	0.248	2½	2.875	2.323	0.0294	0.220
3	3.500	3.068	0.0513	0.383	3	3.500	2.900	0.0548	0.344
3½	4.000	3.548	0.0686	0.513	3½	4.000	3.364	0.0617	0.458
4	4.500	4.026	0.0883	0.660	4	4.500	3.826	0.0798	0.597
5	5.563	5.047	0.139	1.04	5	5.563	4.813	0.126	0.947
6	6.625	6.065	0.200	1.50	6	6.625	5.761	0.181	1.35
8	8.625	7.981	0.3474	2.60	8	8.625	7.625	0.3171	2.38
10	10.75	10.020	0.5475	4.10	10	10.75	9.564	0.4989	3.74
12	12.75	11.938	0.7773	5.82	12	12.75	11.376	0.7058	5.28
14	14.0	13.126	0.9397	7.03	14	14.0	12.500	0.8522	6.38
16	16.0	15.000	1.2272	9.16	16	16.0	14.314	1.1175	8.36
18	18.0	16.876	1.5533	11.61	18	18.0	16.126	1.4183	10.61
20	20.0	18.814	1.9305	14.44	20	20.0	17.938	1.7550	13.13
24	24.0	22.626	2.7920	20.87	24	24.0	21.564	2.536	19.0

Conversion of Inches to millimeters.

in.	mm	in.	mm	in.	mm	in.	mm
1	25.4	26	660.4	51	1295.4	76	1930.4
2	50.8	27	685.8	52	1320.8	77	1955.8
3	76.2	28	711.2	53	1346.2	78	1981.2
4	101.6	29	736.6	54	1371.6	79	2006.6
5	127.0	30	762.0	55	1397.0	80	2032.0
6	152.4	31	787.4	56	1422.4	81	2057.4
7	177.8	32	812.8	57	1447.8	82	2082.8
8	203.2	33	838.2	58	1473.2	83	2108.2
9	228.6	34	863.6	59	1498.6	84	2133.6
10	254.0	35	889.0	60	1524.0	85	2159.0
11	279.4	36	914.4	61	1549.4	86	2184.4
12	304.8	37	939.8	62	1574.8	87	2209.8
13	330.2	38	965.2	63	1600.2	88	2235.2
14	355.6	39	990.6	64	1625.6	89	2260.6
15	381.0	40	1016.0	65	1651.0	90	2286.0
16	406.4	41	1041.4	66	1676.4	91	2311.4
17	431.8	42	1066.8	67	1701.8	92	2336.8
18	457.2	43	1092.2	68	1727.2	93	2362.2
19	482.6	44	1117.6	69	1752.6	94	2387.6
20	508.0	45	1143.0	70	1778.0	95	2413.0
21	533.4	46	1168.4	71	1803.4	96	2438.4
22	558.8	47	1193.8	72	1828.8	97	2463.8
23	584.2	48	1219.2	73	1854.2	98	2489.2
24	609.6	49	1244.6	74	1879.6	99	2514.6
25	635.0	50	1270.0	75	1905.0	100	2540.0

The above table is exact on the basis: 1 in. = 25.4 mm

Conversion of Millimeters to inches.

mm	in.	mm	in.	mm	in.	mm	in.
1	0.039370	26	1.023622	51	2.007874	76	2.992126
2	0.078740	27	1.062992	52	2.047244	77	3.031496
3	0.118110	28	1.102362	53	2.086614	78	3.070866
4	0.157480	29	1.141732	54	2.125984	79	3.110236
5	0.196850	30	1.181102	55	2.165354	80	3.149606
6	0.236220	31	1.220472	56	2.204724	81	3.188976
7	0.275591	32	1.259843	57	2.244094	82	3.228346
8	0.314961	33	1.299213	58	2.283465	83	3.267717
9	0.354331	34	1.338583	59	2.322835	84	3.307087
10	0.393701	35	1.377953	60	2.362205	85	3.346457
11	0.433071	36	1.417323	61	2.401575	86	3.385827
12	0.472441	37	1.456693	62	2.440945	87	3.425197
13	0.511811	38	1.496063	63	2.480315	88	3.464567
14	0.551181	39	1.535433	64	2.519685	89	3.503937
15	0.590551	40	1.574803	65	2.559055	90	3.543307
16	0.629921	41	1.614173	66	2.598425	91	3.582677
17	0.669291	42	1.653543	67	2.637795	92	3.622047
18	0.708661	43	1.692913	68	2.677165	93	3.661417
19	0.748031	44	1.732283	69	2.716535	94	3.700787
20	0.787402	45	1.771654	70	2.755906	95	3.740157
21	0.826772	46	1.811024	71	2.795276	96	3.779528
22	0.866142	47	1.850394	72	2.834646	97	3.818898
23	0.905512	48	1.889764	73	2.874016	98	3.858268
24	0.944882	49	1.929134	74	2.913386	99	3.897638
25	0.984252	50	1.968504	75	2.952756	100	3.937008

The above table is approximate on the basis: 1 in. = 25.4 mm, 1/25.4 = 0.039370078740+

Inch to Metric Equivalents.

Fraction (in.)	Decimal Equivalent		Fraction (in.)	Decimal Equivalent	
	English (in.)	Metric (mm)		Englilsh (in.)	Metric (mm)
1/64	.015625	0.3969	33/64	.515625	13.0969
1/32	.03125	0.7938	17/32	.53125	13.4938
3/64	.046875	1.1906	35/64	.546875	13.8906
1/16	.0625	1.5875	9/16	.5625	14.2875
5/64	.078125	1.9844	37/64	.578125	14.6844
3/32	.09375	2.3813	19/32	.59375	15.0813
7/64	.109375	2.7781	39/64	.609375	15.4781
1/8	.1250	3.1750	5/8	.6250	15.8750
9/64	.140625	3.5719	41/64	.640625	16.2719
5/32	.15625	3.9688	21/32	.65625	16.6688
11/64	.171875	4.3656	43/64	.671875	17.0656
3/16	.1875	4.7625	11/16	.6875	17.4625
13/64	.203125	5.1594	45/64	.703125	17.8594
7/32	.21875	5.5563	23/32	.71875	18.2563
15/64	.234375	5.9531	47/64	.734375	18.6531
1/4	.250	6.3500	3/4	.750	19.0500
17/64	.265625	6.7469	49/64	.765625	19.4469
9/32	.28125	7.1438	25/32	.78125	19.8438
19/64	.296875	7.5406	51/64	.796875	20.2406
5/16	.3125	7.9375	13/16	.8125	20.6375
21/64	.328125	8.3384	53/64	.828125	21.0344
11/32	.34375	8.7313	27/32	.84375	21.4313
23/64	.359375	9.1281	55/64	.859375	21.8281
3/8	.3750	9.5250	7/8	.8750	22.2250
25/64	.390625	9.9219	57/64	.890625	22.6219
13/32	.40625	10.3188	29/32	.90625	23.0188
27/64	.421875	10.7156	59/64	.921875	23.4156
7/16	.4375	11.1125	15/16	.9375	23.8125
29/64	.453125	11.5094	61/64	.953125	24.2094
15/32	.46875	11.9063	31/32	.96875	24.6063
31/64	.484375	12.3031	63/64	.984375	25.0031
1/2	.500	12.7000	1	1.000	25.4000

Metric equivalents.

Length

U.S. to Metric	Metric to U.S.
1 inch = 2.540 centimeters	1 millimeter = .039 inch
1 foot = .305 meter	1 centimeter = .394 inch
1 yard = .914 meter	1 meter = 3.281 feet or 1.094 yards
1 mile = 1.609 kilometers	1 kilometer = .621 mile

Area

1 inch^2 = 6.451 centimeter^2 1 millimeter^2 = .00155 inch^2
1 foot^2 = .093 meter^2 1 centimeter^2 = .155 inch^2
1 yard^2 = .836 meter^2 1 meter^2 = 10.764 foot^2 or 1.196 yard^2
1 acre^2 = 4,046.873 meter^2 1 kilometer^2 = .386 mile^2 or 247.04 acre^2

Volume

1 inch^3 = 16.387 centimeter^3 1 centimeter^3 = 0.61 inch^3
1 foot^3 = .028 meter^3 1 meter^3 = 35.314 foot^3 or 1.308 yard^3
1 yard^3 = .764 meter^3 1 liter = .2642 gallons
1 quart = .946 liter   1 liter = 1.057 quarts
1 gallon = .003785 meter^3 1 meter^3 = 264.02 gallons

Weight

1 ounce = 28.349 grams   1 gram = .035 ounce
1 pound = .454 kilogram   1 kilogram = 2.205 pounds
1 ton = .907 metric ton   1 metric ton = 1.102 tons

Velocity

1 foot/second = .305 meter/second   1 meter/second = 3.281 feet/second
1 mile/hour = .447 meter/second   1 kilometer/hour = .621 mile/second

Acceleration

1 inch/second^2 = .0254 meter/second^2 1 meter/second^2 = 3.278 feet/second^2
1 foot/second^2 = .305 meter/second^2

Force

N (newton) = basic unit of force, kg-m/s^2. A mass of one kilogram (1 kg) exerts a gravitational force of 9.8 N (theoretically 9.80665 N) at mean sea level.

Temperature conversion:Celsius (°C) to Fahrenheit (°F)

Temp. Celsius = 5/9 (Temp. °F − 32 deg.)

Rankine (Fahrenheit Absolute) = Temp. °F + 459.67 deg.

Freezing point of water: Celsius = 0 deg.; Fahr. = 32 deg.

Absolute zero: Celsius = −273.15 deg.; Fahr. = −459.67 deg.

Temp. Fahrenheit = 9/5 × Temp. °C + 32 deg.

Kelvin (celsius Absolute) = Temp. °C + 273.15 deg.

Boiling point of water: Celsius = 100 deg.; Fahr. = 212 deg.

Appendix B

PROPERTIES OF FLAMMABLE LIQUIDS

NFPA 11 defines **flammable liquids** as liquids having a flash point below 100°F (38°C) and having a vapor pressure not exceeding 40 psi (276 kPa) (absolute) at 100°F (38°C).

Class I flammable liquids include liquids having flash points below 100°F (38°C) and may be subdivided into the following subclasses:

- **Class IA Flammable Liquids:** liquids having flash points below 73°F (23°C) and having boiling points below 100°F (38°C).

- **Other Class IA Flammable Liquids:**

acetaldehyde	Ethyl chloride	petroleum ether
collodion	methyl ethyl ether	propylene oxide
ethyl ether	pentane	

- **Class IB Flammable Liquids:** liquids having flash points below 73°F (23°C) and boiling points above 100°F (38°C).

Examples of Class IA Flammable Liquids	Flash Point °F	(°C)	Boiling Point °F	(°C)	Water-Soluble
Acetic aldehyde	−38	(−39)	70	(21)	Yes
Dimethyl sulfide	0	(−18)	99	(37)	Slight
Furan	32	(0)	88	(31)	No

Examples of Class IB Flammable Liquids	Flash Point °F	(°C)	Boiling Point °F	(°C)	Water-Soluble
Ethyl alcohol	55	(13)	173	(78)	Yes
Gasoline	−36	(−38)	100–400	(38–204)	No
Cyclohexane	−4	(−20)	179	(82)	No

- **Other Class IB Flammable Liquids:**

  acetone   ethyl acetate   toluene
  benzene   methyl alcohol
  butyl alcohol   methylcyclohexane

- **Class IC Flammable Liquids:** liquids having flash points at or above 73°F (23°C) and below 100°F (38°C). Class IC flammable liquids include the following:

  amyl acetate   isopropanol   turpentine
  amyl alcohol   methyl alcohol xylene
  dibutyl ether   styrene

Appendix C

PROPERTIES OF COMBUSTIBLE LIQUIDS

NFPA 11 defines **combustible liquids** as any liquid having a flash point at or above 100°F (38°C). Combustible liquids are subdivided into the following classes:

- **Class II Combustible Liquids:** liquids having flash points at or above 100°F (38°C) and below 140°F (60°C).

Examples of Class II Combustible Liquids	Flash Point		Water-Soluble
	°F	(°C)	
Diesel fuel oil (No. 1-D/2-D/4-D)	100–130	(38–54)	NO
Glacial acetic acid	103	(39)	YES
Jet fuel (A & A-1)	110	(43)	NO

- **Other Class II Combustible Liquids:**

acetic acid	fuel oil no. 44	mineral spirits
camphor oil	methyl lactate	varsol
cyclohexane	hydrazine	kerosene

- **Class IIIA Combustible Liquids:** liquids having flash points at or above 140°F (60°C) and below 200°F (93°C).

Examples of Class III A Combustible Liquids	Flash Point		Water-Soluble
	°F	(°C)	
Creosote oil	165	(74)	NO
Butyl carbitol	172	(78)	YES

- **Other Class IIIA Combustible Liquids:**

  aniline   furfuryl alcohol   phenol
  carbolic acid   naphthalenes   pine oil

- **Class IIIB Combustible Liquids:** liquids having flash points at or above 200°F (93°C).

Examples of Class IIIB Combustible Liquids	Flash Point		Water-Soluble
	°F	(°C)	
Fuel oil no. 4	up to 240	(61–116)	NO
Mineral oil	380	(193)	NO
Olive oil	437	(225)	NO

Appendix

D

PIPE DIAMETERS

Table D-1 *Internal diameters for aboveground fire protection piping (in inches).*

	.75"	1"	1.25"	1.5"	2"	2.5"	3"	3.5"	4"	5"	6"	8"
Nominal Diameter												
Steel Pipe Diameter												
Schedule 10		1.097	1.442	1.682	2.157	2.635	3.26	3.76	4.26	5.295	6.357	8.249
Schedule 30												8.071
Schedule 40		1.049	1.38	1.61	2.067	2.469	3.068	3.548	4.026	5.047	6.065	7.981
Copper Pipe Diameter												
Type K	.745	.995	1.245	1.481	1.959	2.435	2.907	3.385	3.857	4.805	5.741	7.583
Type L	.785	1.025	1.265	1.505	1.985	2.465	2.945	3.425	3.905	4.875	5.845	7.725
Type M	.811	1.055	1.291	1.527	2.009	2.495	2.981	3.459	3.935	4.907	5.881	7.785
CPVC Plastic Pipe Diameter												
	.884	1.109	1.400	1.602	2.003	2.423	2.951					

Table D-2   *Internal diameters for underground fire protection piping (in inches).*

	4″	6″	8″	10″	12″	14″	16″	18″	20″	24″
Nominal Diameter										
Ductile Iron Pipe Diameter										
Class 50		6.4	8.51	10.52	12.58	14.64	16.72	18.8	20.88	24.04
Class 51	4.28	6.34	8.45	10.46	12.52	14.58	16.66	18.74	20.82	24.78
Class 52	4.155	6.275	8.385	10.4	12.46	14.52	16.4	18.68	20.76	24.72
Class 54		6.16	8.27	10.28	12.34					
Cast-Iron Pipe Class 150 Diameter										
Unlined	4.1	6.14	8.23	10.22	12.24	14.28	16.32	18.34	20.36	24.34
Enamel-Lined	3.98	6.02	8.11	10.1	12.12	14.09	16.13			
Cement-Lined	3.85	5.89	7.98	9.97	11.99	13.9	15.94			
Plastic Underground Pipe Diameter										
PVC Class 150	4.24	6.09	7.98	9.79	11.65					
PVC Class 200	4.08	5.86	7.68	9.42	11.2					

SELECTED NFPA STANDARDS

The following codes, standards, and recommended practices published by the National Fire Protection Association (Quincy, Massachusetts) contain information pertinent to the design of special hazard and fire alarm systems.

NFPA 11, *Standard for Low, Medium, and High Expansion Foam.*

NFPA 12, *Standard on Carbon Dioxide Extinguishing Systems.*

NFPA 12A, *Standard on Halon 1301 Fire Extinguishing Systems.*

NFPA 13, *Standard for the Installation of Sprinkler Systems.*

NFPA 13D, *Standard on the Installation of Sprinkler Systems in One- and Two-Family Dwellings and Manufactured Homes.*

NFPA 13R, *Standard on the Installation of Sprinkler Systems in Residential Occupancies up to and Including Four Stories in Height.*

NFPA 14, *Standard for the Installation of Standpipe and Hose Systems.*

NFPA 15, *Standard for Water Spray Fixed Systems for Fire Protection.*

NFPA 16, *Standard for the Installation of Foam-water Sprinkler and Foam-water Spray Systems.*

NFPA 17, *Standard for Dry Chemical Extinguishing Systems.*

NFPA 17A, *Standard for Wet Chemical Extinguishing Systems.*

NFPA 20, *Standard for the Installation of Centrifugal Fire Pump.*

NFPA 22, *Standard for the Installation of Water Tanks for Private Fire Protection.*

NFPA 24, *Standard for Private Service Mains and Their Appurtenances.*

NFPA 61, *Standard for the Prevention of Fire and Dust Explosions in Agricultural and Food Processing Facilities.*

NFPA 68, *Guide for Venting of Deflagrations.*

NFPA 69, *Standard on Explosions Prevention Systems.*

NFPA 70, *National Electrical Code.*

NFPA 72, *National Fire Alarm Code.*

NFPA 77, *Recommended Practice on Static Electricity.*

NFPA 101, *Code for Safety to Life in Buildings and Structures.*

NFPA 170, *Standard on Fire Safety Symbols.*

NFPA 214, *Standard on Water Cooling Towers.*

NFPA 409, *Standard on Aircraft Hangars.*

NFPA 495, *Explosive Materials Code.*

NFPA 750, *Standard for the Installation of Water Mist Fire Protection Systems.*

NFPA 2001, *Standard on Clean Agent Fire Extinguishing Systems.*

NFPA STANDARDS CORRELATED TO CORE COMPETENCIES AND LEARNING PERFORMANCE OBJECTIVES

UNIT 1: SPECIAL HAZARD AND FIRE ALARM FUNDAMENTALS

Chapter 1: Fundamental Concepts for Design of Special Hazard and Fire Alarm Systems

NFPA Publications, General overview
  Use of NFPA standards, types of standards
  Core competency: fundamental design concepts

NFPA 170, Standard for Fire Safety Symbols

NFPA 13D, **Standard for the Installation of One- and Two-Family Dwellings and Manufactured Homes**
(2007 Edition)
  Used as an example of automatic sprinkler design
  Core competency: fundamental suppression design concepts

NFPA 550, **Guide to the Fire Safety Concepts Tree** (2002 Edition)
  Use of Fire Safety Concept Tree in special hazard and fire alarm design
  Core competency: fundamental design concepts

NFPA 11, NFPA 12, NFPA 12A, NFPA 13, NFPA 15, and NFPA 16
    Used as examples of prescriptive design
    Core competency: fundamental suppression design concepts

Chapter 3: Special Hazard Suppression Agents and Their Applications

NFPA 1, Uniform Fire Code (2006 Edition)
    Definition of combustion
    Classifications of fire
    Core competency: fundamental suppression design concepts

NFPA 921, Guide for Fire and Explosion Investigations (2004 Edition)
    Definition of smoldering
    Definition of ignition
    Definition of autoignition
    Definition of piloted ignition temperature
    Definition of spontaneous ignition
    Definition of heat transfer
    Definition of flashover
    Definition of explosion
    Core competency: fundamental suppression design concepts

NFPA 15, Standard for Water Spray Fixed Systems for Fire Protection
(2007 Edition)
    Definition of deflagration
    Definition of detonation
    Core competency: fundamental suppression design concepts

NFPA 10, Standard for Portable Fire Extinguishers (2007 edition)
    Design standards for fire extinguisher placement
    Core competency: suppression design fundamentals

NFPA 408, Standard on Aircraft Hand Portable Fire Extinguishers (2004 Edition)
    Example of specialized fire extinguisher occupancies
    Core competency: suppression design fundamentals

NFPA 30A, NFPA 32, NFPA 58, NFPA 86, NFPA 96, NFPA 120, NFPA 122, NFPA 241, NFPA 302, NFPA 385, NFPA 407, NFPA 410, NFPA 418, NFPA 430, NFPA 498, NFPA 1192, NFPA 1194
> Examples of specialized fire extinguisher occupancies
> Core competency: suppression design fundamentals

UNIT 2: WATER-BASED SPECIAL HAZARD SYSTEMS

Chapter 4: Low-Expansion Foam System Design

NFPA 11, Standard for Low- Medium- and High-Expansion Foam (2005 Edition)
> Design of foam systems for vessels and dikes
> Core competency: foam system design and calculation

NFPA 409, Standard on Aircraft Hangars (2004 Edition)
> Design of foam systems for aircraft hangars
> Core competency: foam system design and calculation

NFPA 16, Standard for the Installation of Foam-Water Sprinkler and Foam-Water Spray Systems (2007 Edition)
> Design of foam systems for truck loading racks
> Core competency: foam system design and calculation

Chapter 5: Medium- and High-Expansion Foam System Design

NFPA 11, Standard for Low-, Medium-, and High-Expansion Foam (2005 Edition)
> Design of high expansion foam systems for various occupancies
> Core competency: foam system design and calculation

NFPA 409, Standard on Aircraft Hangars (2004 Edition)
> Design of high expansion foam systems for aircraft hangars
> Core competency: foam system design and calculation

Chapter 6: Water Mist Systems

NFPA 750, Standard for the Installation of Water Mist Fire Protection Systems (2006 Edition)
> Design of water mist systems for various occupancies
> Core competency: special hazard system design and calculation

Chapter 7: Ultra High-Speed Explosion Suppression Systems and Ultra High-Speed Water Spray Systems

NFPA 69, Standard on Explosion Prevention Systems (2002 Edition)
Design of explosion prevention systems for various occupancies
Core competency: special hazard system design and calculation

NFPA 15, Standard for Water Spray Fixed Systems for Fire Protection (2007 Edition)
Design of water spray systems for various occupancies
Core competency: special hazard system design and calculation

NFPA 68, Guide for Venting of Deflagrations (2002 Edition)
Design of explosion protection systems for various occupancies
Core competency: special hazard system design and calculation

UNIT 3: GASEOUS AND PARTICULATE AGENT SPECIAL HAZARD SYSTEMS

Chapter 8: Clean Agent and Halon Replacement Extinguishing System Design

NFPA 2001, Standard on Clean Agent Fire Extinguishing Systems (2004 Edition)
Design of clean agent systems for various occupancies
Core competency: special hazard system design and calculation

NFPA 12A, Standard on Halon 1301 Fire Extinguishing Systems (2004 Edition)
Design of Halon 1301 systems for various occupancies
Core competency: special hazard system design and calculation

NFPA 13, Standard for the Installation of Sprinkler Systems

Chapter 9: Carbon Dioxide System Design

NFPA 12, Standard on Carbon Dioxide Extinguishing Systems (2004 Edition)
Design of carbon dioxide systems for various occupancies
Core competency: special hazard system design and calculation

Chapter 10: Dry Chemical and Wet Chemical Extinguishing System Design

NFPA 17, Standard for Dry Chemical Extinguishing Systems (2002 Edition)
Design of dry chemical systems for various occupancies
Core competency: special hazard system design and calculation
NFPA 17A, Standard for Wet Chemical Extinguishing Systems (2002 Edition)
Design of wet chemical systems for various occupancies
Core competency: special hazard system design and calculation

UNIT 4: SPECIAL HAZARD DETECTION, ALARM, AND RELEASING FIRE ALARM SYSTEMS

Chapter 11: Fire Detection and Alarm Systems

NFPA 72, National Fire Alarm Code (2007 Edition)
   Design of fire detection and alarm systems for various occupancies
   Core competency: fire detection and system design

NFPA 70, National Electric Code (2005 Edition)
   Design of fire detection and alarm systems for various occupancies
   Core competency: fire detection and system design

NFPA 101, Code for Safety to Life in Buildings and Structures

Chapter 12: Fire Alarm System Initiating Devices

NFPA 70, National Electric Code (2005 Edition)
   Design of fire detection and alarm systems for various occupancies
   Core competency: fire detection and system design

NFPA 72, National Fire Alarm Code (2007 Edition)
   Design of fire detection and alarm systems for various occupancies
   Core competency: fire detection and system design

NFPA 92B, Standard for Smoke Management Systems in Malls, Atria, and Large Areas (2005 Edition)
   Design of smoke detection and fire detection and alarm systems for various occupancies
   Core competency: fire detection and system design

Chapter 13: Fire Alarm System Notification Appliances

NFPA 72, National Fire Alarm Code (2007 Edition)
   Design of fire detection and alarm systems for various occupancies
   Core competency: fire detection and system design

Chapter 14: Fire Detector Placement

NFPA 72, National Fire Alarm Code (2007 Edition)
   Design of fire detection and alarm systems for various occupancies
   Core competency: fire detection and system design

Chapter 15: Fire Alarm Circuit Design and Fire Alarm Control Units

NFPA 72, National Fire Alarm Code (2007 Edition)
Design of fire detection and alarm systems for various occupancies
Core competency: fire detection and system design

NFPA 70, National Electric Code (2005 Edition)
Design of fire detection and alarm systems for various occupancies
Core competency: fire detection and system design

GLOSSARY

Abort switch A type of switch that cancels discharge.

Addressable (multiplex) system A system that requires a central processing unit (CPU) or a computer and a software program that specifies and assigns the exact locations or addresses of each initiating device on the fire alarm system.

Addressable detectors Devices that have a preprogrammed electronic address assigned to it that the fire alarm control unit will recognize when it actuates or when it issues a trouble or fault signal.

Addressable manual stations Stations with a discrete preprogrammed electronic identification that the FACU recognizes as to its type, location, and function.

Air-sampling or air-aspirated smoke detection systems Devices that draw air into a sampling chamber for analysis.

Air change The cycle of air movement from HVAC unit to underfloor through equipment and back to the HVAC unit.

Air-regulated fire A scenario in which the ability of the fire to spread is dependent on the quantity of oxygen available.

Alcohol-resistant foam A foam used for the protection of alcohol-based fires.

Americans with Disabilities Act (ADA) Legislation that ensures that persons with disabilities are accommodated.

Annunciator panel A device that provides visible notification of the location of an initiating device by zone and floor.

Aqueous agents Fire suppression agents that are water-based and may involve an additive that enhances the effectiveness of water.

Aqueous film-forming foam (AFFF) A synthetic foam that forms a thin aqueous film that separates the foam from the fuel.

Architectural drawings Drawings that show dimensions of walls, floors, ceilings, and other building features.

Area protection Application of suppressant over the entire floor surface area of a room or enclosure.

Aspirated foam-water nozzles Nozzles possessing an air inlet that allows air to be introduced to the foam solution before it hits the stream deflector.

Atmospheric pressure Sea-level pressure, 14.7 psia (101.283 kPa) or one "G".

Authority having jurisdiction (AHJ) The individual or agency placed in responsible charge of reviewing and approving drawings and completed installations.

Autoignition Initiation of combustion by heat but without spark or flame.

Auxiliary fire alarm system A system of manual pull boxes located throughout a jurisdiction with each box wired to a circuit connected directly to the fire department.

Backdraft A violent expansion of the combustion process caused by the sudden introduction of air to a fire.

Balanced pressure proportioner A device that uses an atmospheric foam concentrate tank, a pump to pressurize the concentrate and force it toward the proportioner, and a proportioner that balances the pumped concentrate pressure to the water supply pressure.

Bar joist A webbed member supported by I-beams.

Bar joist bearing dimension Depth of the top of a joist.

Blackbody An ideal body that would absorb all incident radiation and reflect none.

Bladder tank A tank in which the concentrate is stored within a collapsible bladder.

Boring table Compilation of data obtained by drilling cores of earth at several strategic locations.

Branch lines Sprinkler pipes that have sprinklers or discharge devices installed directly on them.

Breakglass manual stations Manual stations employing a replaceable slender plastic or glass rod that breaks when the handle on the manual station is pulled.

Building sections Cutaway views through a building.

Candela (cd) The standard unit of light intensity measurement.

Carbon dioxide A gaseous fire protection agent, also known by its chemical designation CO_2.

Central station service fire alarm system A fire alarm system that is supervised and constantly monitored by a privately owned company.

Chemical foams A type of foam that depends on the initiation of a chemical reaction within the foam solution to create the air bubbles in the foam; became obsolete with the introduction of AFFF and fluoroprotein foams.

Civil drawings Scaled drawings that coordinate underground utilities entering and leaving a building or group of buildings.

Clean agent An electrically nonconducting, volatile, or gaseous fire extinguishant that does not leave a residue upon evaporation.

Client's value system Values of the receiver of the fire protection system that demand a code-compliant fire protection system at the lowest possible price.

Code of Ethics for Engineers A compilation of rules and obligations published by the National Society of Professional Engineers and used by engineers to resolve ethical conflicts in fire protection system design.

Coded manual stations Stations that send a set of unique and distinctive time-pulsed signals to the fire alarm control unit three or more times, indicating the exact location of the manual station.

Codes Mandatory requirement suitable for adoption into law.

Combination audible/visible notification appliance Possesses both audible and visible notification components.

Combination ultraviolet/infrared (UV/IR) flame detectors Devices that contain sensors for both ultraviolet and infrared wavelengths, and the detector requires that abnormal values in both the UV and the IR spectra be seen before a signal is sent to the FACU.

Combustible liquid A liquid having a flash point at or above 100°F (37.8°C).

Combustion (fire) A chemical process of oxidation that occurs at a rate fast enough to produce heat and usually light in the form of either a glow or flame.

Company manual A written set of company regulations that requires the signature of employees.

Company values Values that require a fire protection professional to behave in a way that would meet the goals of, and bring credit to, the company.

Computer-aided design (CAD) A computerized method of preparing a drawings.

Continuous software interrogation (CSI) A process in which a microprocessor individually interrogates or checks the status of each device or control unit on the signaling line circuit in sequence, then continuously rechecks the devices in order.

Contour lines Lines indicating the elevation of the finished exterior grade.

Contract drawings A set of plans that describe a project in pictorial form.

Critical temperature A temperature beyond which carbon dioxide can exist only in its vapor phase.

Cross bracing Supports that provide structural rigidity between bar joists.

Decibel (dB) A unit of sound intensity at a given distance.

Decommissioning Removing a halon system from service.

Deep-seated fire A fire within an object or enclosure.

Deflagration A reaction in which the flame front moves into the unburned material at less than the speed of sound.

Deluge water mist systems Systems with open nozzles that discharge water mist simultaneously from all nozzles on the system.

Depth of cover Lineal distance from the top of an underground pipe to the finished grade.

Detailed specification In-depth requirements for the design of a fire protection system that allow little latitude for interpretation or alternative design proposals.

Detonation A reaction in which the flame front expands at a rate greater than the speed of sound.

Dike protection systems Systems where the dike area is flooded with foam that will float on top of a flammable liquid that spills within the containment dike.

Diptank A vat used for dipping, coating, or stripping an object in a flammable liquid.

Domestic plumbing piping Piping that supplies water fountains, sinks, and toilets.

Double-action manual stations Devices that require the operator to perform two separate and distinct functions before an alarm is sounded.

Drainboard An object that collects flammable liquid residue that drips from the dipped item onto an inclined surface, allowing the flammable liquid residue to drain back to the diptank.

Drawing Graphic representation of a designer's ideas.

Drop-in agent An agent that allows Halon 1301 to be removed and an equivalent amount of replacement agent inserted.

Dry chemical A power consisting of small particles suspended in a gaseous medium, which permits distribution of the powder to a hazard.

Dry chemical skid A pre-assembled assembly that includes pre-piped dry chemical storage and pressurization facilities.

Dry pipe water mist systems Systems with air-filled piping, having nozzles with an individual actuating device, where each nozzle actuates individually.

End-of-line resistor (ELR) A device that reduces the current of the circuit by creating resistance for the incoming current, allowing less current to leave the resistor than the amount entering it.

Entrainment The process of air being drawn into a fire plume.

Environmental Protection Agency (EPA) U.S. Governmental agency responsible for protecting air and water supplies.

Ethical dilemmas Situations in which pressure is placed upon an individual to do something that is not ethically proper, or when it is not immediately obvious which option would be ethically proper.

Evacuation zone A floor or portion of a floor of a building, or a discrete area or wing of a building, where fire alarm notification can be provided.

"Exit marking" audible notification appliances Installed at the entrance of all building exits and areas of refuge, and emit distinct sound pressure levels capable of directing occupants to the exits.

Expansion ratio Computed by measuring the volume of the foam produced after air is added to the foam solution and comparing that volume to the original volume of foam solution prior to air addition.

Explosion   A rapid release of combustion energy that increases pressure in a vessel, container, or building and results in its eventual rupture.

Explosion isolation   The automatic closing of a valve to limit the pressure rise from spreading to a predetermined area.

Explosion pressure profile   A graph depicting the effect of suppression upon a specific combustible.

Explosion suppression systems   System used for the protection of vessels or other enclosures where overpressurization is the primary concern.

Exposure protection   The wetting of combustibles adjacent to the fire to delay their ignition.

Extended discharge system   A separate system of small pipes and nozzles that provides a rate of discharge after the primary discharge system ceases operation, to compensate for the amount of gas projected to be lost during the required holding period.

Field check   A thorough survey of a proposed fire protection system using a completed fire protection system design as a basis.

Film-forming fluoroprotein foam   A concentrate that uses fluorinated surfactants to produce a fluid aqueous film for suppressing hydrocarbon fuel vapors.

Finish schedule   A contract drawing that lists all rooms in a building and provides details of several room features.

Fire alarm control unit (FACU)   An electronic device to which initiating, notification, and supervisory zones are connected and controlled.

Fire control   A reduction in thermal exposure, threat to occupants, or fire-related characteristics.

Fire extinguishment   Complete suppression of the fire.

Fire-fighting foam   Mixture of the foam solution with air.

Fire protection engineer   A licensed professional engineer who demonstrates sound knowledge and judgment in the application of science and engineering to protect the health, safety, and welfare of the public from the impacts of fire.

Fire protection system design   Design of a fire protection system based on engineering criteria that may not always coincide with criteria found in an accepted national standard.

Fire protection system layout   The performance used by a technician to comply with the requirements of an accepted national standard to execute a drawing.

Fire protection technician   An individual who has achieved NICET Level III of IV certification in the appropriate subfield and who has the knowledge, experience and skills necessary to lay out fire protection systems.

Fire signature   Any fire effect (smoke, heat, light, etc.) that can be sensed by a fire detector.

Fire suppression   A sharp reduction in heat release rate and prevention of regrowth.

Fire tetrahedron   The pictorial representation of the interdependent factors of heat, fuel, oxygen, and uninhibited chemical chain reactions.

Fire triangle   Pictorial representation of the interdependent factors of heat, fuel, and oxygen.

Flame extinguishment   Extinguishing of flammable liquids or gases.

Flammable liquid   A liquid having a flash point below 100°F (37.8°C) and having a vapor pressure not exceeding 40 psi (2068.6 mmHg) (per NFPA 11).

Flashover   A transition phase in the development of a compartment fire in which surfaces exposed to thermal radiation reach ignition temperature more or less simultaneously and fire spreads rapidly throughout the space, resulting in full room involvement or total involvement of the compartment or enclosed space.

Floating roof   A tank roof that floats on the surface of a flammable or combustible liquid.

Fluidization   Mixing of an expellant gas with dry chemical powder for ease of distribution to the hazard.

Fluoroprotein foam A foam that contains fluorochemical additives that make it flow more easily.

Flush-mounted lights Lights mounted below and flush to the ceiling.

Foam proportioner A manufactured product designed to ensure delivery of the precise ratio of foam concentrate to a foam solution.

Foundation plans Plans that show floor and wall structural details and sectional views.

Framing plans Plans that show beam and joist size and elevation.

Fuel-regulated fire A fire that has unlimited access to oxygen and is dependent only on the fuel available.

Gas-sensing detectors Devices that detect the release of a flammable gas before it reaches its ignitable concentration.

General alarm stations Stations that issue a notification alarm immediately upon activation of a general alarm manual station.

Ground fault An IDC fault that can be caused by a bare wire or terminal in contact with a grounded junction box or some other source of grounding.

Guides Informative but nonbinding NFPA documents.

Halocarbon agents Hydrofluorocarbons (HFCs), hydrochlorofluorocarbons (HCFCs), and perfluorocarbons (PFCs).

Heat transfer A process that consists of conduction, convection, and radiation.

High-expansion foam generators Devices that add air to the foam solution spray, creating foam.

High-expansion foam system A system of air-filled bubbles created by the mechanical expansion of a foam solution by air and water, with foam-to-solution ratio of between 200 to 1 and 1000 to 1.

High-pressure water mist systems System in which the pressures encountered by the system piping are 500 psi or greater.

HVAC drawings Heating, ventilating, and air conditioning drawings.

HVAC plenum A space that is kept under negative pressure by fans, allowing stale air to enter the space through grilles on the ceiling.

Hydrocarbon Organic compounds that contain only carbon and hydrogen.

I-beam A solid steel member that looks like an "I" from its end.

Ignition The process of initiating self-sustained combustion.

Incipient fire A small fire in its early stages of combustion.

Inert gas agents Agents that contain one or more non-reactive gases, such as helium, neon, argon, nitrogen, and carbon dioxide.

Inerting Reducing the flammable concentration in an atmosphere to below one-half of its lower flammable limit.

Infrared (IR) flame detectors Devices designed to respond to infrared radiation involving wavelengths from 8500 to 12,000 angstroms.

Initiating device A mechanism that originates a signal that is sent to a fire alarm control unit.

Initiating device circuit (IDC) A circuit to which automatic or manual initiating devices are connected to the fire alarm control unit (FACU), where the signal received by the FACU identifies an alarm condition on the circuit but does not identify the specific device actuated.

Intermediate-pressure water mist systems System in which the pressures encountered by the system piping are between 175 psi and 500 psi.

Inverse square law A law stating that as the distance from the ear to the notification appliance doubles, sound is reduced by 6 dB along the centerline of the notification appliance.

Invert elevation An elevation referencing the bottom of a pipe with respect to the reference elevation.

Ionization detector A device containing a small amount of radioactive material that ionizes the air between a positive and a negative electrode in its sampling chamber to measure conductance.

Laser-based air-sampling smoke detectors Devices that sample air drawn from a network of pipes or tubes and analyze the sample for comparison with a critical baseline.

Legal/ethical conflicts Disagreements related to the absence of congruence between our system of laws and our system of ethical values.

Line-of-sight A visual path above a ceiling created by removing ceiling tiles at regular intervals along a length or dimension of particular interest, such as a fire protection pipe, with the aid of a flashlight and a telescoping elevation pole.

LOAEL Abbreviation for lowest observable adverse effect level.

Local application method A method of fire suppression in which the agent is applied directly onto the point of hazard.

Lower flammable limit (LFL) The minimum concentration of airborne combustibles required for ignition.

Low-expansion foam system A system designed to deliver a foam solution, possessing an expansion ratio of up to 20:1, to a hazard.

Low-pressure water mist systems Systems in which the pressures encountered by the system piping are 175 psi or less.

Manual pull stations Devices for use by building occupants to electronically notify others of a fire.

Manually operated fire alarm devices Mechanisms that require the action of an individual to initiate a fire alarm signal.

Material conversion factor A dimensionless number that increases the basic quantity of carbon dioxide for hazards where the minimum design concentration exceeds 34%.

Metric Conversion Act of 1975 Legislation that created a requirement for conversion to the metric system for all federal projects by 1992.

Montreal Protocol An international agreement intended to sharply restrict the production of chemicals identified as contributing to depletion of the ozone layer.

Multi-sensor detectors A class of detector that increases reliability by combining heat and smoke sensing, heat and carbon monoxide sensing, heat/co/smoke sensing, or ionization plus photoelectric sensing.

Municipal fire alarm system A system of manual pull boxes located throughout a jurisdiction, with each box wired to a circuit that is connected directly to the fire department.

National Fire Protection Association (NFPA) The principal source and publisher of fire protection standards in the United States.

NICET Abbreviation for the National Institute for Certification in Engineering Technologies.

NOAEL Abbreviation for "no observed adverse effect level," a measure of clean agent toxicity to humans, under test conditions.

Nonaqueous agents Agents in which water is not a component.

Nonaspirated sprinklers Foam discharge devices that do not possess an air inlet between the orifice and the deflector.

Non-coded manual stations Devices that do not deliver a distinctive signal and are grouped on a circuit so the control unit can indicate activation.

Notification appliance circuits (NACs) fire alarm circuits to which fire alarm notification appliances are connected.

Notification zone A floor or portion of a floor of a building, or a discrete area or wing of a building, where fire alarm notification can be provided.

Olfactory notification A form of notification that uses odors to arouse sleeping individuals.

Open conductor An IDC Class B fault that can be caused by a cut or broken wire or a loose terminal.

Ozone layer A protective layer of our stratosphere that helps to filter the ultraviolet rays of the sun before they reach Earth.

Performance-based design An engineering approach to fire protection design.

Performance objective Engineering basis for a predetermined end result of a given design.

Performance specification A general specification that provides the minimum information necessary to estimate, design, and install a fire protection system.

Personal values Values obtained from our parents, peers, mentors, television, film, the media, the community, and schools.

Phase diagram A graph that represents the physical state of a specific substance at varying pressures and temperatures.

Photoelectric light-scattering smoke detectors Devices that use a light-emitting diode (LED) and a light sensor that does not receive light from the LED under ambient conditions; light reaches the light sensor when it is reflected to the sensor by smoke particles.

Pilot-actuated ultra high-speed water spray system A preprimed water discharge system with soledoid-operated pilot nozzles, supervised by a pilot piping system.

Piloted ignition temperature Minimum temperature a substance should attain in order to ignite under specific test condition.

Pipe diameter measurement gauge A device used to determine the diameter of an existing pipe into which you are tapping a new fire protection main.

Plan job Design performed using new architectural plans as the basis for design.

Plan reference elevation Sea level, or an elevation chosen specifically for a building or group of buildings.

Plenum space A space above a suspended ceiling that is kept under negative pressure for return air.

Point of contact A person who meets you at the job site, shows you the area to be surveyed, and remains available if any questions arise.

Point protection Application of suppressant directly onto an expected point of hazard.

Preaction water mist systems Systems having nozzles with an individual actuating device, where each nozzle actuates individually, with piping filled with air.

Pre-engineered systems Packaged units in which supply quantity, nozzle selection, pipe size, and detector selection are predetermined for a range of volumes, areas, or applications, and are listed as a unit.

Prescriptive design The direct use of national, local, or manufacturer's standards to design every aspect of a suppression or detection system.

Presignal manual alarm stations Stations that do not actuate a general alarm but, instead, send an advance signal to the FACU when actuated or operated.

Pressure proportioner A device that draws a portion of the incoming water stream and uses it to pressurize the tank holding the foam concentrate.

Pressure sandwich A procedure in which a smoke control system exhausts smoky air from the fire floor with supply air to the fire floor shut down, and all other floors are supplied with fresh air with return air shut down.

Pressure-sensing fire detectors Devices that consist of a plate that depresses when the enclosure pressure increases to a predetermined level and activates a signal to the FACU.

Pressure switches Electrical devices that indicate high or low pressure in a water tank or in a sprinkler system.

Profile plan A plan that specifically shows reference elevations with respect to finished grade.

Proprietary information Unique data that are the sole property of the company.

Proprietary supervising station fire alarm system A system that supervises properties owned by the same company.

Protected premises Buildings or enclosures that are supervised by a fire alarm and detection system.

Protein foam A foam that contains protein-based animal additives.

Pyrolysis A chemical process in which a compound is converted to one or more products in the presence of heat and oxygen.

Rain leaders Rainwater drainage piping.

Rate-by-area method A method of applying carbon dioxide to a two-dimensional surface area based on the capability of listed nozzles to discharge a given amount of carbon dioxide over a fixed area of coverage.

Rate-by-volume method A method of local application of carbon dioxide where an imaginary volume larger than the hazard is created to account for the dissipation and loss of carbon dioxide during discharge.

Rationalization An attempt to justify one's own action that is known to be wrong.

Real protection A method that involves building a dam around the perimeter of a floating roof and filling the seal area with low-expansion foam.

Recessed fixtures Lights whose faces are flat to the ceiling and whose bodies protrude into the ceiling space.

Recommended practices NFPA documents that provide nonmandatory advice.

Reference grid A system of parallel reference lines aligned either to magnetic north or to plant north.

Reflected ceiling plans Plan views of suspended ceilings.

Remote supervising station fire alarm system A system that transmits alarm, supervisory, and trouble signals from one or more protected premises to a distant location where appropriate action is initiated.

Return diffuser Ceiling element used to draw stale air from a room.

Seal protection Filling or covering the seal area with low-expansion foam.

Signaling line circuit (SLC) A circuit over which multiple input and output signals of more than one fire alarm system, transmitter, or device is carried.

Single-action manual stations Devices that activate an alarm and require only one motion by the user.

Smoldering Combustion without flame, usually with incandescence and smoke.

Societal values Values related to the public expectation that all fire protection professionals competently discharge their responsibilities and thereby ensure the safety of all who enter a building that is protected by a fire protection system.

Solenoids Switches used to open and close electrically actuated fire protection valves.

Spark/ember detectors Specialized flame detectors with a photodiode that senses very small amounts of radiant energy, emitted before the full flaming stage.

Specific volume Volume of halon per unit mass.

Specification divisions Broad categories of building component groupings standardized by the Construction Specifications Institute.

Specifications A written description of project requirements.

Specification sections Detailed requirements for each CSI division.

Spontaneous ignition Initiation of combustion of a material by an internal chemical or biological reaction that has produced sufficient heat to ignite the material.

Spot-type analog heat detectors Devices that electronically monitor their status and condition and can sense when they become dirty and require cleaning or replacement.

Spot-type electronic thermistor heat detectors Devices that monitor thermistors with temperature sensing capable of detecting when temperature thresholds have been exceeded.

Spot-type fixed-temperature heat detectors Devices that initiate an alarm when a predetermined temperature is attained by a thermally responsive element.

Spot-type ionization detectors Devices containing a small amount of radioactive material that ionizes the air between a positive and a negative electrode in its sampling chamber to measure conductance.

Spot-type photoelectric light obscuration smoke detectors Devices that project light from a source directly to a sensor that measures reductions in the amount of light received in the presence of smoke.

Spot-type photoelectric smoke detectors Detection devices that project light from a source to a light sensor.

Spot-type rate-compensated heat detectors Devices that compensate for the effects of thermal lag.

Squib-actuated ultra high-speed water spray system A preprimed water-filled piping system controlled by a squib-actuated deluge valve.

Standards Mandatory NFPA requirements that may be used by authorities to approve a fire protection system.

Stratification The predilection or predisposition for smoke in a plume to cease rising when it entrains cool air and cools to the temperature of the surrounding air.

Structural drawings Drawings that provide details related to the floors, roof, and structural elements of a building.

Subsurface injection A system in which foam is discharged below the surface of a flammable or combustible liquid.

Supply diffuser Ceiling element used to distribute fresh air to a room.

Surface application A system designed to roll a thin blanket of foam over the surface area of the fuel.

Surface fire A fire on the exterior of an object.

Survey A thorough investigation of a building and its components for the purpose of taking detailed measurements of the building to serve as reference for a fire protection drawing.

Survey job A project involving an existing building for which plans cannot be obtained.

Survey reference elevation The lowest, flattest, most reliable elevation that can be found.

Survey strategy The approach by which a survey begins—the layout of all structural elements, such as building columns and structural beams, as a reference for all subsequent measurements.

Suspended fixtures Lights hanging from rods or chains.

Tactile appliance A device that provides notification by generating vibrating signals that can be sensed by an individual possessing the appliance.

Tamper switches Electronic devices that indicate a sprinkler valve closure.

Tank farm An enclosure containing vertical cylindrical tanks, horizontal cylindrical tanks, or spherical tanks, which store flammable or combustible liquids, surrounded by a containment dike.

Telescoping elevation measurement pole A device used to measure elevations in the 30- to 50-foot range.

Temperature control Performance objective aimed at reducing room temperatures during combustion to allow safe egress and reduced damage.

Thermal lag The temperature differential between a detector and the heated air surrounding it.

Thermocouple A device consisting of two dissimilar metals joined at their ends.

Tinder Easily ignitable objects used to start a fire.

Total flooding method A method of fire suppression that involves completely filling a room or enclosure volume with a fire protection agent.

Travel distance The actual walking distance from one point to another.

Triple point A point where carbon dioxide exists in all three states simultaneously.

Two-phase flow Liquid and vapor halon phases flowing in pipes simultaneously.

Type I discharge outlet A discharge device designed to deliver foam onto the liquid surface in a gentle fashion.

Type II discharge outlet A discharge device designed to deliver foam less gently than a type I outlet while keeping submergence and agitation to a minimum.

Ultimate stress The stress at which a vessel is likely to rupture.

Ultra high-speed water spray systems Systems used for the protection of explosive hazards, where water is the media used, and where overpressurization is not the primary concern.

Ultraviolet (UV) flame detectors Devices designed to detect radiation falling within ultraviolet wavelengths below 4000 angstroms.

Values of the fire protection profession Values summarized in the Code of Ethics for NICET-Certified Engineering Technicians and Technologists and the National Society of Professional Engineers (NSPE) *Code of Ethics for Engineers.*

Venturi proportioner A device that utilizes the negative pressure created by water flowing past an open orifice to draw foam concentrate into the water stream.

Volume factor A value used to determine the amount of carbon dioxide required to be injected into a room at the minimum design concentration of 34%.

Water-flow switches Electronic devices that indicate water movement in a sprinkler pipe.

Water-level switches Devices that indicate the level of water in a fire protection water storage tank.

Water mist A water spray with water droplets of less than 1000 microns at the minimum operation pressure of the discharge nozzle.

Water mist system A distribution system connected solely to a water supply or alternatively to a water supply and an atomizing media (air or nitrogen), that is equipped with one or more nozzles capable of delivering water mist intended to control, suppress, or extinguish fires, as defined by NFPA 750.

Water oscillating monitors (WOM) Monitors installed at the floor of an aircraft hangar to provide protection below the wing area.

Wet chemical A solution of water and chemical to form an extinguishing agent.

Wet pipe water mist systems Systems having nozzles with an individual actuating device, where each nozzle actuates individually, and with piping filled with water.

Whistle-blowing An external action in which an individual exposes a situation perceived to be unethical.

Wire-to-wire short A condition in which two bare wires touch each other, creating an incomplete circuit.

Yield stress The stress at which a vessel is likely to deform.

Zoned application systems Systems in which a volume is protected by several distinct suppression zones, each with its own detection system.

ACRONYMS

AC	alternating current
ADA	Americans with Disabilities Act
ADAAG	Americans with Disabilities Act Accessibility Guidelines
AFFF	aqueous film-forming foam
AHJ	authority having jurisdiction
ANSI	American National Standards Institute
ARC	alcohol-resistant concentrates
ASHRAE	American Society of Heating, Refrigerating, and Air-Conditioning Engineers
ASMET	Atria Smoke Management Engineering Tools
CAD	computer-aided design
CNS	central nervous system
COE	Corps of Engineers
CPU	central processing unit
CSI	Construction Specifications Institute
CSI	continuous software interrogation
DC	direct current
ELR	end-of-line resistor
EPA	Environmental Protection Agency
FACU	fire alarm control unit
FFFP	film-forming fluoroprotein foam
FM	Factory Mutual
HCFC	hydrochlorofluorocarbon
HFC	hydrofluorocarbon
HPR	highly protected risk
HVAC	heating, ventilating, and air conditioning
IDC	initiating device circuits
IR	infrared
LED	light-emitting diode
LFL	lower flammable limit
LNG	liquified natural gas
MEC	minimum explosive concentration test
MIE	minimum ignition energy test
MOC	maximum allowable oxygen concentration test
MSDS	Material Safety Data Sheets
NAC	notification appliance circuits
NASA	National Aeronautics and Space Administration
NEC	National Electric Code
NFPA	National Fire Protection Association
NICET	National Institute for Certification in Engineering Technologies
NIST	National Institute of Standards and Technology
NOAEL	no observed adverse effect level
NSPE	National Society of Professional Engineers
PFC	perfluorocarbon
PSI	pounds per square inch
PSIA	pounds per square inch absolute
RTI	response time index
SCBA	self-contained breathing apparatus
SLC	signaling line circuits
UL	Underwriters Laboratories
UV	ultraviolet
VAC	volts alternating current
VDC	volts direct current
WOM	water oscillating monitors

ADDITIONAL RESOURCES

Alpert, R. L., "Incentive for Use of Misting Sprays as a Fire Suppression Flooding Agent," in *Proceedings of the Water Mist Fire Suppression Workshop*, National Institute of Standards and Technology, Gaithersburg, Maryland, March 1–2, 1993.

Bachalo, W. D., "Advances in Spray Drop Size and Velocity Measurement Capabilities for the Characterization of Fire Protection Systems," in *Proceedings of the National Institute of Standards and Technology Water Mist Fire Suppression Workshop*, Gaithersburg, Maryland, March 1–2, 1993.

Benedetti, Robert P., *Flammable and Combustible Liquids Code Handbook*, National Fire Protection Association, Quincy, Massachusetts, 1996.

Bryan, John L., *Fire Suppression and Detection Systems*, Macmillan, 1993. New York, NY.

Bukowski, Richard W., and O'Laughlin, Robert J., *Fire Alarm Signaling Systems*, National Fire Protection Association/Society of Fire Protection Engineers, Quincy, Massachusetts, 1994.

Butz, J. R., "Application of Fine Mists to Hydrogen Deflagrations," in *Proceedings of the Halon Alternatives Technical Working Conference*, Albuquerque, New Mexico, May 11–13, 1993.

Cholin, John M., "Fire Alarm Systems Inspection, Testing, and Maintenance," Chapter 5-5 in *Fire Protection Handbook*, 18th Ed., National Fire Protection Association, Quincy, Massachusetts, 1997.

Chow, W. K., "On the Evaporation Effect of a Sprinkler Water Spray," *Fire Technology*, vol. 25, no. 4, pp. 364–373, November 1989.

Conroy, Mark T., "Fire Extinguisher Use and Maintenance," Chapter 6-23 in *Fire Protection Handbook*, 18th Ed., National Fire Protection Association, Quincy, Massachusetts, 1997.

Conway, Donald J., *Human Response to Tall Buildings*, Dowden, Hutchinson, and Ross, Inc., Stroudsburg, Pennsylvania, 1977.

De George, R. T., Business Ethics, Macmillan, New York, 1982.

DiNenno, Philip J., "Direct Halon Replacement Agents and Systems," Chapter 6-19 in *Fire Protection Handbook*, 18th Ed., National Fire Protection Association, Quincy Massachusetts, 1997.

DiNenno, Philip J., "Halon Replacement Clean Agent Total Flooding Systems," Chapter 4-7 in *SFPE Handbook of Fire Protection Engineering*, 2nd Ed., SFPE/NFPA, Quincy, Massachusetts, 1995.

Drysdale, Dougal, *An Introduction to Fire Dynamics*, John Wiley and Sons, New York, NY, 1985.

Evans, D., and Pfenning, D., "Water Sprays Suppress Gas-Well Blowout Fires," *Oil and Gas Journal*, pp. 80–86, April 29, 1985.

Ewing, C. T., Hughes, J. T., and Carhart, H. W., "The Extinction of Hydrocarbon Flames Based on the Heat-Absorption Processes that Occur in Theory," *Fire and Materials*, vol. 8, no. 3, 1984.

Fire Protection Equipment Directory, Underwriters Laboratories, Chicago, Illinois, 2007.

Fire Protection Handbook, 20th Ed., National Fire Protection Association, Quincy, Massachusetts, 2007.

Fire Technology, John M. Watts, editor, National Fire Protection Association, Quincy, Massachusetts.

Fleming, R. P., "New Interest in Water Mist," *NFSA Sprinkler Quarterly*, no. 84, p. 22, Fall 1993.

Foam Systems Design and Applications, Ansul Fire Protection, Marinette, Wisconsin, 2007.

Friedman, Raymond, *Principles of Fire Protection Chemistry*, National Fire Protection Association, Quincy, Massachusetts, 1989.

Friedman, Raymond, "Theory of Fire Extinguishment," Chapter 1–8 in *Fire Protection Handbook*, 18th Ed., National Fire Protection Association, Quincy, Massachusetts, 1997.

Gagnon, Robert M., "Water Mist Fire Suppression Systems—Theory and Applications," *Fire Protection Contractor Magazine*, pp. 36–42, May 1994; pp. 42–47, June 1994.

Gagnon, R. M., and McCuen, R., *The Engineering Student's Guidebook for Professional Development*, University of Maryland, College Park, Maryland, 1999.

Gagnon, Robert M., *Design of Water-Based Fire Protection Systems*, Delmar Publishers, Albany, New York, 1997.

Gagnon, Robert M., "Ultra High Speed Suppression Systems for Explosive Hazards," in *Fire Protection Handbook*, 20th Ed., National Fire Protection Association, Quincy, Massachusetts, 2007.

Gagnon, R. M., and Kirby, R. H, *A Designer's Guide to Fire Alarm Systems,* National Fire Protection Association, Quincy, Massachusetts, 2003.

Gagnon, R. M., editor and contributing author, *Designer's Guide to Automatic Sprinkler Systems*, Society of Fire Protection Engineers and The National Fire Protection Association, Quincy, Massachusetts, 2005.

Gameiro, V. M., "Fine Water Spray Fire Suppression Alternative to Halon 1301 in Gas Turbine Enclosures," in *Proceedings of the Halon Alternatives Technical Working Conference*, Albuquerque, New Mexico, May 11–13, 1993.

Gott, Joseph E., *Analysis of High Bay Hangar Facilities for Fire Detector Sensitivity and Placement*, NIST Technical Note 1423, National Institute of Standards and Technology, Gaithersburg, Maryland, 1997.

Hague, David R., "Dry Chemical Agents and Application Systems," Chapter 6–21 in *Fire Protection Handbook*, 18th Ed., National Fire Protection Association, Quincy, Massachusetts, 1997.

Hayes, W. D., Jr., *Literature Survey on Drop Size Data, Measuring Equipment, and a Discussion of the Significance of Drop Size in Fire Extinguishment,*

Report Number NBSIR 85-3100-1, National Institute of Standards and Technology, Gaithersburg, Maryland, July 1985.

Hickey, Harry E., "Foam System Calculations," Chapter 4–5 in *SFPE Handbook of Fire Protection Engineering*, 2nd Ed., SFPE/NFPA, Quincy, Massachusetts, 1995.

Hill, R. G., Marker, T. M., and Sarkos, C. P., "Evaluation and Optimization of On-Board Water Spray Fire Suppression Systems in Aircraft," in *Proceedings of Water Mist Fire Suppression Workshop*, National Institute of Standards and Technology, Gaithersburg, Maryland, March 1–2, 1993.

Hills, A. T., Simpson, T., and Smith, D. P., "Water Mist Fire Protection Systems for Telecommunication Switchgear and Other Electronic Facilities," in *Proceedings of the Water Mist Fire Suppression Workshop*, National Institute of Standards and Technology, Gaithersburg, Maryland, March 1–2, 1993.

Jackman, L. A., Glockling, J. L. D., and Nolan, P. F., "Water Sprays: Characteristics and Effectiveness," in *Proceedings of the Halon Alternatives Technical Working Conference*, Albuquerque, New Mexico, May 11–13, 1993.

Johnson, Peter F., "Special Systems and Extinguishing Techniques," Chapter 6-25 in *Fire Protection Handbook*, 18th Ed., National Fire Protection Association, Quincy, Massachusetts, 1997.

Journal of Applied Fire Science, Paul R. DeCicco, Editor, Baywood Publishing Company Inc., Amityville, New York.

Klote, John H., "Smoke Control," Chapter 4-12 in *SFPE Handbook of Fire Protection Engineering*, 2nd Ed., SFPE/NFPA, Quincy, Massachusetts, 1995.

Lawson, J. R., Walton, W. D., and Evans, D. D., *Measurement of Droplet Size in Sprinkler Sprays*, National Institute of Standards and Technology, Gaithersburg, Maryland, 1988.

Lee, S., and Sichel, M., "Evaporation of Liquid Droplets in a Confined Medium," in *Chemical and Physical Processes in Combustion*, 20th Fall

Technical Meeting, combined technical meetings of the Combustion Institute/Eastern States Section and NBS/CFR Annual Conference on Fire Research, Gaithersburg, Maryland, 1987.

Lev, Y., "Cooling Sprays for Hot Surfaces." *Fire Prevention*, no. 222, pp. 42–47, 1989.

Levine, R. S., *Navy Safety Center Data on the Effects of Fire Protection Systems on Electrical Equipment*, National Institute of Standards and Technology, Gaithersburg, Maryland, 1991.

Lindenberg, Michael R., *Engineering Unit Conversions*, Professional Publications, San Carlos, California, 1988.

Liu, S. T., *Analytical and Experimental Study of Evaporative Cooling and Room Fire Suppression by Corridor Sprinkler System*, National Institute for Standards and Technology, Gaithersburg, Maryland, 1977.

Lugar, J. R., *Water Mist Fire Protection*, David W. Taylor, Naval Ship Research and Development Center, Bethesda, Maryland, 1979.

Manicas, Peter T., Kruger, and Arthur N., *Essentials of Logic*, American Book Company, New York, NY, 1968.

Marlatt, F. P., "Maryland Fire and Rescue Institute Hosts High-Pressure Sprinkler Testing Program," *MFRI Bulletin*, vol. 24, no. 5, November 1993.

Marttila, P., "Water Mist in Total Flooding Applications," in *Proceedings of Halon Alternatives Technical Working Conference*, Albuquerque, New Mexico, May 11–13, 1993.

Mawhinney, J. R., "Characteristics of Water Mists for Fire Suppression in Enclosures," in *Proceedings of the Halon Alternatives Technical Working Conference*, Albuquerque, New Mexico, May 11–13, 1993.

Mawhinney, J. R., "Engineering Criteria for Water Mist Fire Suppression Systems," in *Proceedings of the Water Mist Fire Suppression Workshop*, National Institute of Standards and Technology, Gaithersburg, Maryland, March 1–2, 1993.

Mawhinney, J. R., *Fine Water Spray Suppression Project*, National Fire Laboratory, Institute for Research in Construction, National Research Council of Canada, May 5, 1992.

Mawhinney, J. R., "Water Mist Suppression Systems May Solve an Array of Fire Protection Problems," *NFPA Journal*, vol. 88, no. 3, May/June 1994.

Maybee, Walter, Editor, *804 Newsletter*, Santa Fe, New Mexico.

McCarthy, Shawn P., *Engineer Your Way to Success*, National Society of Professional Engineers, Alexandria, Virginia, 1989.

McCuen, Richard H., "The Ethical Dimensions of Professionalism," *Engineering Issues*, vol. 105 (E11), pp. 89–105, April 1979.

Milke, James A., "Smoke Management in Covered Malls and Atria," Chapter 4-13 in *SFPE Handbook of Fire Protection Engineering*, 2nd Ed., SFPE/NFPA, Quincy, Massachusetts, 1995.

The Montreal Protocol on Substances That Deplete the Ozone Layer, Final Act, United Nations Environment Program, HMSO, CM977, September 1987.

Moore, Wayne D., "Automatic Fire Detectors," Chapter 5–2 in *Fire Protection Handbook*, 18th Ed., National Fire Protection Association, Quincy, Massachusetts, 1997.

Moore, Wayne D., "Fire Alarm System Interfaces," Chapter 5-4 in *Fire Protection Handbook*, 18th Ed., National Fire Protection Association, Quincy, Massachusetts, 1997.

National Foam Engineering Manual, Chubb National Foam, Lionville, Pennsylvania, 1997.

O'Hern, T. J., and Rader, D. J., "Practical Application of In Situ Aerosol Measurement," in *Proceedings of the Halon Alternatives Technical Working Conference*, Albuquerque, New Mexico, May 11–13, 1993.

Papavergos, P. G., *Fine Water Sprays for Fire Protection—A Halon Replacement Option*, British Petroleum Ventures, BP Research, Sunbury Research Center, 1990.

Parker, Sybil P., *McGraw-Hill Dictionary of Science and Engineering*, McGraw-Hill Book Company, New York, NY, 1984.

Peterson, Marshall E., "The Role of Extinguishers in Fire Protection," Chapter 6–24 in *Fire Protection Handbook*, 18th Ed., National Fire Protection Association, Quincy, Massachusetts, 1997.

Pfenning, D., and Evans, D., "Suppression of Gas Well Blowout Fires Using Water Sprays—Large and Small Scale Studies," in *Proceedings of Production Session*, pp. 1–12, American Petroleum Institute, San Antonio, Texas, September 11–13, 1984.

Quintiere, James G., *Principles of Fire Behavior*, Delmar Publishers, Albany, New York, 1997.

Reischl, U. "Water Fog Stream Heat Radiation," *Fire Technology*, vol. 15, pp. 262–270, November 1979.

Rosander, M., and Giselsson, K., "Making the Best Use of Water for Fire Extinguishing Purposes," *Fire*, pp. 43–46, October 1984.

Ross, S. D., *Moral Decision: An Introduction to Ethics*, Freeman, San Francisco, California, 1972.

Scheffey, Joseph L., "Foam Agents and AFFF System Design, Chapter 4-4 in *SFPE Handbook of Fire Protection Engineering*, 2nd Ed., SFPE/NFPA, Quincy, Massachusetts, 1995.

Scheffey, Joseph L., "Foam Extinguishing Agents and Systems," Chapter 6-22 in *Fire Protection Handbook*, 18th Ed., National Fire Protection Association, Quincy, Massachusetts, 1997.

Schifiliti, Robert P., "Notification Appliances," Chapter 5-3 in *Fire Protection Handbook*, 18th Ed., National Fire Protection Association, Quincy, Massachusetts, 1997.

Schifiliti, Robert P., Meacham, Brian J., and Custer, Richard L. P., "Design of Detection Systems," Chapter 4-7 in *SFPE Handbook of fire Protection Engineering*, 2nd Ed., SFPE/NFPA, Quincy, Massachusetts, 1995.

Schuchard, Walter F., "Household Fire Warning Equipment," Chapter 5–6 in *Fire Protection Handbook*, 18th Ed., National Fire Protection Association, Quincy, Massachusetts, 1997.

SFPE, *The SFPE Handbook of Fire Protection Engineering*, Society of Fire Protection Engineers and the National Fire Protection Association, Quincy, Massachusetts, 1995.

Simplex Time Recorder Company, *Fire Alarm Basics*, Pace Series-Product Application, and Concept Education, Simplex Time Recorder Company, Gardner, Massachusetts, 1996.

Simplex Time Recorder Company, *Technical Reference Manual, Volumes 1 and 2*, Simplex Time Recorder Company, Gardner, Massachusetts, 1996.

Spring, D. J., Simpson, T., Smith, D. P., and Ball, D. N., "New Applications of Aqueous Agents for Fire Suppression," in *Proceedings of the Halon Alternatives Technical Working Conference*, Albuquerque, New Mexico, May 11–13, 1993.

Sunar, D. G., *Getting Started as a Consulting Engineer*, Professional Publications, San Carlos, California, 1986.

Sunar, D. G., *How to Become a Professional Engineer*, 2nd Ed., Professional Publications, San Carlos, California, 1989.

Tapscott, Robert E., "Combustible Metal Extinguishing Agents and Application Techniques," Chapter 6-26 in *Fire Protection Handbook*, 18th Ed., National Fire Protection Association, Quincy, Massachusetts, 1997.

Taylor, Gary M., "Halogenated Agents and Systems," Chapter 6-13 in *Fire Protection Handbook*, 18th Ed., National Fire Protection Association, Quincy, Massachusetts, 1997.

Training Manual on Fire Alarm Systems, National Electrical Manufacturers Association, Washington, DC, 1992.

Turner, A. R. F., "Water Mist in Marine Applications," in *Proceedings of the Water Mist Fire Suppression Workshop*, National Institute of Standards and Technology, Gaithersburg, Maryland, March 1–2, 1993.

Wilson, Dean K., "Fire Alarm Systems," Chapter 5-1 in *Fire Protection Handbook*, 18th Ed., National Fire Protection Association, Quincy, Massachusetts, 1997.

Wysocki, Thomas J., "Carbon Dioxide and Application Systems," Chapter 6-20 in *Fire Protection Handbook*, 18th Ed., National Fire Protection Association, Quincy, Massachusetts, 1997.

INDEX